RECENT ADVANCES
IN
DEVELOPMENTAL BIOLOGY

Recent Advances in Developmental Biology

Vol. 7

Reproduction and Development in Vertebrates

A.P. DIWAN
N.K. DHAKAD

ANMOL PUBLICATIONS PVT LTD
New Delhi-110 002

ANMOL PUBLICATIONS PVT LTD
4374/4B, Ansari Road
Daryaganj, New Delhi-110 002

Recent Advances in Developmental Biology

First Edition 1995
ISBN 81-7488-039-9

PRINTED IN INDIA

Published by J.L. Kumar for Anmol Publications Pvt. Ltd. and
Printed at Mehra Offset Press, Delhi.

Preface

The fundamentals of Developmental Biology are presented within the framework of scientific discovery. Researches in Embryology have made almost incredible strides in the past few decades. Consequently, the existing concepts of the developmental biology have expanded. There has been a revolution indeed in this direction.

The text integrates the descriptive, experimental and biochemical approches into a conceptual framework for the analysis of development. All important points are illustrated diagrammatically. The title is not intended to be comprehensive, nor could it be at this length, but concentrates as putting across the basic principles of the subject as briefly and lucidly as possible. It does this with the aid of careful selected examples, some recent and other classic of the field and with numerous illustrations. The aim is to enthuse the reader with this active and exciting area of research and to lay a solid foundation on which further study of its various facts may be based.

Authors

Contents

1

Anatomy of the Reproductive System

All the living creatures develope from the pre-existing organisms and performs all the activities for their growth and maturity. But as none of the living creatures is immortal, so the individuals do something for the continuity of their race. And the activity for the continuation of the race is called *reproduction.* In all vertebrates, special organs are developed for the purpose, which are called the *reproductive organs* or *sex organs.* In a large number of sexually reproducing animals there are two distinctly different kinds of individuals, males and females. The males produce sperms and the females produce eggs. The term egg also refer to the familiar hen's egg which contain yolk, albumen and a shell in addition to the female gamete, the real egg. Sometimes the same individuals possesses the capcity to function as both male and female. Such an animal is termed *hermophrodite* or *bisexual.* Hermophroditic animals are rare among vertebrates. A common example is hagfish.

All the vertebrates reproduce sexually, by the fusion of sperm cells with egg cells. Not only does this union of gametes bring together in a single fertilized egg the hereditary factors form both parents, but it induces the egg, otherwise inert, to start division and the formation of an embryo.

The reproductive system of vertebrates consists of the *reproductive organs* or *gonads,* in which the germ cells become differentiated, and the reproductive ducts, through which the germ cells leave the body. During the development of the individual, the reproductive and urinary system arise in close association so that structuraly they form what is known as the *urino-genital* system. The functions of the two parts of this sytem are entirly unrelated, and we shall be concerned here with only the reproductive or genital system.

The gonads of vertebrates produce gametes and, in most cases, steroid hormones. These hormones serve various functions, such as inducing or facilitating sexual behaviour and/or parental behaviour, preparing the reproductive tract for receiving the gametes, caring for the zygote, and other functions.

Here we present a general outline of the anatomy of the male and female reproductive systems of representatives of the different classes of the vertebrates.

MALE REPRODUCTIVE SYSTEM

In most vertebrates, the testes are paired, but in some, e.g., the cyclostomes, the left and right testes are fused, and in some teleost species only one testis develops as in *Notopterus notopterus*. In most classes of vertebrates the sperm produced in the testes are conducted to the outside by a duct system : in cyclostomes and the Salmonidae, however, such a duct system is lacking. In vertebrate species that do have a sperm duct system, this system may be separate from the urinary Wolffian duct system, as is the case of teleosts. In all other vertebrates, sperm are transported in tubules than originate as pronephric and mesonephric tubules.

Cyclostomes

In lampreys, the one testis consists of two gonads fused in the middle. It is located on the midline and is attached to the body wall by a mesentery. The testis consists of lobules, each containing several ampullae lined with germinal epithelium. When the lamprey becomes sexually mature and starts spawning, these ampullae break down, and the sperm are released into the coelom. There is no duct system, and the sperm leave the body through an opening secondarily connecting

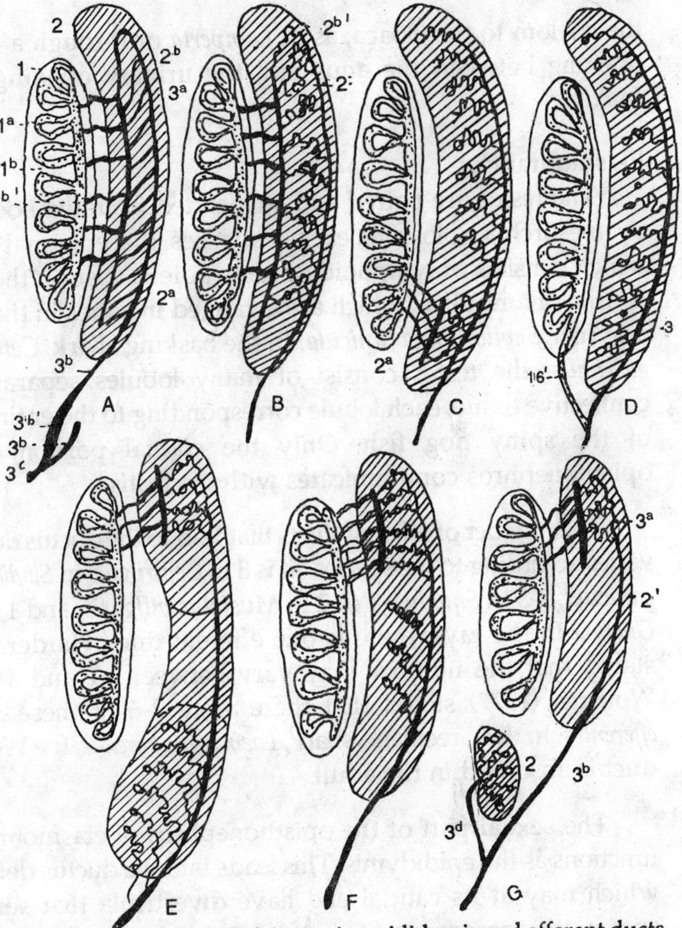

Fig. 1.1. Interrelationships of the testis, epididymis and efferent ducts in vertebrates. *A.* Basic schme. *B.* Acipenser Lepisosteus. Neoceratodus; C. Lepidorsiren. D. Polypterus. E, Selachians. F. Amphibia. G. Amniotes.

1 = testis, 1*a* = seminiterous ampullae or seminiferous tubules, 1b = rete testis, 1b = ductuli efferentes, 2 = opisthonephros 2′ = mesonephros of aminotes, 2″ = metanephros of amniotes, 2^a kidney border canal, 2^b = epididymal ductule 2b = lobuli of kidney, 2c = glomerulus of kidney, 3 = primary urinary duct and its derivatives, 3a= epididymal duct, 3^b = ductus deferens, 3^{b1} = ampulla of ductus deferens, $3^{b''}$ = vesicula seminalis, 3^c = ejaculatory duct, 3^d = ureter.

the coelom to the cloaca, as in *Lamperta* or through a median opening between the anus and the urinary opening, as in *Myxine.*

Elasmobranchs

The testes are paired and suspended from the body wall by mesorchia. Close to each testis lies the epigonal organ, which consists of lymphoid or hemapoietic tissue. "the testes consist of ampullae, which are arranged in zones in the spiny dog fish, *Scyliorhinus canicula.* In the basking shark, *Cetorhinus maximus,* the testes consist of many lobules separated by connective tissue, each lobule corresponding to the entire testis of the spiny dog fish. Only the cranial portion of the opisthonephros communicates with the testis.

The number of efferent ducts that lead to the ductus deferens varies according to the species : it is 8 in *Centrina*, 5 in *Scyliorhinus canicula*, 2-3 in *Squalus*, 1 or 2 in *Mustelus vulgaris*, and 1 or 2 in *Galeus*. In the rays, there is one efferent duct. Harder (1975) stated that the number may vary between 1 and 18, and Wourms (1977) stated that there were 2-6 of these *ductuli efferentes*. In the Greenland shark, *Laemargus borealis*, the Wolffian duct is absorbed in the adult.

The sexual part of the opisthonephros of elasmobranchs functions as the epididymis. This leads into the ductus deferens, which may at its caudal end have diverticula that serve as sperm storage organs (sperm sacs), or the ductus deferens may have a wide ampulla. The cranial part of the kidney forms a Leydig gland which secretes the seminal fluid. This Leydig gland is not homologous with the Leydig cells (which are sometimes called Leydig gland) of the higher vertebrates. The sperm may be packed into spermatophores (e.g., in the basking shark).

The copulatory organs of the alasmobranchs are very elaborate. They consist of claspers and a siphon sac in sharks or a clasper gland in skates and rays. The claspers are modifications of the pelvic fin. Wourums (1977) stated that the clasper may be considered part of the pelvic fin rolled up to form a tube and with edges of the fin overlapping. The siphon sacs are paired structures in the pelvic area between the skin and the abdominal wall. They have muscular tissue and are lined by a secretory epithelium, which in skates, may be very glandular.

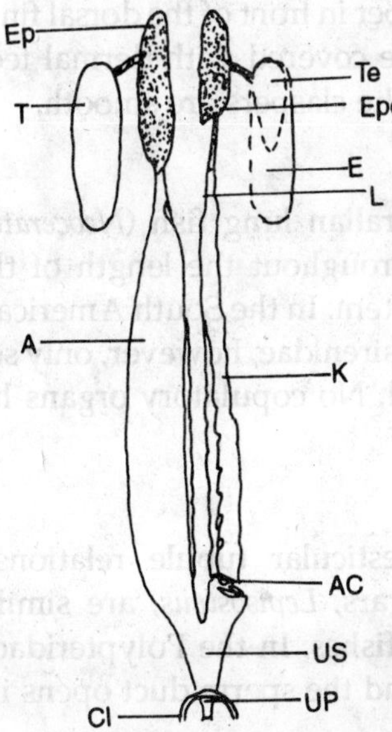

Fig. 1.2. Reproductive organs of the male basking shark (*Cetorhinus maximums*). Most of the ampulla on the right side of the drawing has been removed. A = ampulla ductus deferentis. A.C = cut end of the ampulla, Cl = cut wall of the cloaca. E = epigonal organ, Ep = epididymis, Ep. C = cut end of the epididymis, K = kidney, L = Leydig's gland. T = testis and epigonal organ, Te = position of the testis (dotted line), UP = urogenital papilla, US = urogenital sinus.

Holocephall

The urogenital tracts of male Holocephali and elasmobranchs are very similar. The ductuli efferentes are more numerous in the Holocephali than in the elasmobranchs and may extend into the mesorchium. Several ureters are present, some which open into the posterior part of the nephric duct or ampullae. The ductus deferens of the Holocephali has a complex system of chambered ampullae with the epithelium differentiated into regions.

The chimeras of Holocephali have one pair of pelvic fin claspers homologous with the claspers of the elasmobranchs and a single clasper in front of the dorsal fin. In the Holocephali, both claspers are covered with dermal teeth, whereas in the elasmobranchs the claspers are smooth.

Dipnoi

In the Australian lung fish (*Neoceratodus forsteri*), some renal tubules throughout the length of the kidney from an efferent duct system. In the South American and African lung fishes, the Leptosirenidae, however, only some posterior renal tubules are used. No copulatory organs have been found in the Dipnoi.

Chondrostei

The renal-testicular tubule relationship in sturgeons, *Acipenser*, and gars, *Lepisosteus*, are similar to those in the Austrilian lung fishes. In the Polypteridae, the duct systems are separated, and the sperm duct opens in the urinary duct.

Amoult (1964) and Holden (1971) have presented evidence that the anal fin is used as a copulatory chamber in *Polypterus senegalus*. Holden (1971) speculates that this copulatory mechanism is required to conserve sperm, because the testes of these fish are quite small, even during the breeding season.

The testes of sturgeons are lobed and are suspended by mesorchia. The male sturgeon has a well-developed Mullerian duct, but its function has not been determined.

Teleosts

There is no connection between testes and kidney in teleosts. In some species, the sperm duct has its own opening to the exterior, whereas in others there is no sperm duct, e.g., in the

Salmonidae and Anguillidae, and the sperm are released into peritoneal cavity and then leave via peritoneal canals.

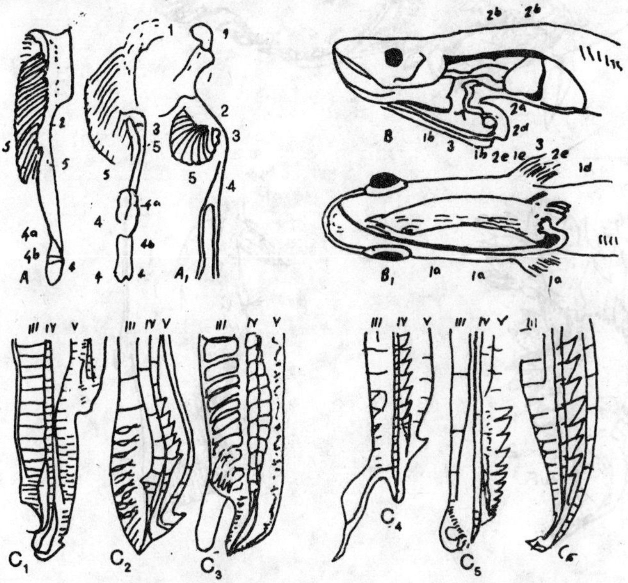

Fig. 1.3. A "Mixopterygia" (Clasper) of the cartilagenous fishes, A_1 = felt pelvic fin or a shark (*Chlamydoselachus*), dorsal view, A_2 = left pelvic fin of Dasyatis, dorsal view, A_3 = right pelvic fin of Chimaera, (Holocephali), ventral view, 1 = pelvic plate, 1' = tenaculum, 2 = basal cartilage of metaptergii, 3 = connecting cartilage, 4 = appendix, 4a = marginal cartilage, 4^b = terminal cartilage, 4^c = ventral covering cartilage, 5 = radii, 5′ = modified radius.

B, Priapum of Phallostethidae, B_1 = *Phallostethus*, lateral view, B_2 = *Solenophallus* in ventral view. 1a = Ctenactinium (fin ray of right pelvic fin), $1^{a'}$ = desmactnium —fin ray of left pelvic fin, 1^b = pelvic bone of the right side, 1^c= toxactinium, 1^d = 4-6 fin rays of the right pelvic fin, $1^{d''}$ = fin rays of the left side, 2^a = testis, 2^b = epididymis, 2^c = ductus deferens, 2^d = vesicle of ductus deferens, 2^e = genital papilla, 3 = anus, C, Gonopodia in the Poecillidae, C_1 = *Xiphophorus*, C_2 = *Gambusia*, C_3 = *Furcipenis*, C_4 = *Cnesterodon*, C_5 = *Giradinus*, C_6 = *Priapichthys*, III, IV and V are anal fin rays that have been modified into gonopodia.

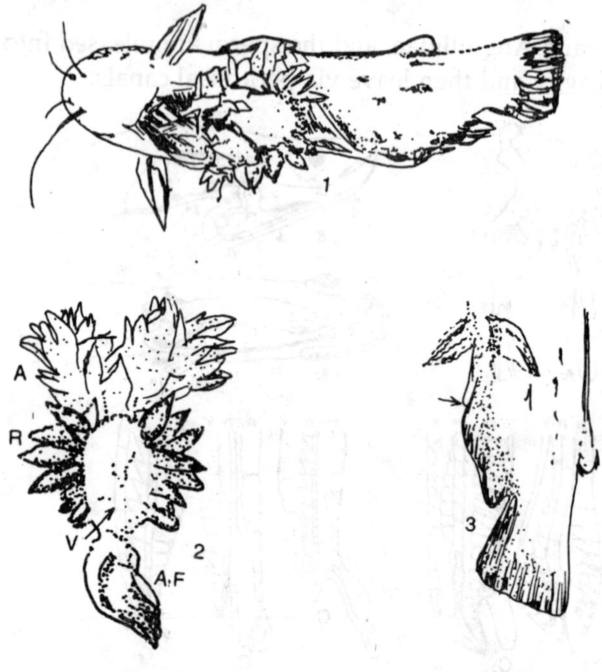

Fig. 1.4. Genital organs of *Trachycorystes*. 1 T. striatulus, adult male, 2. T. *Striaaatulus*. A = sperm-producing lobe of testis, B = gelatin-producing lobe of testis. A.f. = anal fin with pseudopenis, 3. Tail of juvenila male with incompletely developed pseudopenis (arrow).

The testes of most species of teleosts are paired, but sometimes only one testis is present; e.g., *Notopterus notopterus* has only a left testis. The testes are often oval or cylindrical, although in *Perca* and *Cyprinus* they are Y-shaped, whereas in the lctaluridae. *Glyptosternum* and *Trachycorystes* and testes are branched, fingerlike organs. In this type of gonad, spermatogenesis takes place only in the anterior part. The posterior lobes secrete a fluid, which, in *Trachycorystes*, seals off the vagina after mating, in the lctaluridae and *Glyps tosternum*, the function of this secretion is not known. The testes are suspended by mesorchia, but when the testes fuse, the mesolchium becomes attached to the mesentery of the intestine.

Acinar and Tabular Testes

Teleost testes have been classified on the basis of either the presence of acini (lobules) or tubules or on the basis of the distribution of spermatogonia in the tubules. The testes may be of acinar type which is reportedly found in the Clupeidae. Cyprinidae, Salmonidae, and Esocidae; the tubular type is found in Perciformers. Grier et al. (1980, after a critical review of the literature, concluded that the criteria for classifying teleost testes as either lobular or tubular were not well established and that the classification of Leydig cells, Sertoli cells, and lobule boundary cells had led to misconceptions when this dual classification was used. They proposed a classification based on the intratubular distribution of spermatogonia.

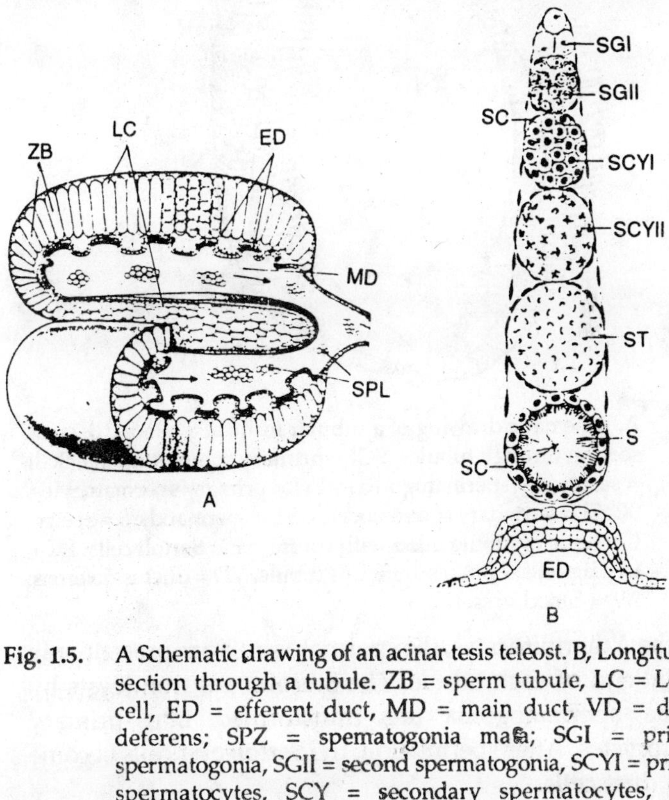

Fig. 1.5. A Schematic drawing of an acinar tesis teleost. B, Longitudinal section through a tubule. ZB = sperm tubule, LC = Leydig cell, ED = efferent duct, MD = main duct, VD = ductus deferens; SPZ = spematogonia maša; SGI = primary spermatogonia, SGII = second spermatogonia, SCYI = primary spermatocytes, SCY = secondary spermatocytes, ST = spermatids, S = sperm, SC = Sertoli cell.

Intratubular Distribution of Spermatogonia

In the Salmoniformes and Perciformes, tubules form either anastomosing or branching networks, and spermatogonia are present along the entire length of each tubule. The spermatogonia and their associated Sertoli cells from nearly solid cords within the tubules when the testes are regressed. When spermatogenesis begins, the Sertoli cells from the border of the cysts within which spermatogenesis occurs.

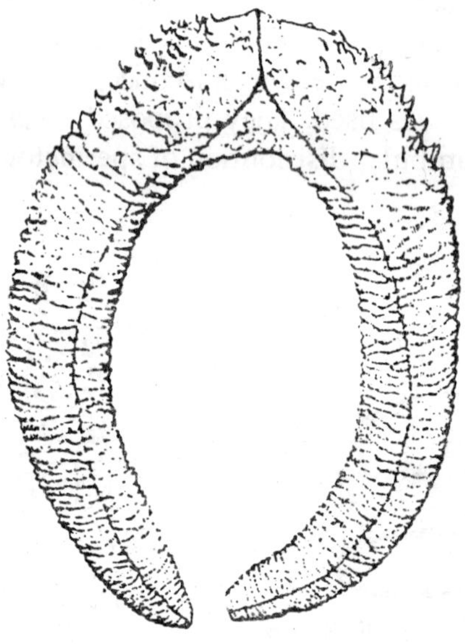

Fig. 1.6. A. schematic drawing of a tubular testis of a teleost. B. cross section through tubules, SGI = primary spermatogonia, SGII = secondary spermatogonia. SCYI = primary spermatocytes. SCYII = secondary spermatocytes. ST = spermatids, S = sperm. L = lume of tubule filled with sperm, SC = Sertoli cells. LC = Leydig cells. ZB = seminiferous tubule. VD = ductus deferens. BV = blood vessel.

In the Atheriniformes, the spermatogonia are present only the distal end of each tubule. The Steroli cells from cysts by the time spermatogonia are transformed into primary spermatocystes. After spermiation, the Sertoli cyst cells become efferent duct cells.

The identity and homologies of Leydig, Sertoli, and lobule boundary cells has caused controversies. These have been reviewed by Grier et al. (1980) and Grier (1981). It appears from the available microscopic and ultrastructural evidence that the Leydig cells may be interstitial cells or they may be distributed around the efferent ducts at the periphery of the testes. The Sertoli cells are tubular cells, which have several functions, i.e., nourishment of germ cells, production of steroids, phagocytosis of the residual bodies of spermatids, and in some cases, participaiton in the formation of efferent duct cells. The Sertoli cells have erroneously been called lobule boundary cells in earlier investigations. The true lobule boundary cells of teleost testes lie outside the basement membrane and they may be myoid in nature, as they are in mammalian testes. The ultrastructural evidence argues against the steroidal capacity for these cells.

The posterior lobe of some fishes, e.g., *Clarias* spp. and *Heteropneustes* spp., both genera belonging to the Siluridae, secrete materials that participate in the formation of spermatophores (sperm packets).

In Atheriniformes, which have internal fertilization and in which naked sperm bundles (spermatozeugmata) are produced, the Sertoli cell-spermatid or Sertoli cell-sperm evolves in slightly different ways in different species. In the Poeciludae, sperm nuclei become embedded within the cytoplasmic recesses of the Sertoli cells, whereas in the Goodeidae, the flagella of the spermatid become associated with the Sertoli cyst cells. Spermatophore formation, at least in *Horaichthys setnai*, involves the Sertoli cells in the sense that their secretory product coalesce around the mature sperm.

Grier (1981) has pointed out the coevolution of testes and anal fin in the various groups of teleosts, which has resulted in modifications of those into different structures for the transfer of sperm into the female reproductive tract. Generally, species that have a gonopodium have unmodified testes and do not produce either spermatozeugmata or spermatophores, although in the most primitive representative of the Anablepidae, *Anableps dowi*, partial spermatozeugmata are formed Grier (1981) has speculated that the ancestral "type" from which the Anablepidae have evolved may have had a testicular structure in which spermatozeugmata were formed.

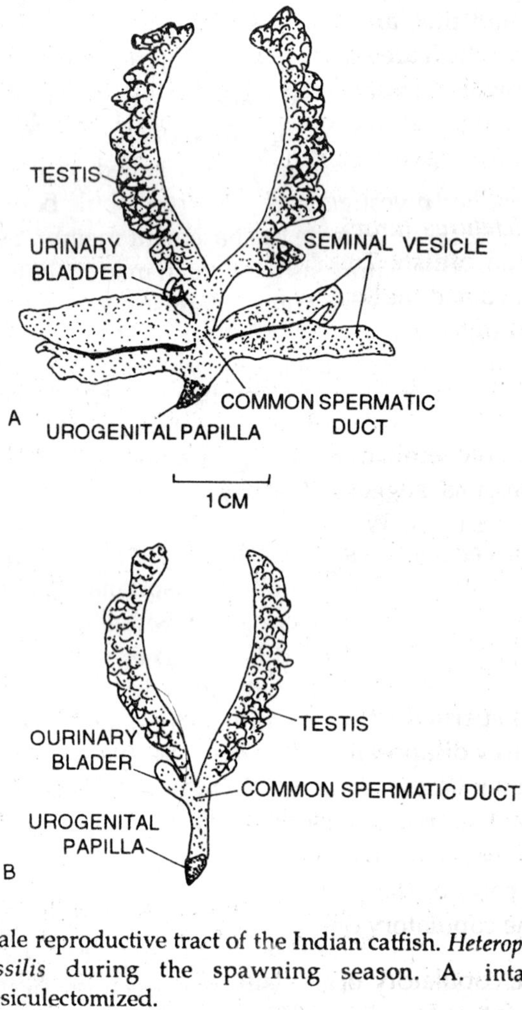

Fig. 1.7. Male reproductive tract of the Indian catfish. *Heteropneustes fossilis* during the spawning season. A. intact. B. vesiculectomized.

The sperm duct, the *ductus deferens,* does not produce secretions, but there are sometime accessory sexual organs, such as the seminal vesicles of the Inidan catfish. *Heteropneustes fossilis,* which do produce secretions. The term *seminal vesicles* has been used for different structures. For instance in the frillfin goby. *Bathygobius soporator,* the seminal vesicle do contain sperm and thus are truly seminal vesicle. So-called seminal vesicles, which are separate glandular structures, are found in the following teleost fish: the toad fish. Opsanus tau; the gobies, *Gobius niger, G. minutus, G. paganellus;* the loach, *Cobitis forsilis:* the goatfish, *Mullus barbatus*; the northen pike, *Esox lucius*; the mudsucker, *Gillichthys mirabilis*; the Indian catfish. *Heteropneustes fossili. Ictalurus furcatus, I. catus, I. nebulosus,* and the South American catfish. *Trachycorystes striatulus.* In the lctaluridae mentioned and the South American catfish, the seminal vesicles are part of, or are derived from, the testes. The secretions of the seminal vesicles reach the exterior via the sperm duct and genital papilla.

That the seminal vesicles of the mudsucker do not respond to androgens suggests that these seminal vesicles are not derivatives of the Wolffian duct system. We cannot, however, use such a failure to respond to androgens as definitive evidence. For instance, in the musk shrew (*Suncus murinus*), the uterus does not respond to estrogen, although it is clearly a Mullerian derivative.

The sperm ducts of the Salmonidae and Asguillidae have secondarily disappeared. The copulatory organ of teleosts varies from being absent in many groups, to the bizarre priapus of the Phallostethidea (phallus = penis, stethus = chest) and Neosthethidae, in which vetiges of the pelvic fin, the pelvic girdle, parts of the pectoral girdle, and the first pair of ribs from the copulatory organ.

The copulatory organ may thus be variously formed. It may be formed by the genital papilla, which may be elongated e.g., in *Zenarchopterus*, some Cottoidei, and in some species of the Blennoidei. Some brotulid fish, e.g., of the genus *Typhlias, Stygicola, Dinematichthys, Brosmophycis,* and the species *Lucifuga subterranea* and *Dipulus caecus,* have a penislike structure. The morphology of this structure was studied by Turner (1946), who concluded that it was of an elaborately modified genital sinus.

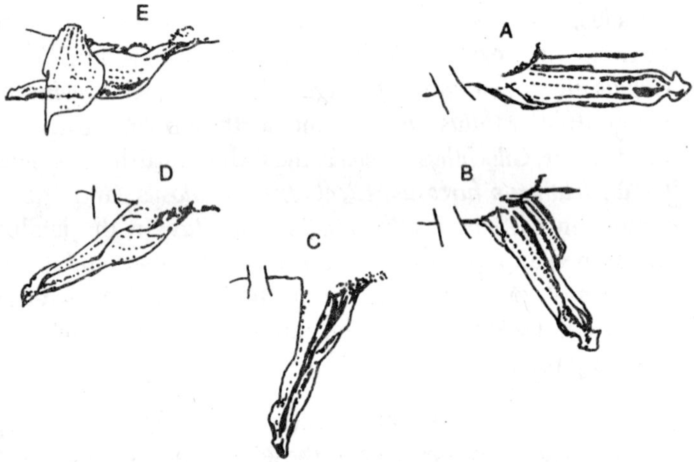

Fig. 1.8. Changes in the external contours of the gonopodium of *Xiphophorus helleri* Hecket during erection, showing the manner in which a groove is formed along one side of the fin. In the resting phase, *A,* the much smaller rays 6, 7, 8 and 9 are depressed on the base of rays (Ray 5 are not shown in figure). As the gonopodium begins its swing downward and out to one side (toward the reader) B and C rays 4 and 5 fall sideways and the tissue, which relates the smaller posterior rays, is distended. In *C,* the beginning of a groove along the side of the gonopodium facing the observer is revealed by the deeply shaded interior areas. In *D,* one edge of ray 5 is approaching the opposite edge of ray 3 preparatory to forming a closed channel, and the posterior rays are further separated in this region. Note position of spines on ray 3 in relation to the tip of the fin. Note also now both terminal holdfast elements are now facing in same direction. In *E,* the gonopodium is fully erected and the opening at the base of the gonopodial groove is facing upward and slightly away from the reader. In their completely distended condition in

E, the posterior rays together serve as a cup to receive discharged spermatophores. The position of the pelvic fin that is swung forward with the gonopodium is indicated in *E*. The base of this fin is indicated in outline in A to D.

The anal fin may be modified to form a copulatory organ called a gonopodium. In the clinid fish, *Starksia*, the copulatory organ is formed by the anal fin, which is attached to the genital papilla. The copulatory organ of the Jenynsiidae and Anablepidae is peculiar in that in an individual, it can move in one direction only, left in some individuals, right in others. (The females correspondingly have either a left or a right genital opening). Turner (1948) found that the copulatory organ of *Jenynsia lineata* is formed by elongation of anal fin rays 3, 4, 6, 7 and 8 (1, 2 and 5 are resorbed) and a thickening with bilateral asymmetry of rays 6 and 7, so that the fine becomes converted into a tube into which the sperm duct opens. The anal fin of the Embiotocidae is enlarged and fleshy, so that it looks tubercular.

In the cottid fish, *Oligocottus snyderi*, the first two rays of the anal fin are separate, and the first ray, which is much larger than the others, is prehensile and can be bent around the female, so that it can act as a clasper as well as in intromittent organ. In *Tomeurus gracilis*, the gonopodium, which extends beyond the tip of the head, is separated from the anal fin and is displaced under the pectoral fin. It is, however, not an intromittent organ, for the spermatophores are deposited near but not in the female's genitial opening.

In the scorpaenid fish, *Sebastiscus marmorathus*, the mesonephric duct and urethra jut out behind the anus and forms the copulatory organ. In the Phallostethidae and Neosthetidae, the pectoral girdle, first pair of ribs, vestiges of the pelvic fins and of the pelvic girdle together help from the copulatory organ.

It is not clear what structures are involved in the formation of the "copulatory organ" of *Trachycorystes striatulus*. Such an

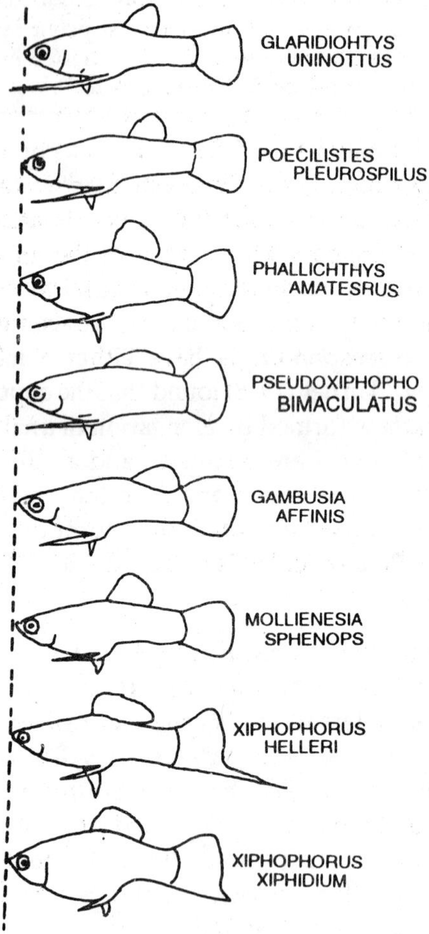

Fig. 1.9. Comparison of the males of eight representative species of poeciliid fishes with their gonopodia swung forward into the erected position. The fish are drawn to the same size (standard length) to illustrate differences in the lengths and relative positions of their gonopodia. The left ventral fin is cut off at the base.

intromittent organ occurs also in the genera *Asterophysis, Pseudoauchenipterus, Auchenipterichthys,* and *Tatia,* and probably in *Ceratocheilus.* Mohsen (1961) described a pear-shaped pseudopenis for *Skiffia lermae* (Goodeidae), into which the ducti deferentes and the urinary duct drain. It is not clear what the anatomical origin of this pseudopenis is and whether it is an intromittent organ.

In certain species of the genus *Corynopoma*, spoon-shaped appendages are found on the left side of the body. These organs are not intromittent organs, although they may function in establishing sexual contact. In *C. landonia*, these appendages develop from scales, whereas in *C. aliata* and *C. riisei*, they develop from the edge of the gill cover.

AMPHIBIANS

The testes are paired, and the sperm reach the exterior via the ductus deferens, which opens in the cloaca. Near the testes lie the fat bodies, which are essential for normal spermatogenesis, since their removal leads to a decrease in testicular weight in *Rana esculens* and *Triturus viridescens.*

It has been seen that the connections between the kidney and testes are similar to those found in elasmobranchs. A more eritical examination, however, shows that in most Anura, the Wolffian duct serves as the urinary duct and the semen duct, except that in the genus *Alytes*, the two duct systems are completely separate. In the urodeles, the urinary and gonadal systems are not separate. in *Megalobatrachus, Triturus (Diemictylus)*, and *Hynobius*, the mesonephric tubules drain directly into the cloaca and not into the Wolffian duct.

The ductus deferens of some anurans have a bottle-shaped dialation (ampulla) at the caudal portion, which functions as a sperm storage organ. This has been found, e.g., in the toad (*Bufo bufy japonicus*), in the tree frog (*Hyla arborea japonica*), in *Rana nigramaculata*, in the bullfrog (*Rana catesbeiana*), in *Rhacophorus buergeri* and *Rh. japonicus.* In other species, e.g., *Rana japonica, R. ornati ventris, R. tagoi, R. tsushimensis,* and

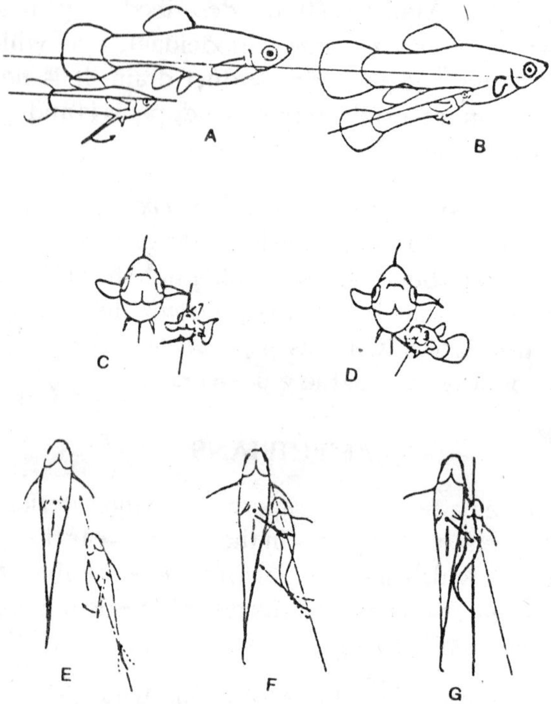

Fig. 1.10. Spawning *Fundulus heteroclitus:* note interlocked anal and pelvic fins.

Fig. 1.11. Orientation of male and female poeciliid fishes during copulation.

okinavana, there is knoblike outgrowth of the outer wall of the Wolffian duct, which functions as a true seminal vesicle for storing sperm. In still other frogs, e.g., *Rhacophorus arboreus*, the part of the Wolffian duct lateral to the kidney is coiled and serves as a sperm storage organ.

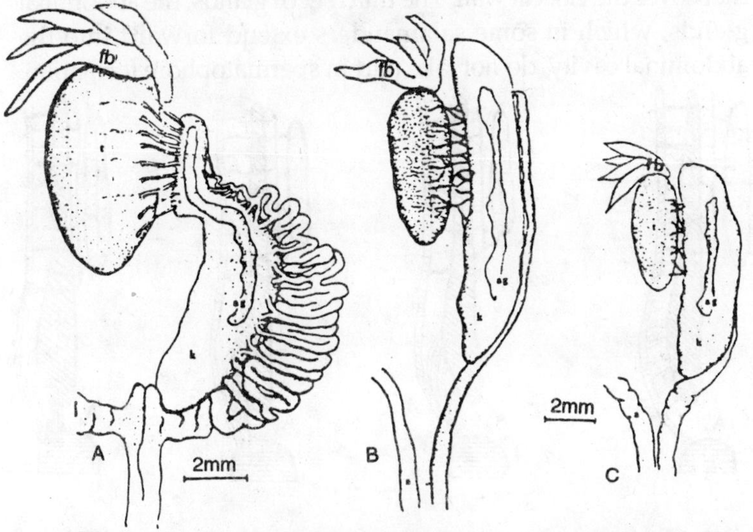

Fig. 1.12. Ventral view of urogenital organs. A, *Rhacophorus aboreus;* B. *Rnacophorus buergeri;* C. *Rhacophorus jeponicus.* a = ampulla, ag = adrenal gland, k = kidney, t = testis, fb = fat bodies.

The ductus deferens opens into the cloaca. Copulatory organs are rare in amphibia. In the primitive limbless Caecilia or Gymnophiona, fertilization is internal and the males have a modified cloaca, which they evert and introduce into the cloaca of the female. Among the Anura, fertilization is internal in the "tailed" frog, *Ascaphus truei*, which has a premanently everted cloaca that serves as a copulatory organ, and in the viviparous toad of East Africa (*Nectophrynoides vivipara*), in which no copulatory organ has been found.

In the urodeles, fertilization is internal, but the sperm are not introducted into the cloaca by the male; they are deposited instead on the ground in spermatophores, which are subsequently picked up by the female by means of the cloacal lips. The formaiton of the spermatophore of the salamander requires two of the three sets of cloacal glands, e.g., the pelvic gland located in the roof of the cloaca and the cloacal glands that cover the cloacal wall. The third set of glands, the abdominal glands, which in some salamanders extend forward into the abdominal cavity, do not take part in spermatophore formation.

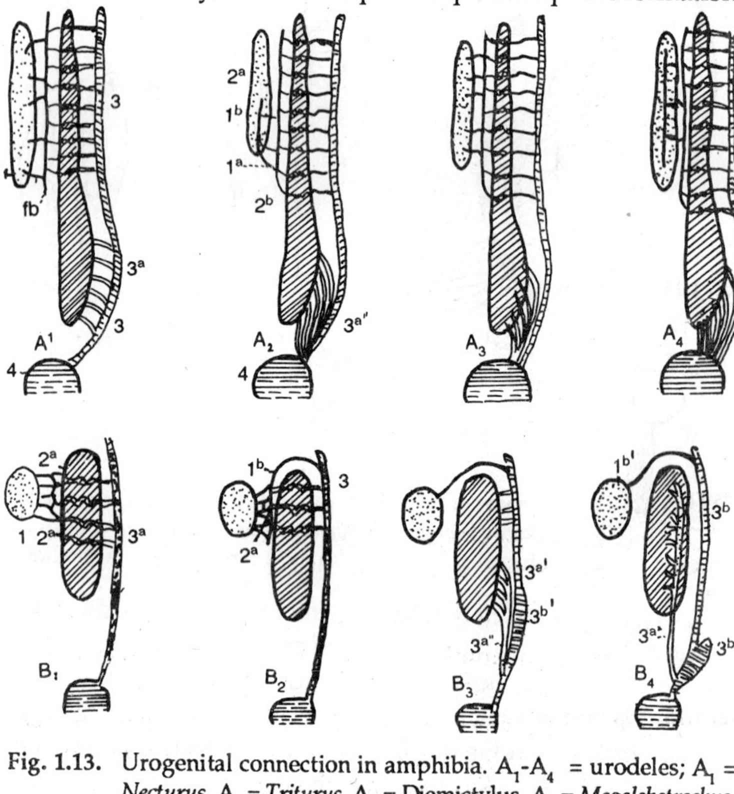

Fig. 1.13. Urogenital connection in amphibia. A_1-A_4 = urodeles; A_1 = *Necturus*, A_2 = *Triturus*, A_3 = Diemictylus, A_4 = *Megalobatrachus*. B_1-B_4 = Anura : B_1 = *Rana*, B_2 = Bombina, B3 = *Discoglossus*, B_4 = *Alytes*, 1 = testis, 1^b = rete testis, 1b = ductuli efferentes, 2^a = kidney border canal, $2^{b''}$ = kidney tubules, 3 = primary urinary duct, 3^a = collecting tubes separated, 3^a = collecting duct changed into secondary urinary duct, 3^b = ductus deferens, $3^{b''}$ = ampulla of ductus deferens, $3^{b'}$ = seminal vesicle, 4 = cloaca.

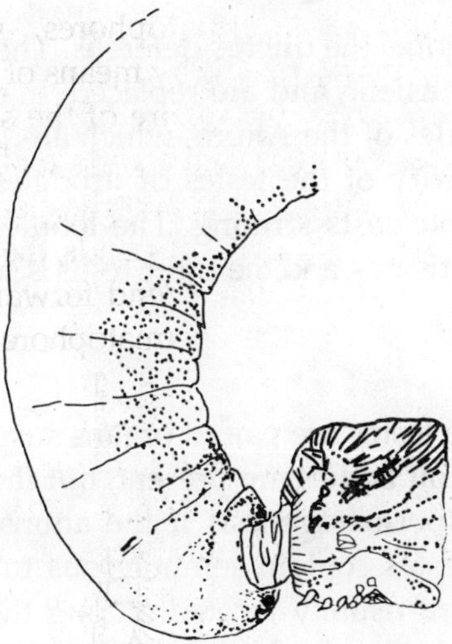

Fig. 1.14. Intoromittent organ of *Scolecomorphus* ulugurensis.

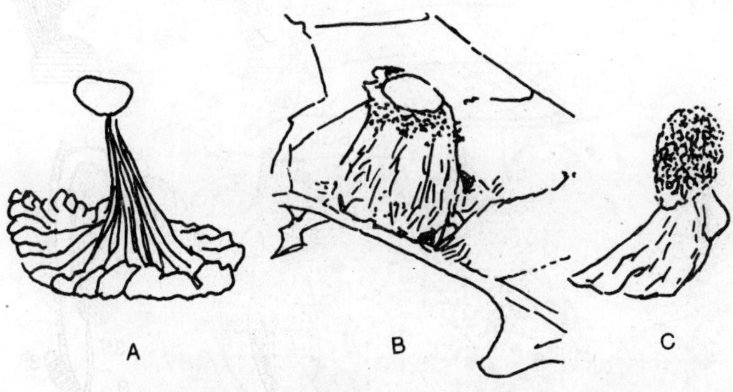

Fig. 1.15. Spermatophores of common salamanders. *A, Triturus viridescens* (after Smith), *B, Desmognathus fuscus* (after Noble and Webar), C. *Eurycea bislineata.*

The testes of the Anura are generally compact, and consist of convoluted seminiferous tubules. The interstitial tissue, which contains the Leydig cells, lies among the tubules. Spermatogenesis is uniform in all the tubules. The testes of the urodeles differ from those of the Anura; they consist of lobules connected by narrow bridges of tissue, each of which drains into a short duct that joins the ductus deferens. These lobules of the urodeles are transient, and are replaced, in contrast to the seminiferous tubules of the Anura, which are permanent structures. The similarity of the testes of urodeles with the acinar testes of some teleosts striking. The lobules are held together by connective tissues, and the entire testis is surrounded by a fibrous coal.

Reptiles

In the reptiles, the testes are paired organs suspended by mesorchia. In lizards, fat bodies are present, but these do not appear essential for spermatogenesis if the animal receives sufficient food. The testes contain seminiferous tubules and Leydig cells, which are usually located among the tubules. However, in all teiid lizards investigated (17 species of *Cnemidophorus* and one of *Ameiva*), the Leydig cells form a circumtesticular tunic.

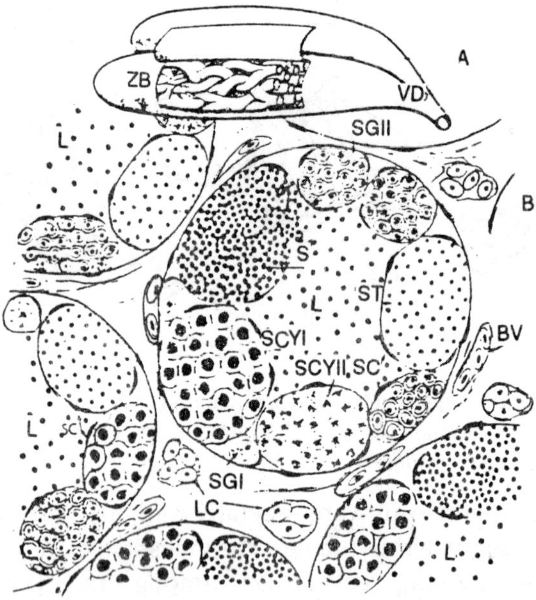

Fig. 1.16. Everted hemipenis of a rattlesnake (*Crotalus atrox*).

As in all amniotes, the Wolffian duct has lost its connection with the mesonephros and has become the ductus deferens. The genital region of the mesonephros functions as the epididymis. The arrangement of the ductuli efferentes varies among reptiles. In snakes, several seminiferous tubules drain into one ductule, and the ductuli are arranged along the length of the testes; in lizards (Sauria), they are reduced to a single marginal canal. In the Anguidae, they are intermediate; several of the ductuli efferentes anastomose and drain into the ductuli epididymides. In the Cholomidae, there are many efferent ductuli that anastomose and form a network, the *rete testis*, which differs from the mammalian rete testis in being located outside the testis..

The epididymis is large in lizards (Sauria), worm lizards (Anguidae), and turtles (Chelonia); in snakes the size of the epididymis varies with the species. In reptiles, the kidney has a sexual segment, the size of which varies, there being a greater variation among lizards than among snakes. In snakes and *Varanus*, this sexual segment corresponds to the medial region of the distal convoluted tubules; in other Squamata, the collecting tubules form the main part of the sexual segment. In lizards and worm lizards, the sexual segment includes part or all of the ureter and, in lizards, increases under the influence of androgens.

The functions of this sexual segment and its secretions have not been determined, but the following functions have been proposed, for the secretions : (1) They may help separate semen from urine by blocking renal tubules and the ureter during copulation. (2) They may assist in complete emptying of the ampullae and seminal grooves. (3) They may form the copulatory plug, which in garden snakes (*Thamnophis sirtalis*) prevents other conspecifics from copulating with the mated female. (4) They may contain nutrients for the sperm. (5) They may act as pheromones; evidence for this is the fact that male garter snakes do not court recently mated females.

In males, under the influence of androgens, a part of each

kidney tubule thickens. The cells of these tubules are very active during the breeding season, and their secretion may be important in either the transport or the nourishment of the sperm.

Male lizards and snakes have a paired hemipenis, the two arms of which are caudal extensions of the cloaca, each of which can be everted independently of the other into the cloaca of the female. During this copulation, the sperm run along an external groove, and the spermatic, sulcus of the hemipenis into the female's cloaca. Spines of ridges help to maintain intromission. The penis of chelonians and crocodiles is a modification of the floor of the cloaca. It consists of a median groove with ridges of erectile tissue along each side. The caudal end of the penis is a raised gland, which acts as an entering wedge prior to the filling of the erectile blood sinuses; consequently intromission in turtles also involves an eversion of the cloaca. After testosterone propionate treatment of young crocodiles (*Crocodylus palustris*), there is no change in the sexual segment of the kidney tubules, a slight response of the Wolffian duct, but an enormous development of the penis.

BIRDS

The testes of birds, which can be very large (e.g. 360 mg in a 30 g grass parakeet, *Melopsittacus undulatus*) are suspended by mesorchia. There are no tubuli recti, but the rete testis connects the seminiferous tubules to the ductuli efferentes. The epididymis is small compared with the epididymis of mammals. It is homologous with the caput epididymidis of mammals, and different regions of the ductus deferens are homologous with the corpus and cauda epididymidis of mammals.

The ductus deferens, under the influence of either endogenous or exogenous androgens, becomes coiled. It has no glandular areas that might correspond to the accessory glands of the mammalian reproductive system. In some species there is an enlargement of the ductus deferens, called the

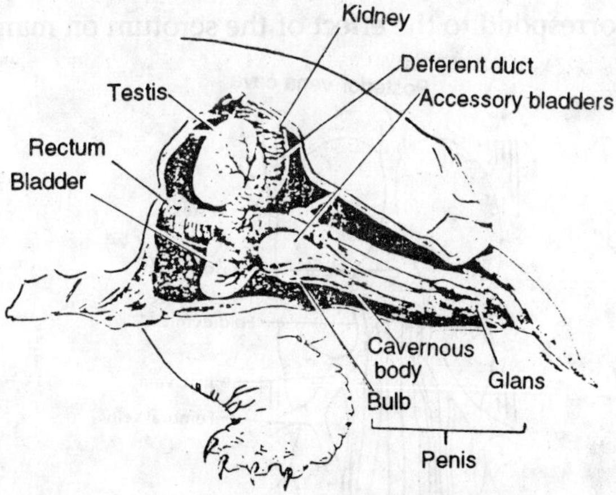

Fig. 1.17. General system of a male turtle (*Pseudemys* sp). The cloaca has been opened dorsally to expose the entrance of the rectum and the dorsal groove of the penis.

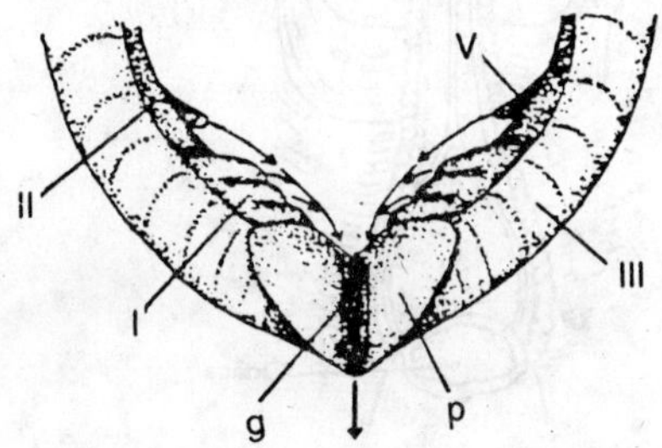

Fig. 1.18. Copulatory organ of a rooster. II = second fold of cloaca, I = swollen lymph fold, g = longitudinal groove, *p* = erected phallus, III = third fold of cloaca, v = papillary process of vas deferens.

"seminal vesicle" or "seminal glomus," which is also highly coiled and in which sperm are stored. Wolfson (1954) demonstrated that in a number of passerine species (*Junco hyemalis, Zonotrichia albicollis, Z. leucophrys,* and *Melospiza melodia*), the temperature of the seminal vesicle was 0.8-4.70 C below the deep body temperature. He proposed that such low temperatures might be beneficial for the sperm and might correspond to the effect of the scrotum on mammalian testes.

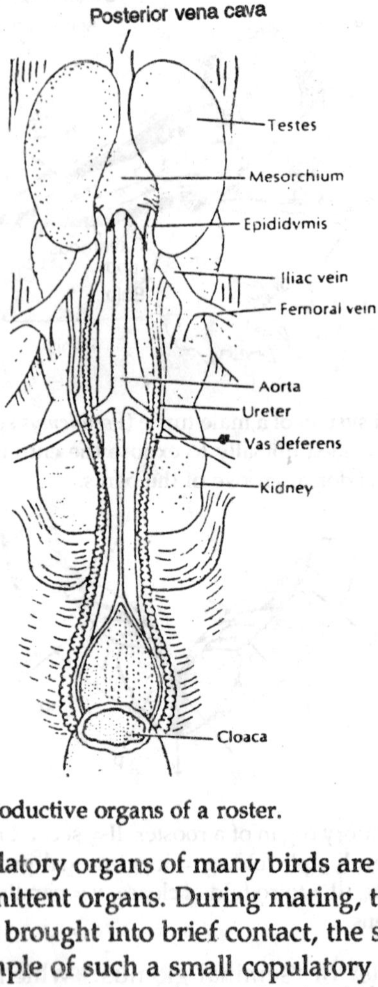

Fig. 1.19. Reproductive organs of a roster.

The copulatory organs of many birds are small, and they are not intromittent organs. During mating, the cloaca of the two sexes are brought into brief contact, the so-called cloacal kiss. An example of such a small copulatory organ is that of the rooster. Greese, swans, ducks, the Antidae and ostriches, cassowarries, emus and rheas have an intromittent organ, which is called a penis but which is actually a pseudopenis.

The male Japanese quail has foam glands, which produce a foam that presumable aids in the deposition or transport of sperm. These glands, which respond to androgen treatment

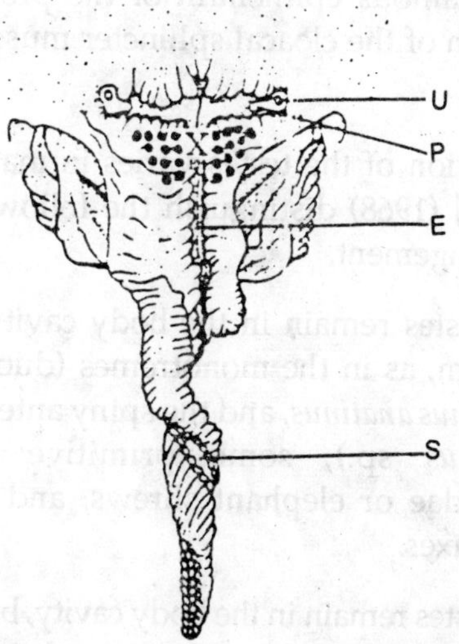

Fig. 1.20. **Copulatory organ of a drake. *Anas* sp. U = Opening of ureter, P = papilla of ductus deferens, E = ejaculatory groove, S = seminal groove, X = collecting point of fluid Ejaculatory groove region is marked with large dots.**

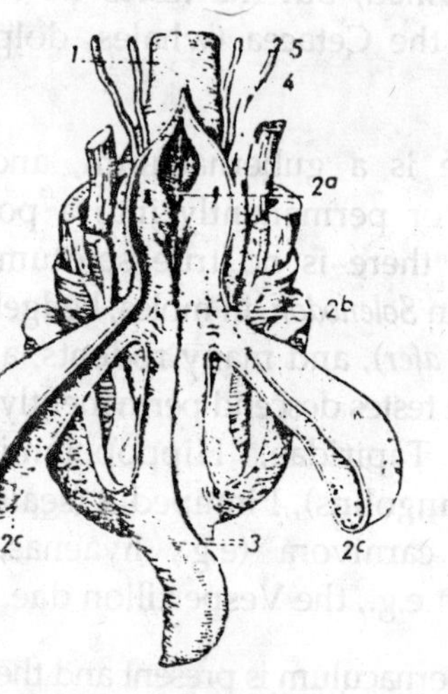

Fig. 1.21. **Copulatory organ of *Rhea americana*, 1 = colon, 2^a = utodeum, 2^b = protodeum, 2^c = bursa of Fabricius, 3 = penis, 4 = ductus deferens, 5 = ureter.**

by an increase in secretory activity, are located between the stratified squamous epithelium of the proctodeum and the fibrous sheath of the cloacal sphincter muscle.

Mammals

The location of the testes varies in mammals. Giersberg and Rietsehel (1968) distinguish the following five different types of arrangement.

1. The testes remain in the body cavity, and there is no gubernaculum, as in the monotremes (duck-billed platypus, *Ornithorhynchus anatinus,* and the spiny anteaters. *Tachyglossus* sp., *Zaglossus* sp.), some primitive insectivora (the Macroscelididae or elephant shrews, and *Tenrec ecaudatus*), and the Hyraxes.

2. The testes remain in the body cavity, but a gubernaculum that later regresses is formed, as in the Sirenia (seacows), elephants, and the Edentata (sloths and anteaters).

3. The testes move cauded, a gubernaculum and inguinal canal are formed, but the testes do not enter the inguinal canal, as in the Cetacea (whales, dolphins, porpoises) and armadillos.

4. There is a gubernaculum, and the testes descend temporarily or permanently into a pouch of the cremaster muscle, but there is no true scrotum. The testes descend temporarily in *Solenodon,* in moles, hedgehogs, shrews, aardvark (*Orycteropus afer*), and many rodents, and in the lagomorphs (rabbits). The testes descend permanently in the Phasocolomidae (wombats), Tapiridae, Hippopotamidae, Rhinocerotidae, Manidae (pangolins), Pinnipedia (seals seal lions, walruses), some land carnivora (e.g., hyaenas, *Crocuta*) and some Chiroptera, * e.g., the Vespertilion dae.

5. A gubernaculum is present and the testes descend through the inguinal canal either temporarily (usually during the breeding season) or permanently into the scrotum.

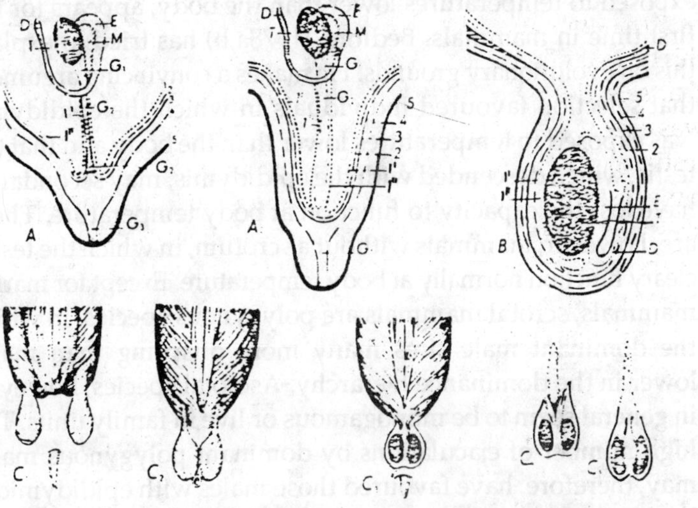

Fig. 1.22. Descent of the testes. A_1 = with an inguinal cone, A_2 = without an inguinal cone, B = scheme of testes in scrotum after complete descent, sagital section, C_1-C_5 = extra abdominal position of the testes; C1 = original position cremaster pouch separate, penis cloacal; C_2 = Tupaja, scrotum with prepenial cremaster pouch; C_3 = Methatheria with prepenial scrotum; C_4 = Eutheria with prepenial scrotum; C_5 = Eutheria with postpenial scrotum and pendulous penis. 1 = peritoneum; 1' = parietal layer of 1; 1" = visceral layer of 1; 2 = transverse fascia and transverse muscle; 3 = musculus obliquus internis abdominis; 4 = M, abaquus externus abdominis; 5 = skin and hypodermis. G = gubernaculum of the testis; G_1 = ligament of the testis; G_2 = inguinal ligament; G_2 = inguinal cone; G_3 = ligament of scrotum; T = testis; E = epididymis; D = ductus deferens; M = mesorchium.

In most species in which the testes descend either into a cremaster muscle pouch or into a scrotum, the descent is permanent. In some species, the testes can return to the abdominal cavity or to the inguinal canal. Examples are : *Orycteropus* (aardvarks), some rodents (*Rattus, Sciurus, Tamias,* i.e., chipmunks) and Chiroptera, and some primates (*Loris and Perodicticus*).

The phenomenon of testes in a scrotum, in which they are

exposed to temperatures lower than the body, appears for the first time in mammals. Bedford (1978a,b) has tried to explain this on evolutionary grounds. He makes a convincing argument that selection favoured individuals in which the epididymis was exposed to temperatures lower than the body and that the testis, which descended with the epididymis, may secondarily have lost the capacity to function at body temperature. There are, however, mammals without a scrotum, in which the testes cleary function normally at body temperature. Except for marine mammals, scrotal mammals are polygynous species, in which the dominant male sires many more offspring than males lower in the dominance hierarchy. Ascrotal species, however, in general seem to be monogamous or live in family units. The high number of ejaculations by dominant polygynous males may, therefore, have favoured those males with epididymides in which sperm could be stored successfully. Such storage would probably be more successful at lower temperatures because metabolic rates would be lower; thus eventually, individuals with epididymides (and as a consequence testes) a temperatures lower than the body would have a selective advantage.

The apparent contradiction of polygynous ascrotal marine mammals may, in fact, be only an apparent contradiction. In the southern elephant seal (*Mirounga leonina*), the temperature of the testis is 6° C below body temperature, apparently as a result of blood flow from the rear flippers. In some seals, e.g. *Callorhinus* and *Zalophus*, the testes are located in a scrotal pouch.

Many mammalian species have one or more well-developed accessory reproductive glands. These glands are :

1. *Ampullary glands*. These are derived from the Wolffian duct system and have a simple cuboidal, or columnar epithelium; their secretions are high in ergothioneine.

2. *Seminal vesicles*. More properly called vesicular glands,

these are also derived from the Wolffian duct system. They have a simple or pseudostratified epithelium and secrete fructose, critic acid, proteins and ascorbic acid.

3. *Paraurethral glands.* Also called glands of Littre, these are probably synonymous with the glands named paraprostate glands by McKeever (1970) and Holtz (1972). These glands originate from the epithelium of the urethra and are thus part of the urogenital sinus. They are usually small structures, the ducts of which drain into the urethra. According to Holtz (1972), there are two different types of paraprostate glands in the rabbit : (a) glands consisting of small clusters of fluid-filled vesicles, which, under microscopic examination, resemble the bulbourethral glands, and (b) small, compact, whitish glands, which are histologically identical with the prostate.

4. *Proprostate glands.* These are found in the rabbit, and appear to be part of the prostate; however, they are separated from the prostate by a connective tissue system. These glands originate, as does the prostate, from endoderm of the urogenital sinus. Holtz (1972) emphasizes that some names given to these glands are not appropriate, e.g., coagulating gland (because the glands are not the source of the gel mass in the ejaculate), and prostate and glandula vesicularis (because they are not Wolffian duct derivatives). The proprostate is an acinar gland, rich in smooth muslce and with a columnar epithelium. It secretes a white granular substance, which gives the gland a while colour and has two large excretory ducts that lead to small urethral diverticula.

5. *Prostate gland.* This originates as part of the urogenital sinus and has simple or pseudostratified columnar epithelium. The secretions contain enzymes, such as diastase, glucoronidase, fibrinolysin, acid phosphatase as well as citic acid. The gland has ducts that drain into the urethra.

6. *Urethral gland.* This gland is found in some Chiroptera between the prostate and the openings of the ducts at the

bulbourethral glands. It lies around the urethra, is covered by the muscles of the urethra, and its secretions drain via short ducts into the urethra.

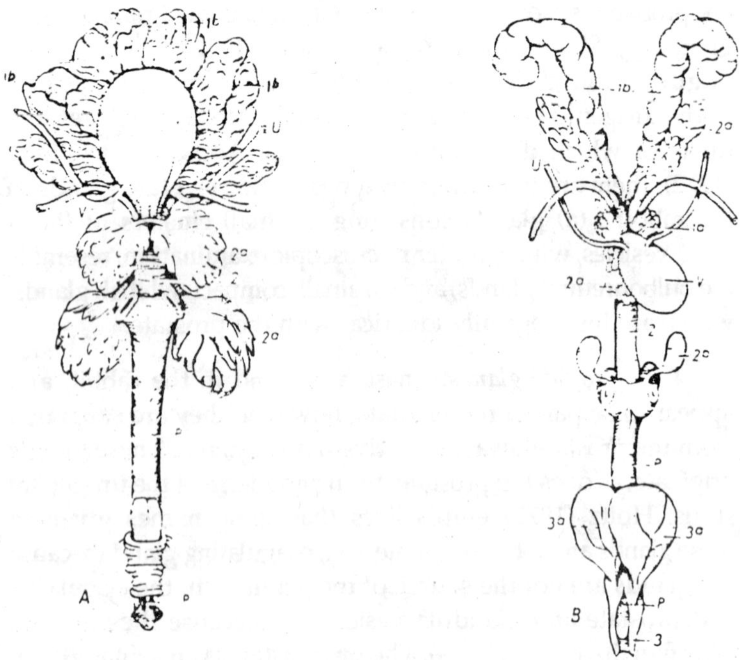

Fig. 1.23. Accessory sexual glands of the male. A. Hedgehog (*Ennaceus europaeus*); B, mouse (*Mus musculus*) 1 = ductus deferens; 1^a = empullary gland; 1^b = seminal vesicle; 1^b-$1^{b''}$ = parts of the seminal vesicles; 2 = urethra; 2^a = prostate gland; $2^{a'}$-$2^{a''}$ = parts of the prostate gland; 2^b = bulbourethral gland (note that the urethral glands in the wall of the urethra cannot be seen); 3 = preputium. 3^a = preputial glands, P = penis, P′ = glans penis; U = ureter (urinary duct); V = urinary bladder.

7. *Bulbourethral glands* (Cowper's glands). These glands are also part of the urogenital sinus. They have a simple columnar epithelium which secretes a viscous fluid containing sialoprotein, and have ducts that drain into the urethra.

8. *Bulbar glands*. Also called the *interior bulbourethral glands*, these glands seem to be present in some of the Sciuridae. They

are glandular developments of the bulbourethral glands inside the corpus spongiosum. The secretions drain via a glandular duct (ductus penis) at the ventral side of the urethra.

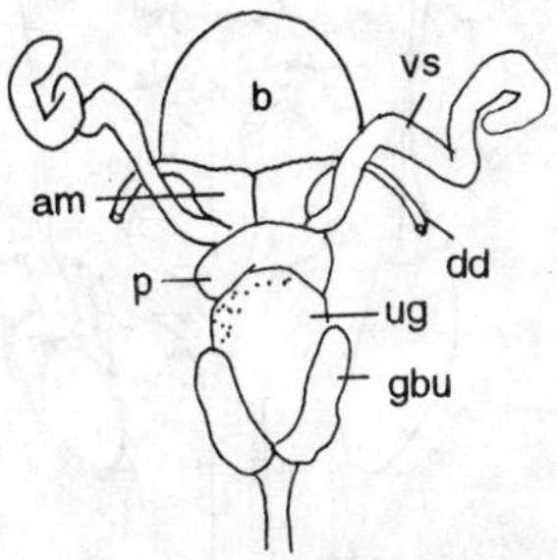

Fig. 1.24. Reproductive organs of one of the Chiroptera, the fruit bat (*pteropus giganteus*) dorsal view, am = ampuliary glands, b = utrinary bladder, dd = ductus deferens gbu = bulbourethral gland p = prostate, ug = urethral gland vs = seminal vesicle.

9. *Prostatic utricle.* Also called *uterus masculinus,* this is usually vestigial or even absent but can, according to Prasad (1974), be well developed. It is derived from the Mullerian duct, and the ducts drain into the urethra.

Since this is a derivative of the Mullerian duct system, it might be expected to be sensitive to estrogens. As estrogens have been suspected of inducing cancer of derivatives of the Mullerian duct system in women, it may be prudent to consider seriously the wisdom of the use of estrogens in treating cancer of the prostate.

10. *Preputial glands.* These are modified sebaceous glands with a stratified, squamous epithelium and with ducts draining at the surface of the glands. They are enormously developed in the beaver.

11. *Inguinal glands.* These are sebaceous glands and sweat glands found on the surface of the inguinal canal and in skin folds of the genital region. Holtz (1972) suggests the name *perineal glands* for these in the rabbit.

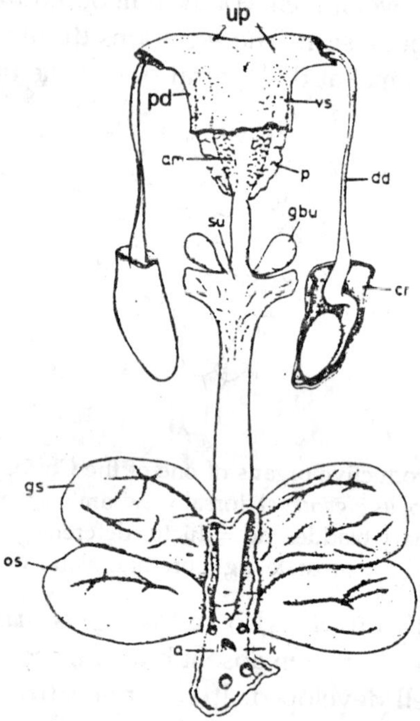

Fig. 1.25. Reproductive organs of the beaver (*Castor fiber*). The rectum is opened in order to show the penis and ducts of the glands. a = anus, am = ampulla of ductus deferens, cr = cremaster pouch, dd = ductus deferens, gbu = bulbourethral gland, gs = preputial glands, k = cloaca, os = oil sac, p = prostate, pd = plica Douglasi, su = urogenital sinus, vs = seminal vesicle, up = prostatic utricle.

The presence of these various glands in some mammalian species. The effect of the presence of these accessory sex glands on fertility has been studied in the boar (*Sus scrofa*), which produces a large amount of fluid in the ejaculate (about 200 cc) 26 per cent of which is contributed by the seminal vesicles, 56 per cent by the prostate, and 19 per cent by the bulbourethral glands. These large quantitites of secretions make this species a very suitable animal for studying these glands. Removal of the seminal vesicles, or the bulbourethral glands, or both, had no effect on fertility. The effect of removal of the prostate was not reported for the boar, but apparently removal of the prostate in humans does not affect fertility.

In rats, removal of the seminal vesicles and the prostate and coagulating gland reduces fertility severely, but removal of the following glands or combination of glands has no effect on fertility; bulbourethral glands, seminal vesicles, prostate and coagulating gland, bubbourethral glands and seminal vesicles, bulbourethral gland and coagulating gland, seminal vesicles and prostate.

Prototheria. The penis of the Protheria, like that of the reptiles and birds, is located on the floor of the cloaca, and lies in a preputial fold of the ventral wall of the cloaca. The penis conducts the semen but not the urine, to the outside, because the semen-urine duct gives off a branch to the cloaca before it enters the penis. The ductus deferens drains into the urogenital sinus, dorsal to the ureters, not ventral, as in the Metatheria and Eutheria. The penis contains fibrovascular tissue which is homologous with the corpus cavernosum penis of the Metatheria and Eutheria. The glans of the ducst-billed platypus is bifurcated.

Metatheria. The penis is located in a cloacal fold, which is surrounded, together with the anus, by a cloacal sphincter muscle, so that it it is not visible generally except during erection. However, *Phascogale* (marsupial pocket mice) and *Dasyurus* (Austrilian "native cats") the penis is free and not in a cloacal fold. The penis of the oppossum *Didelphis virginiana* is bifurcated as is the penis of *Dasyurus* and *Phascolomis*, although in these last two genera, the bifurcaiton is not as marked. The penis of the Macropodidae and Tarsipes is definitely not forked.

In *Peamelidae* (bandicoots), the urogenital sinus conducts only the semen to the outside; the urine is voided via the cloaca. In *Didelphis* and *Caenolestes* (bat opossum), the distal opening of the urinary duct lies at the base of glans and the ductus deferens ends in the tip of the penis. In *Myrmecobius* (numbats). *Thylacinus* (pouched wolf) and *Dasyurus*, the urethra divides into two branches, one branch going to one tip of the bifurcated glans, the other to the other tip.

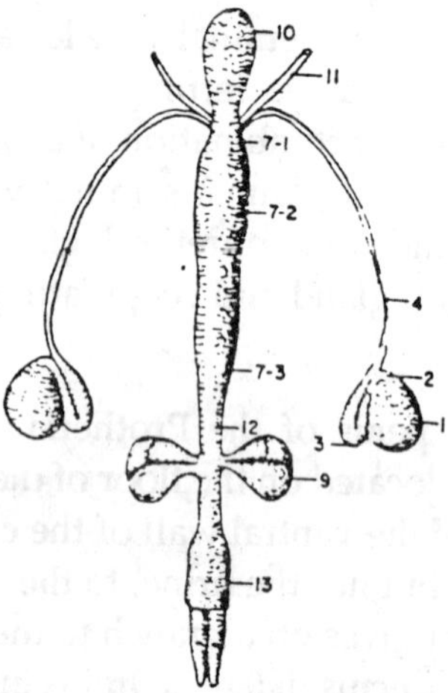

Fig. 1.26. Reproductive organs of the opposum. Note the bifurcated penis. 1 = testis, 2 = caput epididymis, 3 = cauda epididymis, 4 = ductus deferens, 5 = ampullary gland, 6 = vesicular gland, 7 = prostate gland, 8 = anterior lobe of prostate, 9 = bulbourethral gland, 10 = bladder, 11 = ureter, 12 = urethra, 13 = penis.

Eutheria. The perineum divides the cloaca into the dorsocaudal rectum and the ventrocranial urogenital sinus, so that the penis lies on the ventral side of the anus. The distance between anus and penis can be small, as in beavers (*Castor* sp.) and hares (*Lepus europeaus*). Prasad (1974) states that in beavers, the urogenital openings and the anus open into a cloaca, a situation also found in the Prototheria, Metaheria, Edentata, and in the Insectivora, and *Neomys* sp. In other species, the penis lies, with its preputial fold, near the umbilicus. The double skil fold, which surrounds the penis, partly or completely forms the prepuce, which in the horse consists of two circular preputial folds which are telescoped so that the penis can be pulled inside the folds. The pars intrapreputialis penis, corresponds to the glans penis in humans.

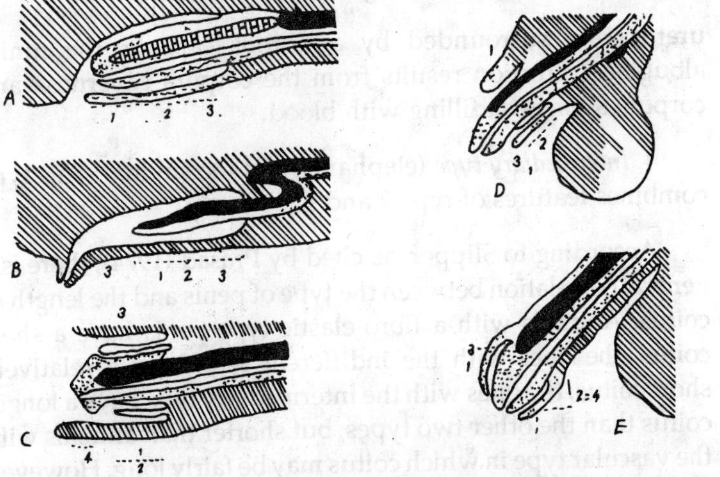

Fig. 1.27. **Medial view of the penis of some mammals.**

A, seal (*Phoca*); B, bull (*Bos taurus*); C. horse (*Equus caballus*); D, long-ta ed monkey (*Cercopithecus* sp.); E, human (*homo sapiens*), Black = corpus cavemosum penis stippled = corpus apongiosum penis and corpus cavernosum glandis; heavy black lines = cs penis in A. 1 = preputium, 2 = pars intrapreputias, penis, 3 = preputial cavity, 4 = glands penis.

Several types of penes have been distinguished :

1. *Indifferent type* (edentates and rodents). The penis is short, and the corpus spongiosum penis is not covered by a tunica. The corpus cavernosum can be either fibrous tissue or trabecular erectile tissue.

2. *Fibro-elastic type* (ruminants and whales). Usually the penis is long and fibrous and has a sigmoid flexure in the resting stage. There is little erectile tissue in the corpus cavernosum, the erection consisting of the straightening out of the sigmoid flexure. The corpus spongiosum is surrounded by a well-developed tunica.

3. *Vascular type* (Perissodactyla, Carnivora, Primates, and some Insectivora). The penis, if not in erection, is long and flexible. The corpus cavemosum penis and corpus spongiosum

urethra are surrounded by a fibrous capsule, the tunica albuginea. Erection results from the corpora cavernosa and corpora spongiosa filling with blood.

4. *Intermediary type* (elephants, and the sirenia). This type combines features of type 2 and 3.

According to Slipper, as cited by Prasad (1974), there is a general correlation between the type of penis and the length of coitus. Animals with a fibro-elastic type penis have a short coitus, the ones with the indifferent type have a relatively short coitus, the ones with the intermediary type have a longer coitus than the other two types, but shorter than animals with the vascular type in which coitus may be fairly long. However, coitus in swine which have an intermediate type penis, may last as long as 30 minutes!

In most species of Chiroptera, Rodentia, Carnivora, Insectivora, and many primates (but not humans) there is a penilea bone or baculum present. The male reproductive systems of some mammalian species are illustrated, in situ.

FEMALE REPRODUCTIVE SYSTEM

In most vertebrates, the female has two ovaries, but there are exceptions to this, and among birds, most species have only one functional ovary, although some species have a high incidence of two functional ovaries.

The female genital duct system of vertebrates parallels that of the males to the extent that in cyclostomes there is no genital duct system, in the Salmonidae the funnel is a secondary formation, and in the teleosts, the oviduct is not a Mullerian duct system but is formed from part of the ovary and mesentery.

Cyclostomes

The lampreys have one very large ovary, which originates from the fusion of the two gonad primordia. The ovary contains many follicles, (24,000-236,000), each of which consists of a fibrous theca externa, a glandular theca interna, and a granulosa layer. The follicles are all in the same stage of development at any particular time, and as Dodd (1972) has pointed out, not oogonial nests remain at the time of metamorphosis. The unique feature may be related to the fact that lampreys spawn and die shortly thereafter. The follicles form neither corpora atretica nor corpora lutea, although atresia does occur in *Petromyzon marinus*. Atresia consists of reabsorption of the yolk and follicle cells, so that only stroma cells are left.

TABLE 1.1

Vertebrates with one Functional Ovary

Species	*Explanation for one-ovary condition*
Lampreys	Fusion of two gonad
Hagfishes	One gonad fails to develop
Perches, *Perca*	Fusion of two gonads
Pike perch, *Lucia-Stizostedion sp.*	Fusion of two gonads
Stone loach, *Noemacheilus sp.*	Fusion of two gonads
European bittering, *Rhodeus amarus*	Fusion of two gonads
Japanese ricefish, *Oryzias latipes*	One gonad fails to develop
Guppy, *Pocilia reticulata*	One gonad fails to develop
Sharks	
Scyliorhinus	Left ovary becomes atrophic
Pristiophorus	Left ovary becomes atrophic
Carcharhinus	Left ovary becomes atrophic
Galeus	Left ovary becomes atrophic
Mustelus	Left ovary becomes atrophic
Sphyrna	Left ovary becomes atrophic
Wiviparous ray, *Urolophus*	Left ovary functional
Ovoviviparous ray, *Dasyatis*	Right ovary absent
Blind worm snakes, *Typhlopidae*	Left ovary and oviduct absent; reason unknown
Birds	Left ovary functional in most species; right ovary regresses embryos

Table cont...

Duckbill platypus, *Omithorhynchus anatinus*	Left ovary functional
Bats	
Miniopterus nata ensis	Left ovary functional
Miniopterus schrelbersis	Right ovary functional
Rninolophus	Right ovary functional
Tadarida cyancephala	Right ovary functional
Molossus ater	Right ovary functional
Mountain viscacha Lagidium ceruanumi	Right ovary functional, but after its removal left ovary becomes functional
Water buck, *Kobus defassa*	Left ovary functional

The single large ovary of the Myxiniidae (hagfishes) reuslts from the failure of development of one of the two gonad primordia. There are few oocytes (1-21), and these are arranged in a row along the ovary. Each oocyte is enclosed in a tough shell which is secreted by the follicle. One can distinguish four layers in the follicular wall, an inner, simple, follicular epithelium : two layers of connective tissue; and an outer squamous layer. In the hagfishes there are corpora atretica and "corpora lutea," which originate either from follicular cells or from fluid-filled cysts. There are no gonoducts in either lampreys or hagfishes, and the gametes are shed in the body cavity and leave via an orifice secondarily connecting the coelom with the cloaca, in *Lampetra,* and through a single median orifice between anus and urinary opening in *Myxine.*

TABLE 1.2

Avian Species in Which 50 Percent or More of the Individuals Examined have two Functional Ovaries

Species	*Number + examined*	*Incidence of two ovaries*
Apteryx australis (kiwi)	2	2/2
Podiceps cristatus (great crested grebe)	4	2/4
Fulmarus glacialis (furmar)	2	2/2

Cathartes aura (turkey vulture)	5	2/5
Circus aeruginosus (marsh harrier)	6	4/6
C.approximans (swamp harrier)	17	16/17
C.pygarus	5	4/5
C.cyaneus cyaneus	13	13/13
C.cyaneus hudsonius (marsh hawk)	31	22/31
C.macrourus	7	5/7
Accipiter gentilis atricapillus	14	13/14
A.cooperi (Cooper's hawk)	3	3/3
A. fasciatus	5	5/5
A. nisus	30	20/30
Ieracidea berigora	2	2/2
buteo buteo (common buzzard)	11	6/11
E.jamacensis (red-tailed hawk)	15	12/15
B. platypterus	5	4/5
Falco columbarius (pigeon hawk)	4	2/4
F. vespertinus	9	6/9
F. tinnuculus (European kestrel)	20	12/20
Eclectus roratus	4	2/4
Bubo bubu (eagle owl)	7	7/7
Strix aluco (tawney owl)	4	4/4

*Data are not included if findings on one bird only were listed and if the number of birds examined was not given.

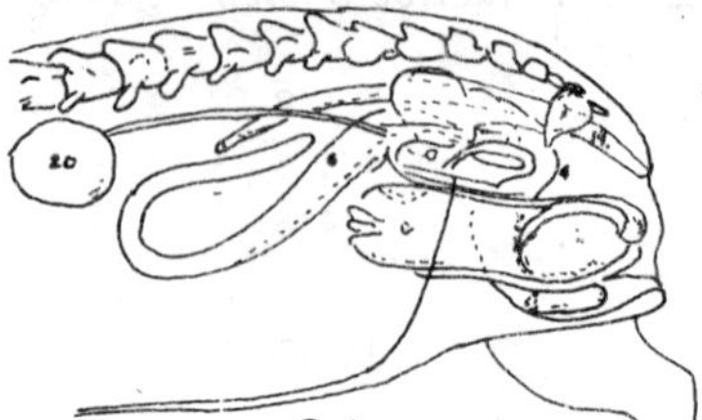
Guinea Pig

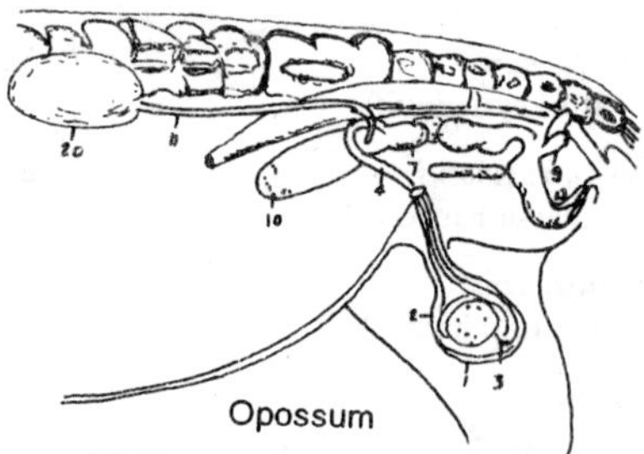
Opossum

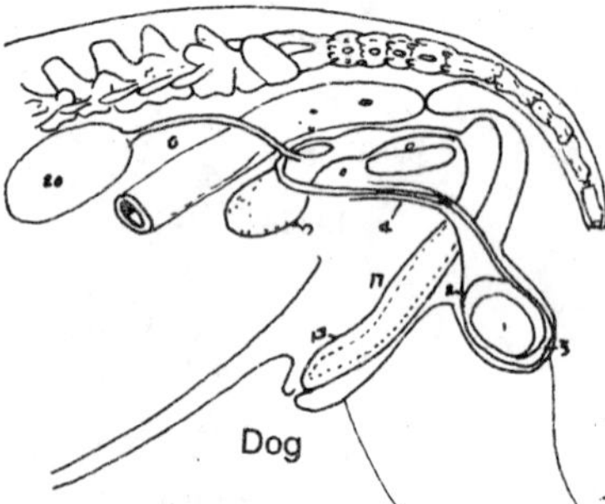
Dog

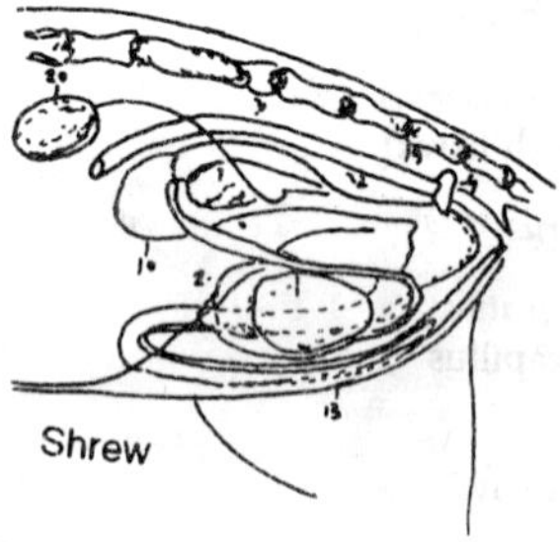
Shrew

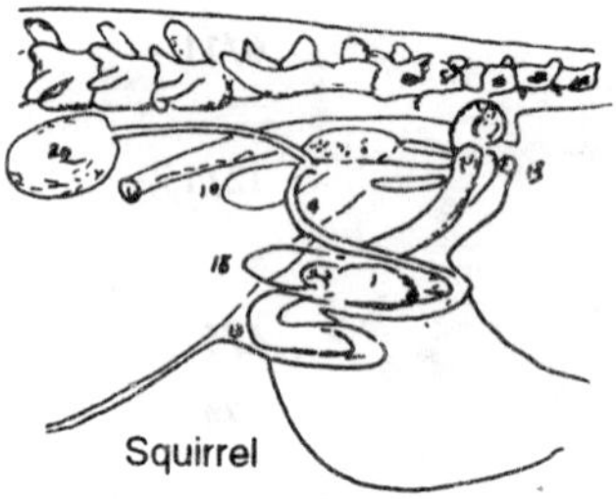
Squirrel

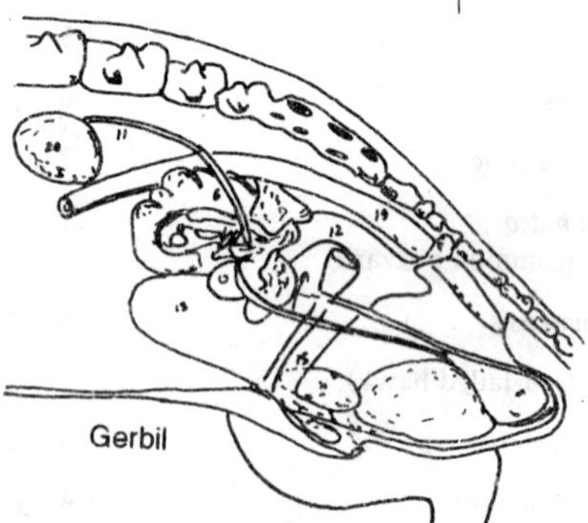
Gerbil

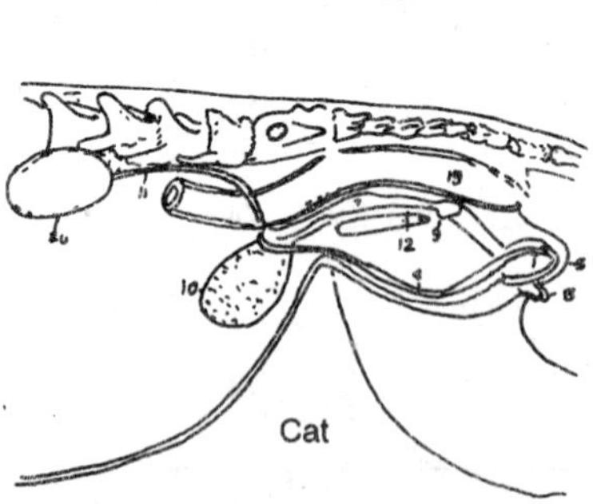
Cat

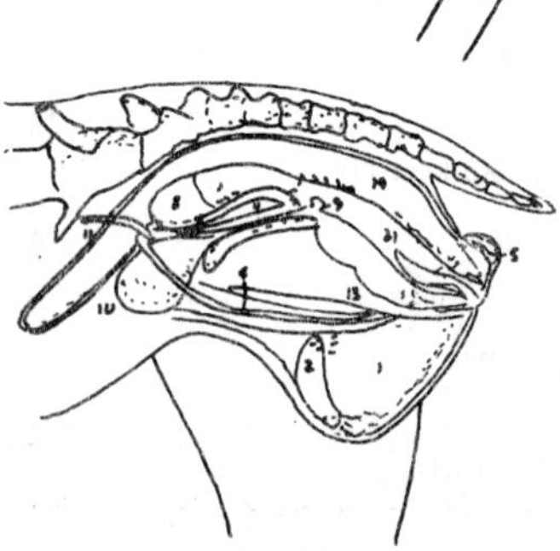
Rabbit (Cottoniai)

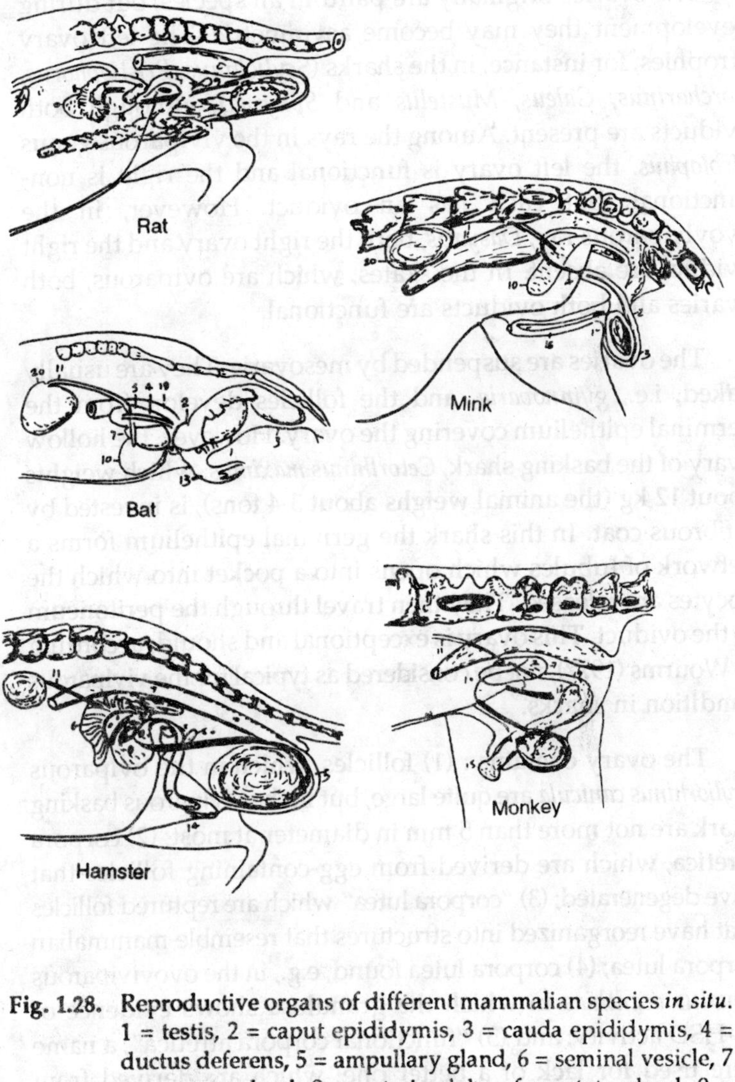

Fig. 1.28. **Reproductive organs of different mammalian species *in situ*. 1 = testis, 2 = caput epididymis, 3 = cauda epididymis, 4 = ductus deferens, 5 = ampullary gland, 6 = seminal vesicle, 7 = prostate gland, 8 = anterior lobe of prostate gland, 9 = bulbourethral gland, 10 = bladder, 11 = ureter, 12 = urethra, 13 = penis, 14 = preputial gland, 15 = bulbar gland, 16 = paraprostate gland, 17 = baculum, 18 = fat body, 19 = rectum, 20 = kidney.**

Elasmobranchs

The ovaries originally are paird in all species, but during development they may become asymmetric. The left ovary atrophies, for instance, in the sharks (*Scyliohinus, Pristiophorus, Corcharinus, Galeus, Mustelus* and *Sphyrna*), although both oviducts are present. Among the rays in the viviparous genus *Urolophus*. the left ovary is functional and the right is non-functional, but each has an oviduct. However, in the ovoviviparous ray, *Dasyatis*, both the right ovary and the right oviduct are absent. In the skates, which are oviparous, both ovaries and both oviducts are functional.

The ovaries are suspended by mesovaria. They are usually naked, i.e., *gymnovaria*, and the follicles develop from the germinal epithelium covering the ovary. However, the hollow ovary of the basking shark, *Cetorlhinus maximus*, which weights about 12 kg (the animal weighs about 3-4 tons), is invested by a fibrous coat. In this shark the germinal epithelium forms a network of tubules which opens into a pocket into which the oocytes are released. They then travel through the peritoneum to the oviduct. This ovary is exceptional and should, according to Wourms (1977), not be considered as typical for the viviparous condition in sharks.

The ovary contains : (1) follicles, which in the oviparous *Scyliorhinus canicula* are quite large, but in the viviparous basking shark are not more than 5 mm in diameter at most; (2) corpora atretica, which are derived from egg-containing follicles that have degenerated; (3) "corpora lutea" which are reptured follicles that have reorganized into structures that resemble mammalian corpora lutea; (4) corpora lutea found, e.g., in the ovoviviparous *Squalus acanthias* in which the granulosa shows evidence of 38-HSD activity; and (5) "functional corpora atretica", a name here used for lack of a better one, which are derived from atretic follicles but show luteinization and the presence of steroids. Functional corpora aretica are found in ovoviviparous electric rays. *Torpedo marmorata* and *T. ocellata*. The reptured follicles, "corpora lutea," do not show evidence of steroidogenesis in these rays.

The (Mullerian) oviduct has four regions :

1. *Funnel* (anterior ostium tubae). This collects the ovulated oocytes. It is formed either by fusion of the anterior end of the left and right oviduct, as in the basking shark, or by asymmetric development of one primitive funnel, as in *Scyliorhinus.*

2. *Shell gland or nidamental gland.* This structure is well developed in oviparous and ovoviviparous species, but may be vestigial in viviparous ones. It is a tubular gland which secrets albumin, mucus, and in those species producing an egg case, egg case proteins. In *Scyliorhinus*, sperm are stored in this part of the oviduct.

3. *Isthmus.* This connects the shell gland to the uterus.

4. *Uterus.* This may play a role in the nutrition of the embryos. The left and right uterus may fuse posteriorly and open into a common vagina, or they may open separately into the cloaca. The common vagina, if present, also opens into the cloaca. In some cases a hymen is present at the posterior end of the oviduct.

Holocephali

In the chimaera. *Hydrolagus colliei*, two ovaries and two oviducts are present. In general, the reproductive system resembles that of the elasmobranchs, except that in the Holocephali, the two oviducts always have one common funnel, and each oviduct has a separate opening into the urogenital sinus lateral to the excretory duct, whereas in the elasmobranchs the oviducts open via either a common vagina or separate gavinas into the cloaca. The uterus contributes to the formation of the egg case in *H. colliei*, but does not do so in elasmobranchs.

Chondrostei

The Acipenseridae (sturgeons) have a vestigial connection

between ovary and kidney. The ovary starts out as a ribbonlike structure, as it does in teleosts, but it rolls up laterally and dorsally, so that it forms a hollow organ. The Mullerian Oviduct is continuous with the ovary and opens into the cloaca next to the Wolffian duct (Baer, 1964 : Harder, 1975). The oocytes, after release from the follicle, pass through the Mullerian oviduct to the cloaca.

Teleosts

The ovaria start as ribbonlike structures, but they roll up in a lateral and dorsal direction, so that the side at which ovulations occur is turned inward and surrounds a tube, the oviduct. The oviduct is called *entovarial,* when the free side of the ovary is against ovarian tissue, but *parovarial,* when the

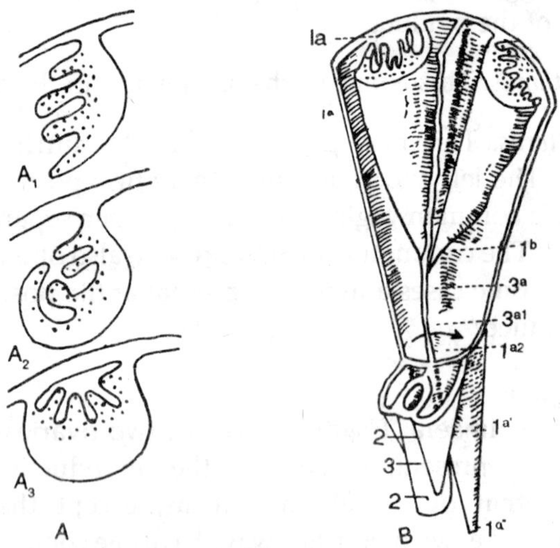

Fig. 1.29. Ovaries and oviducts of teleosts A_1-A_3 = cross section through ovaries : A_1 = ribbonlike ovary (original condition), A_2 = ovary rolled up on itself (entovarial); A_3 = ovary rolled against body wall (parovanal). B = Ovary and oviducts of salmonidae. 1^a = ovarum, $1^{a\prime}$ = lamella of ovary, 1^b = oviduct, 1c = genital funnel, $1^{c\prime}$ = opening of genital funnel, 1^{c*} = genital pore, 2 = coelom cavity, 2' = coelom pore, 3 = intestine, 3^a = mesentery, $3^{a\prime}$ = opening of 3^a via $1^{c\prime}$ (arrow).

free side of the ovary is against the wall of the peritoneal cavity. The oviducts are thus extensions of these peritoneal folds, which enclose the ovary during its development.

The ovaries are usually paired, hollow organs suspended by mesovaria. The left ovary of the smelt (*Osmerus eperlanus*) is located anteriorly and the right ovary posteriorly. Only one gonad primordium develops in the Japanese rice fish (*Oryzias latipes*) and the guppy (*Poecilia reticulata*). In the perches (*Perca* spp.), the ovaries are fused completely, whereas in the pike perch (*Lucia,* formerly *Stizostedion, perca*), they are fused only caudally. The ovaries are fused medially in the stone loach (*Noemacheilus* sp.), and in the European bitterling (*Rhodens amarus*). Oocytes and embryos develop in the wall of the ovary.

The teleost ovary may contain follicles, corpora atertica, pre- and postovulatory "corpora lutea" and pre- and postovulatory corpora lutea. The follicular wall may, in addition, from (1) a calyx from the theca closing the ovarian wall after rupture of the follicle, (2) a calyx nutricius from the follicular wall that secretes nutritious material (embryotrophe) for the embryo either into the ovary or into the follicle containing the embryo.

The oviducts, which, as noted, are *not* Mullerian ducts but peritoneal folds that have grown backward to form oviducts, are absent in some species. In the Salmonidae (salmons and trouts), the anterior part of the oviduct has disappeared, and the oocytes fall into the body cavity and then reach the funnel-like remnant of the posterior part of the oviduct. In the Galaxiidae, Glyodontiidae, Notopteridae, Osteoglossidae, and Anguillidae, and in *Misgurnus,* the oviducts have completely disappeared, and the gametes leave the body cavity via a genital pore, which is not, however, homologous with the abdominal pore found in many sharks.

The genital opening may be in various locations. The midgullar position of the female's genital opening in the

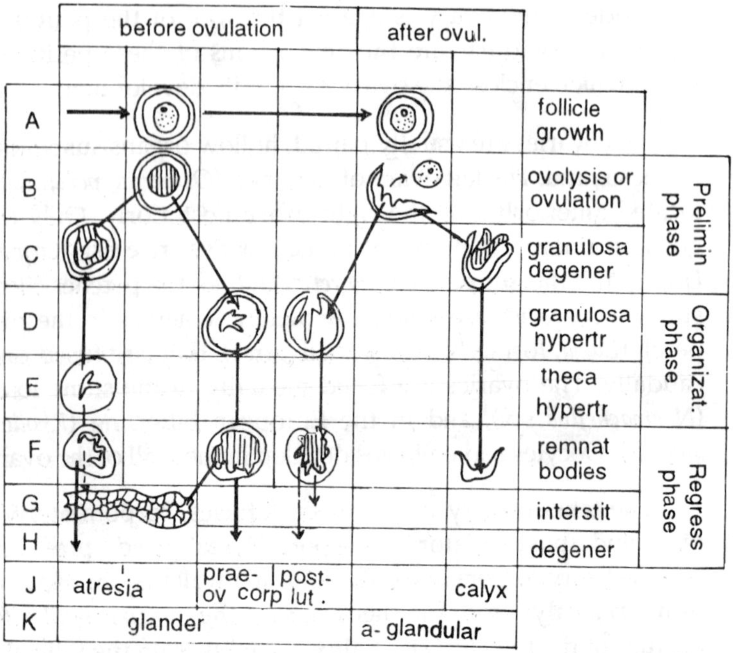

Fig. 1.30. Fate of the ovarian follicle of the European bittering. *Rhodeus amarus*

Phallosthide is adapted to the specialized penis of the male. Teleosts have some specialized structures and associated behaviours, e.g., the ovipositor of the bitterling (*Rhodeus amarus*), a skinlike structure derived from the genital papilla. In this species, the growth of the oviposterior and the nuptial colour of the male require the presence of a fresh water mussel, *Unio pictorum* or *Anodonta intermedia* (Banarescu, 1973). Males select the female with the largest oviposterior. At spawning, the females thrust their ovipositor into the *excurrent* syphon of the mussel and thus have to push their oocytes upsteram. Breder and Rosen (1966) state that, exceptionally, the female lays her oocytes in the current syphon. The male subsequently spawns near the *incurrent* syphon, or the sperm would never enter the mussel. The oocytes are fertilized inside the mussel and they dvelop in its gill chambers. After about a month the embryos hatch and the young bitterlings leave the mussel. The larvae of the mussel attach themselves to the gills and fins of the young fish and live their as parasites for a time.

The Jenynsiidae and Anblepidae have their genital opening either on the left or right, corresponding to the males having a left or right gonopodium. Among the Jenynsiidae, three-fifths of the males are dextral and two-fifths sinistral, whereas for the females the ratio is just the reverse.

Amphibians

The ovaries are paired, lobed, hollow organs consisting mainly of a cortext. The outside of the ovary is covered with germinal epithelium; the cavity is lined with cells of medullary origin.

Anura. The ovaries of the Anura, which are largely oviparous, have follicles and scars from ruptured follicles, corpora atretica and "functional corpora atretica," that is, structures formed from non-ovulated follicles, but resembling corpora lutea. In the viviparous toad *Nectophrynoides occidentalis*, Xavier (1974) identified corpora lutea formed from ruptured follicles. These corpora lutea contained cholesterol, showed a positive reaction for 3β-HSD, and were capable of synthesizing progesterone from pregnenolone. The quantity of progesterone synthesized paralleled the size of the corpora lutea, and both decreased as pseudopregnancy (ovulation without mating) progressed beyond four months.

Ruptured follicles appear to have a steroid secretory function in *Rana esculenta* and *R. cyanophlyctis*, which are oviparous, and in some ovoviviparous species. The females of some species carry the fertilized eggs either in a pouch, e.g., *Gastrotheca marsupiata* and *G. pygmaea*, or in separate chambers, e.g., *Pipa pipa* and *Cryptobatrachus evansi*. The developing embryos probably derive nutrients from the mother.

Urodeles. The ovaries resemble those of the Anura, with

follicles, corpora lutea, and corpora atretica, which originate from follicles in which the yolk has been resorbed. According to Lofts (1974), there is little or no evidence that these structures secrete steroids. Corpora lutea, which result from a reorganization of the ruptured follicle into a progestin-secreting structure, are found in *Triturus cristatus* and *Salamandra salamandra*.

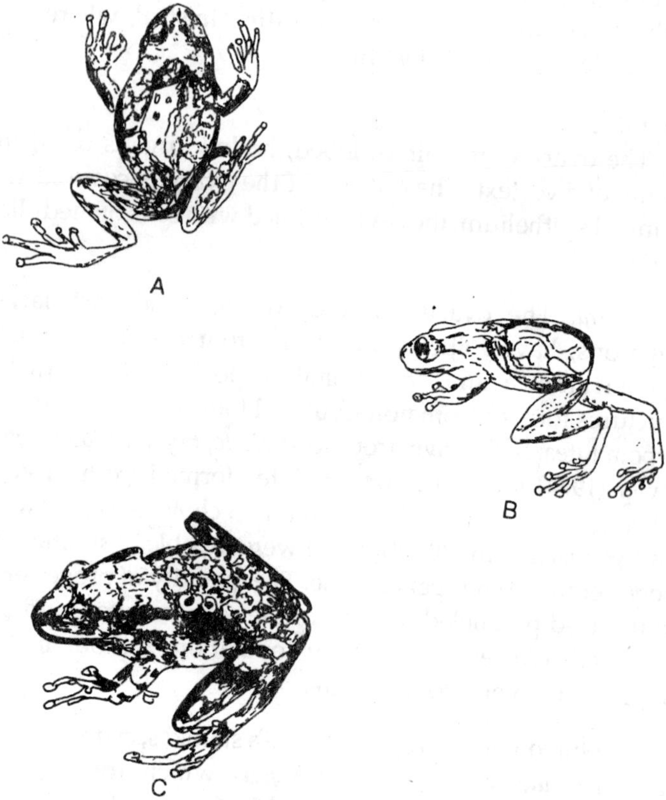

Fig. 1.31. The brood pouch in females of different species of amphibians. *A, Gastrotheca marsupiata; B, Gastrotheca pygmaea*, eggs removed from the brood pouch; C, *Cryptobatrachus evansi*.

The (Mullerian) oviducts of amphibia have three regions: the pars recta, the pars convoluta, and the pars uterina. The oviducts end in the cloaca, which in the urodeles, shows some specialization. In salamanders, three sets of glands are associated with the oviduct : (1) *pelvic glands,* which are located in the roof of the cloaca, and serve as reservoirs for the storage of spermatozoa; (2) *cloacal glands,* present in all ambystomids, salamandrids, and the primitive plethodontids, which may play a role in formation of the egg case; and (3) *abdominal glands,* present in *Ambystome, Necturus,* and *Eurycea,* which are apparently rudimentary.

In the urodeles with internal fertilization, the female cloaca may have well-developed cloacal lips, with which the spermatophores can be picked up from the ground.

Reptiles

The ovaries are paired, hollow organs, suspended by mesorchia and with squamous epithelium lining the ovarian cavity. The follicles project from short stalks, and the walls, which surround the large yolk, consist of a granulosa layer, a theca interna, and a theca externa. Corpora lutea are formed after ovulation as the result of the luteinization and proliferation and/ or hypertrophy of the granulosa cells and supporing connective tissue from the thecal layers. In oviparous species, the corpora lutea regress soon after oviposition. In the lizard, *Anolis carolinensis,* follicles that have become atretic form corpora atretica. The corpora atretica seem to determine the response of the ovarian follicles to exogenous and probably endogenous gonadotrophins. After ovulation, the oocytes may according to Bellairs (1970), wander through the body cavity and enter either of the two oviducts.

Abdominal fat bodies have been found in all temperate zone reptiles that have been examined and in some anoline lizards of warmer regions. The lipids from these bodies pass to the liver and are incorporated into yolk precursons. In the lizard, *Uta stansburiana,* removal of the fat bodies inhibits follicular development or, when the follicles are developed, causes atresia of the developed follicles. There appears to be a reciprocal relationship between ovary and fat bodies, because ovariectomy prevents rapid fat mobilization from the fat bodies.

The oviduct, which the Mullerian ducts, are convoluted. They consist of (1) an infundibulum or funnel : (2) a convoluted segment with a layer of longitudinal and a layer of circular muscle, and a mucosa with many alveolar glands and ciliated epithelium; (3) an isthmus; (4) a uterine segment, which has a strong musculature and is lined with ciliated and non-ciliated columnar epithelium; and (5) a vagina which has mucous glands, and traverses the cloaca.

The cloacal openings for the paired oviducts are separate in *Lacerta viridis*, *Sceloporus* sp., *Hemidacytylus* sp., *Phrynosoma* sp., but united in *Gerrhonotus* sp. and *Cnemidophorus* sp. In the blind, libless lizards (*Anniella*) both ovaries are functioned and of the same size, but only the right oviduct is functional. The left oviduct is smaller than the right oviduct and has numerous glands, the secretions of which probably pass out via the cloaca. In blind worm snakes (Typhlopidae), the left oviduct is absent.

In some reptiles the sperm retain their fertilizing capacity for a considerable length of time after insemination. It is not surprising, therefore, that there are structures for storing sperm in the oviduct. In snakes, turtles and lizards, sperm are retained in "seminal vesicles," which are specialized alveolar glands at the base of the infundibulum. From these glands, ducts lead into the oviductal lumen. In the snake, *Thamnophis elegans terrestris*, the sperm are retained in the posterior two-thirds of the oviduct, and the sperm do not reach the "seminal vesicles" until February—March, if the snake mates in the fall. If insemination occurs in the spring, however, the sperm reach the "seminal vesicles" in April. In the adder, *Vipera aspis*, the sperm remain in the vagina after fall mating and proceed to the "seminal vesicles" in the spring just before they fertilize the oocytes.

Iguanids have "seminal vesicles" in the anterior segment of the vagina, whereas gekkonids have these structures between uterus and infundibulum. The seminal receptacles, found in the caudal part of the albuminous region of the oviduct of the box turtle (*Terrapene carolina*), are morphologically unspecialized glands.

Birds

Most avian species have only a left ovary and a left oviduct, but some species have been found to have two functional ovaries, and in most of these species, it is the left oviduct that is functional. Oocytes ovulated from the right ovary will, in the absence of a right oviduct, drop into the body cavity and will be absorbed, since a mesentery separates the right ovary from the left oviduct, so that an oocyte from the right ovary cannot be picked up by the left oviduct.

The ovary is attached to the body wall by a mesovarium. It has many follicles embedded in a sparse stroma of connective tissue. During the laying season, there is a series of large follicles, which show a hierarchy of size.

The follicular wall consists of a granulosa layer, a theca interna, and a theca externa. The large follicles are filled with yellow yolk; some of the smaller follicles are filled with either yellow or white yolk, depending on the stage of development. On the wall of the large follicles is a clear band, the stigma, which macroscopically is devoid of blood vessels and along which the follicle ruptures.

In addition to the large and developing follicles, one finds atretic and ruptured follicles. In atretic follicles, the yolk is reasbored, and the granulosa cells become dissociated from the follicular wall and move into the follicles. Atretic follicles are not always present in laying chickens, and ruptured follicles rapidly disappear after ovulation. In laying chickens, they are difficult to find five days after their rupture. There are no corpora lutea in birds.

Usually one (the left) oviduct, which is a Mullerian duct, is present. By inbreeding, however, Morgan and Kohlmeyer (1957) were able to obtain chickens strains with two oviducts (and one ovary). The oviduct has five anatomical and functional regions.

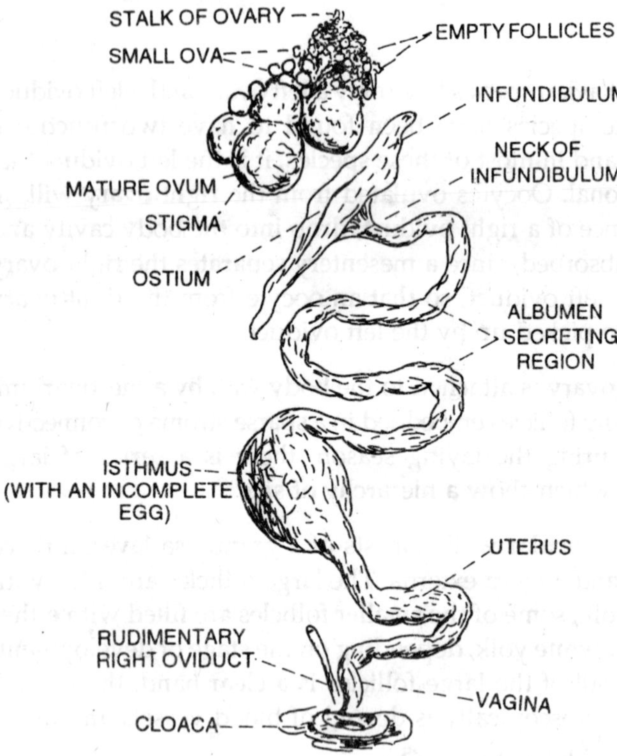

Fig. 1.32. Reproductive organs of the hen.

1. *Infundibulum.* This region consists of a funnel and a tubular part. In the sexually mature chickens, the infundibulum is about 7-9 cm long. It has two muscle layers— a circular and a longitudinal layer—and a few muscle cells scattered in bundles. The wall of the funnel is thin and has low mucosal folds. The tubular part has a thicker wall and higher mucosal folds. This part of the oviduct secretes thick albumen around the yolk during its passage, a passage that takes 15-30 minutes.

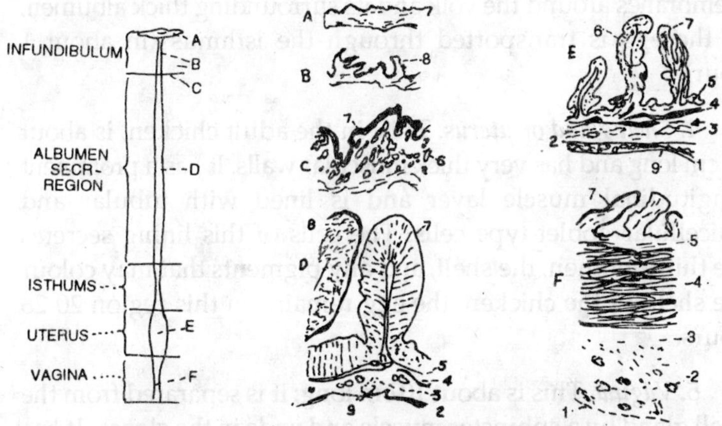

Fig. 1.33. **Transverse sections through different parts of the oviduct of the adult chicken. A = lips of infundibulum, B = neck of infundibulum, C = transition region from infundibulum to magnum, D = magnum of albumen secreting region, E = walls of the uterus, F = walls of the vagina, 1 = Peritoneal membrane, 2 = longitudinal muscle fibres, 3 = connective tissue, 4 = circular muscle layer, 5 = inner layer of connective tissue, 6 = thick layer of convoluted tubular glands, 7 = epithelium of the duct, 8 = ducts of the convoluted glands, 9 = blood vessels.**

2. *Magnum or albumen-secreting region.* This, in the adult chicken, is about 32-34 cm long, white, and highly coiled. It has thick walls with longitudinal circular layers of muscle cells and a longitudinally folded mucous membrane lined with high, tubular, glandular, ciliated epithelium. The glandular epithelium gives this region of the wall its thickness. Thick albumen is secreted in this region around the yolk as it transverses the magnum (in about 2-3 hours).

3. *Isthmus.* This region is about 10 cm long in the adult chicken and is separated from the magnum by a clear translucent band of non-glandular tissue. The isthmus has a thicker circular muscle layer than the infundibulum and the magnum, and its glandular development is less marked than that of magnum. The glands are tubular and secrete the proteinaceous egg

membranes around the yolk and its surrounding thick albumen, as the egg is transported through the isthmus (in about 1 hour).

4. *Shell gland or uterus*. This, in the adult chicken, is about 11 cm long and has very thick muscular walls. It has a prominent longitudinal muscle layer and is lined with tubular and unicellular goblet-type cells. The cells of this lining secretes the thin albumen, the shell, and any pigments that may colour the shell. In the chicken, the egg remains in this region 20-26 hours.

5. *Vagina*. This is about 10 cm long; it is separated from the shell gland by a sphincter muscle and ends in the cloaca. It has a longitudinal muscle layer that is easily distinguished and a circular muscle layer that is well developed. The mucous membrane has low narrow folds and is lined with ciliated and non-ciliated cells. There are no glands and no secretion in the vagina proper; the egg is "finished" in the shell gland.

In addition to the glans just mentioned in connection with egg formation, there are two areas where spermatozoa are stored and can retain their fertilizing capacity (about 7 - 14 days in chickens and 40 - 50 days in turkeys). These two regions are at the extreme ends of the oviduct, i.e., at the infundibulum and at the utero-vaginal junction.

The ovary receives its blood from the gonadrenal artery, which is a branch of the dorsal aorta that sends a branch, the ovarian-oviductal artery, to the ovary. The blood is drained by a cranial ovarian-oviductal and a caudal ovarian vein, each of which drain into the vena cava. The venous system of the ovary is very well-developed.

The innervation of the ovary is complex; adrenergic and cholinergic fibres have been found. The adrenergic supply originates from the 5th, 6th, 7th, sometimes the 4th thoracic, and the 1st and 2nd lumbar ganglia of the sympathetic chain. The origin of the parasympathetic fibres is not known.

Mammals

Ovary. The ovaries are usually paired, but in some species (Table 1.2) one may be inactive; that is, either it does not ovulate or it ovulates only after the active ovary has been removed. For example, if the right ovayr of the mountain viscacha (*Lagidium peruanum*) is removed, the left one becomes functional. The right ovary is also the functional one in the following bats: *Miniopterus, schreibersi, Rhinolophus* sp., *Tadarida cyanophala and Molossus ater*. The left ovary is dominant in the bat, *Miniopterus natalensis*, and in the waterbuck, *Kobus defassa*.

Each ovary is suspended from the dorsal body wall by the mesovarium or secondarily by the broad ligament. In many species, the ovary lies in a membranous sac called the *ovarian bursa*, which is a fold of the mesosalpinx, the mesentery of the oviduct. The function of bursa has not been determined.

The ovary consists of the stroma, the germinal epithelium, the germ cells and their associated structures, interstitial cells, and various vestigial structures. The germinal epithelium, which covers the ovary, may be either directly on the ovarian stroma (e.g., in shrews) or it may be on a fibrous ovarian tunica albuginea. It may consist of squamous or low columnar cells. The germ cells and their associated structures include oocytes, follicles in different stages of development, connective tissue, corpora lutea, old corpora lutea called *corpora albicantia*, and interstitial tissue.

1. *Follicles*. These can be characterized as small, medium and large follicles, according to Pedersen and Peters (1968). The main distinctions, are made on the basis of oocyte size, number of layers of follicular cells, and the presence of follicular fluid. The follicular wall has a granulosa layer, a theca externa, and a theca interna. The oocyte is surrounded by an acellular layer, the zone pellucida, which separates it from the surrounding follicular cells. In the large follicles, the oocyte is surrounded by granulosa cells, which form a structure, the cumulus oophorus, which connects it to the follicular wall.

Immediately surrounding the zona pellucida is a layer of columnar or pseudostratified columnar cells, which is called the *corona radiata.*

TABLE 1.3

Occurrence of ovarian bursa in mammals

	Anatomy of bursa				
Animals	1	2	3	4	5
Prototheria					
Tachyglossus	+				
Ornithorhynchus				+	
Mernatheria					
Phalangeridae				+	
Phascolomidae		+			
Wacropodidae				+*	
Eurneria					
Insectivora					
Tenrecinae	+				
Erinaceinae	+				
Soricinae	+				
Talpidae	+				
Chiroptera	+				
Primates					
Lemuridae			+		
Hominidae			+	+	+
Edentata					
Bradypodidae	+				
Dasypodidae		+			
Lagomorpha					
Lepus europaeus			+		
Oryctolagus cuniculus		+			
Rodentia					
Sciuridae		+			
Castoridae		+			

Circetidae	+				
Muridae	+				
Gliridae	+				
Dipodidae					+
Hystricidae		+			
Cavidae			+		
Dasyproctidae					+
Cetacea					+
Carnivora					
Canidae	+				
Ursidae	+				
Procyonidae	+				
Mustelidae	+				
Viverridae	+	+			
Hyaenidae		+			
Felidae		+	+		+
Pinnipedia		+			
Hyracoidea					
Procaviidae		+			
Perissodactyla					
Equidae				+	+
Artiodactyla					
Suidae		+			
Camelidae	+				
Cervidae			+	+	
Bovidae		+			

*Ovary end used by infundibulum.

1 = Complete bursa, with participation of infundibulum.

2 = Incomplete bursa, with participation of infundibulum wide opening to pertoneal cavity.

3 = Incomplete bursa without participation of infundibulum.

4 = Peritoneal pocket only; ovary lies outside this pocket.

5 = No peritoneal pocket.

2. *Atretic follicles.* These are degenerating follicles. The first sign of atresia is the pykonosis of the epithelial follicular

cells, which are adjacent to the lumen of the large follicles and in layers just external to the corona radiata of secondary follicles. Subsequently the cumulus oophorous shrinks, the granulosa cells disappear, and fibroblasts and leukocytes invade the atretic follicle.

3. *Corpora lutea.* These originate from ruptured follicles and are formed by enlargement and multiplication of the granulosa cells. The corpora lutea consist of a solid mass of polyhedral cells. Mossman and Duke (1973a) proposed the term "luteal gland" in preference to corpus luteum, but the use of the term *corpus luteum* is such long standing that substitution of a new term seems to have little chance of adoption. The corpora lutea of the Madagascar hedgehog (*Setifer setosus*) develop from a follicle in which no antrum or cavity is formed. The oocytes is fertilized inside the follicle, and the zygote is extruded by contractions of the follicular sheath and a shift in position of the granulosa cells, so that the rygote becomes exposed on the surface of the follicle.

4. *Accessory corpora lutea.* These structures can have two different origins : (1) granulosa cells of atretic follicles, as in the North America porcupine (*Erithizon dorsatum*) and in the primates, *Tupaia* sp., and *Urogale* sp., or (2) the epithelium of medullary cords, as in juvenile easter chipmunks (*Tamias striatus*), in which medullary follicles ovulate and form large corpora lutea.

5. *Corpora albicantia.* Thes are small masses of connective tissue which are the remains of degenerated corpora lutea.

6. *Thecal glands.* These consist of a layer or zone of glandular cells adjacent to the granulosa cells of mature follicles.

7. *Interstitial glands.* These may originate from the ovarian stroma, medullary cord cells, and the theca interna of atretic follicles. Mossman and Duke (1973b) stress the inaccuracy of calling these structures "luteinized follicles."

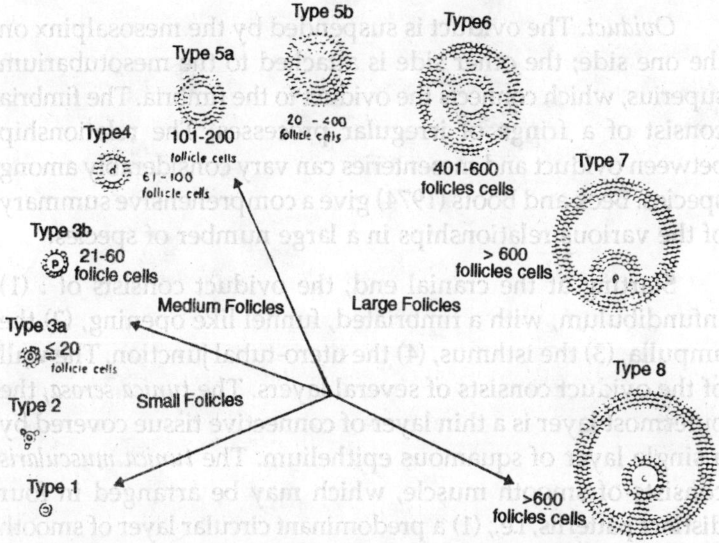

Fig. 1.34. Classification of ovarian follicles of the mouse.

Some structure in the ovary are vestigial. The *rete ovarii* plays a role in oogenesis. It is a regular network of spaces lined with a simple low columnar epithelium. It is well developed in the tarsier, *Tarsius* sp. (a primate) and in pangolins, *Manisa* sp. The *epoophoron* consists of vestiges of efferent ductules and in homologous with the epididymis of the male. *Gartner's duct* or the duct of the epoophoron is a vestige of the ductus deferens. The *paraoophoron* consists of the more rostral and the more caudal mesonephric tubules—or ductuli aberrantes. It is homologous with the (also vestigial) paradidymis of the male.

The ovary receives blood from the ovarian artery, a branch of the dorsal aorta, and blood leaves the ovary via the ovarian vein, which drains into the caudal vena cava. Some branches of the ovarian artery and vein anastomose with ovarian branches of the uterine blood vessels.

The ovary receives fibers from the vagus nerve and also adrenergic fibers. It has been seen that adrenergic mechanisms are involved in ovulation.

Oviduct. The oviduct is suspended by the mesosalpinx on the one side; the other side is attached to the mesotubarium superius, which connects the oviduct to the fimbria. The fimbria consist of a fringe of irregular processes. The relationship between oviduct and mesenteries can vary considerably among species. Beck and Boots (1974) give a comprehensive summary of the various relationships in a large number of species.

Starting at the cranial end, the oviduct consists of : (1) infundibulum, with a fimbriated, funnel like opening, (2) the ampulla, (3) the isthmus, (4) the utero-tubal junction. The wall of the oviduct consists of several layers. The *tunica serosa*, the outermost layer is a thin layer of connective tissue covered by a single layer of squamous epithelium. The *tunica muscularis* consists of smooth muscle, which may be arranged in four distinct patterns, i.e., (1) a predominant circular layer of smooth muscle without distinguishable longitudinal layers, (2) thin longitudinal layers on both sides of a thicker circular layer, (3) a thick circular layer with an outer longitudinal muscle layer, and (4) a circular layer with an inner longitudinal muscle layer. The *tunica mucosa*, the innermost layer, consists of a lamina propria of connective tissue and a lamina epithelialis, which has ciliated cells and three types of non-ciliated cells : secretory cells, wedge-type cells called "peg cells," and basal cells along the base of the epithelium.

In species such as pigs, mice, rats, humans and rhesus monkeys, the presence of ciliated cells depends on the stage of reproductive cycle; however, the occurrence of cyclic changes in cilia is controversial.

The anatomy of the utero-tubal junction varies among species. Beck and Boots (1974) described ten different morphological types of utero-tubal junctions. Type 1-3 lack mucosal projections of oviductal or uterina origin. Type 4-10 have specialized mucosal projections of oviductal or uterine origin. The oocytes or the zygotes are transported rapidly through the utero-tubal junction, which may play a role in sperm transport.

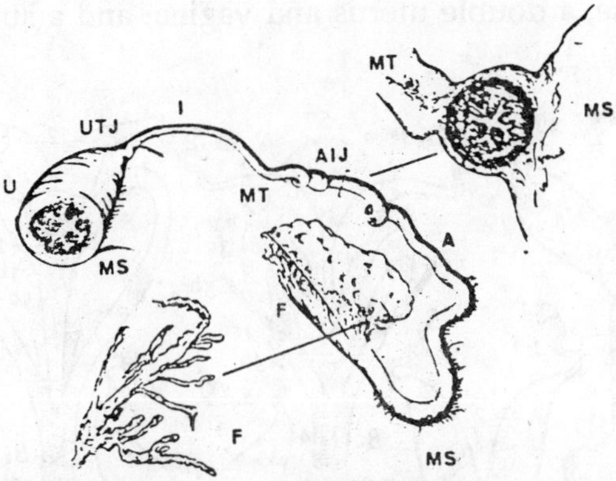

Fig. 1.35. Diagram of the rabbit oviduct. A = ampula, AIJ = ampullary-isthmic junction, F = fimbrae, I = isthmus, MS = mesosalpinx, MT = mesotubarium, U = uterus, UTJ = uterotubal junction.

Uterus. The uterus is suspended by the mesometrium. There are different types of uteri. In the prototheria, the uteria are paired and open into the urogenital sinus; no vagina is present. In the metatheria, or marsupials, there are 2 vaginae, 2 cervices, 2 uteria. Sharman (1976) has illustrated some of the vaiations of this general plan and two such variations. The kangaroo (*Macropus* sp.) has an open midline vagina after the first parturition, so that it resembles the Eutheria midline vagina, but it has been formed in a different manner.

Among the Eutheria one finds four kinds of uteri : (1) The duplex uterus found in the rabbit has 1 vagina, 2 cervices, and 2 uterine horns. (2) The bicornuate uterus found in the pig, has 1 vagina, 1 cervix, and 2 uterine horns. (3) the bipartite uterus found in cattle, sheep, and horses has 1 vagina, 1 cervix and either a small or a prominent uterine body, and 2 separate uterine horns, (4) the simplex uterus found in most primates, including humans, has 1 vagina, 1 cervix and 1 uterine body. In humans, various anomalies have been found; for example, all the three different types of uteri listed for the Eutheria, a doubt uterus and vagina has a result of failure of the Mullerian ducts to fuse, a double uterus and vagina, and a subseptate uterus.

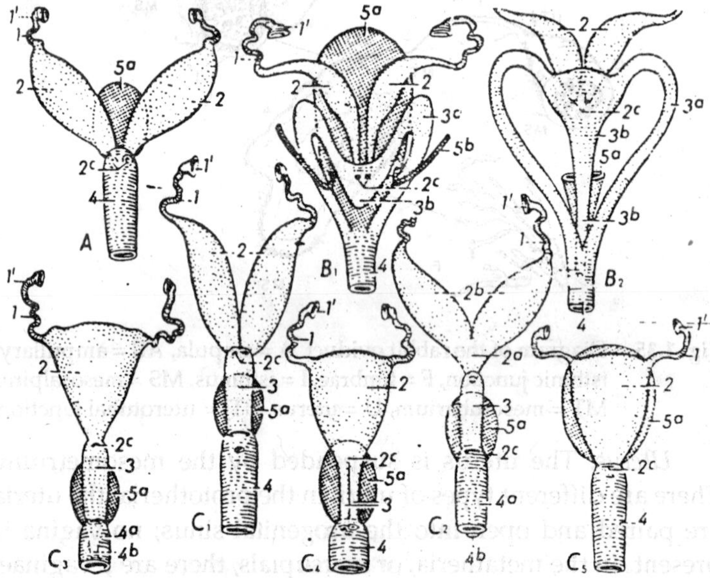

Fig. 1.36. A = Prototheria : paired uterin, no vagina, B_1 = Metatheria: *Didelphis,* paired uteri, paired vaginae, single sinus vaginalis which ends bindly caudally, B_2 = Metatheria, *Macropus,* as B1, but the sinus viginalis is secondary connection with the urogenital sinus, C_1 = Eutheria, duplex uterus; C_2 = bicomuate uterus; C_3 = simplex uterus, C_4 = uterus simplex with paired vagina as result of sagittal septum (*Bradypus*); C_5 = simplex uterus which ends into urogenital sinus because vagina is lacking (*Dasypus*).

1 = oviduct, 1' = opening of oviduct, 2 = uterus, 2^a = body of uterus, 2^b = uterine horn, 2^c + cervix, 3 = vagina, 3^a = paired vaginae, 3^b = unpaired vaginae, 4 = urogenital sinus, 4^a = vestibule of vagina, 4^b = urethra, 5^a = urinary bladder, 5^b = ureter.

Small variations are found in the four different types of uteri. For example, the sloth *(Bradypus* sp.) has a simplex uterus, one cervix, but a paired vagina as a result of sagittal septum; the armadillo (*Dasypus* sp.) has a simplex uterus with one cervix opening into the urogenital sinus.

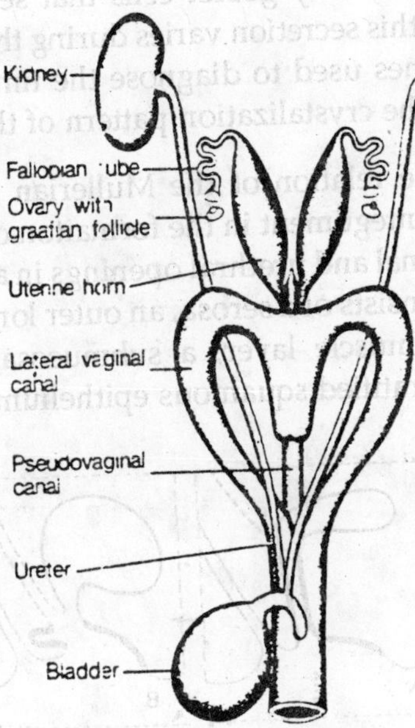

Fig. 1.37. Peculiar anatomy of the female reproductive tract in marsupials. Spermatozoa ascend by way of lateral vaginal canals, and the fetus is born through the central pseudovaginal canal, which is occluded at other times.

Sharman (1976) has pointed out that the peculiar vagina of marsupials is the result of the position of the ureters of marsupials, which grow forward *between* the Wolffian and Mullerian ducts. This prevents the fusion of the vaginae in the midline (because of the intrusion of the ureters). The lateral vaginae serve for transported of spermatozoa, but the birth of the young occurs via the midline birth canal, which connects the uterine opening *directly* with the arogenital sinus.

Cervix. The cervix is a sphincter at the posterior end of the uterus. Its wall is thick, and the inner wall has ridges. It consists of a serosa, an outer longitudinal and an inner circular muscle, a submucosa, a mucosa consisting of columnar epithelium at the cranial end and squamous epithelium at the caudal end, and many goblet cells that secrete mucus. The consistency of this secretion varies during the menstrual cycle and is sometimes used to diagnose the time of ovulation in women, as in the crystalization pattern of the mucus.

Vagina. The relation of the Mullerian ducts, urogenital sinus, and the integument in the formaiton of the vagina. The location of vacinal and urethral openings in a few species. The vaginal wall consists of a serosa, an outer longitudinal and an inner circular muscle layer, a submucosa, and a mucosa consisting of stratified squamous epithelium.

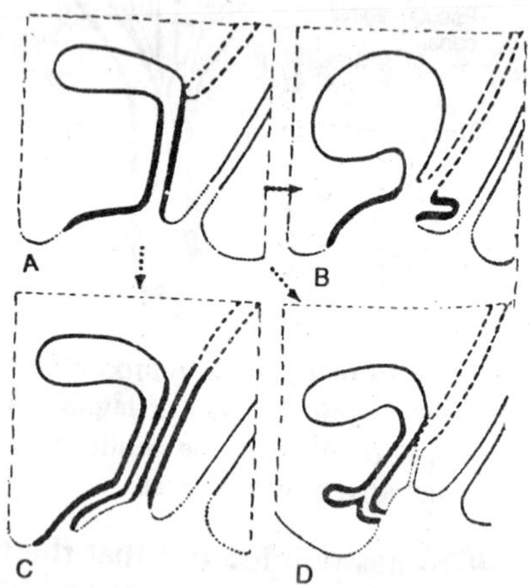

Fig. 1.38. Contributions of Mullerian ducts, urogenital sinus, and integument in formation of the eutherian vagina. A. Hypothetical basic plan; B. Human; C. Mouse (*Mus* sp.) D. Mole (*Talpa* sp.). Dotted line = integument (ectoderm) thin line = intestine and urinary bladder; heavy line = urogenital sinus entoderm; dashed line = Mullerian duct mesoderm.

In some laboratory species, especially the rat, mouse, and hamster, the stage of the estrous cycle is correlated with differences in the histological appearance of the vaginal epithelium, so that vaginal smears can be used to diagnose the stage of the estrous cycle. During proestrus epithelial cells predominate; during estrus there are some remnants of epithelial cells and many large cornified cells; during metestrus there are many cornified and epithelial cells and leukocytes; during diestrus few epithelial cells remain and few to many leukocytes are found.

Prostate. Although the prostate is mostly a male secondary sexual gland, a female prostate is well developed in some species. In laboratory rats, the "Witschi rat," developed by selective breedings, has a well-developed prostate.

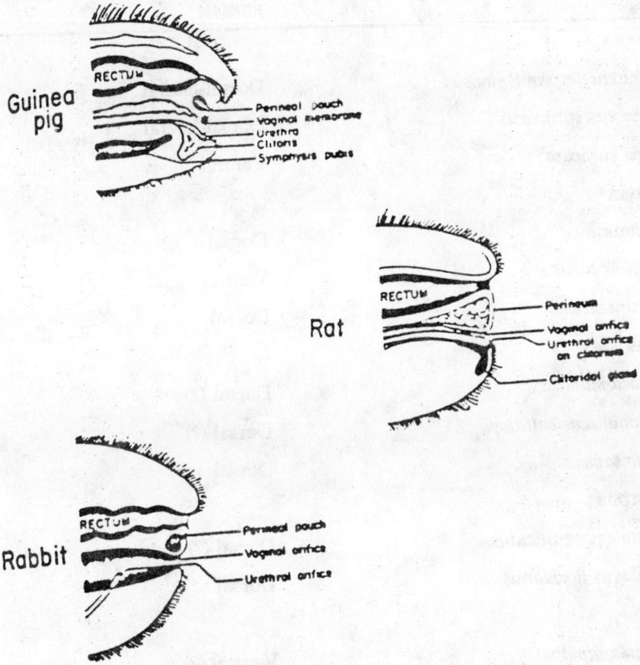

Fig. 1.39. Location of the urethral and vaginal orifices in the abdomen of the guinea pig (anestrous condition, vagina closed by an epthelial membrane). rat. and rabbit.

External Genitalia. The external genitalia consist of the labia majora, labia minora, and the clitoris, but only the human has true labia. Generally, the clitoris is poorly developed unless the fetus has been exposed to larger amounts of androgens than normal, e.g., in the adrenogenital syndrome, in the freemartin, or after experimental androgen treatment. However, in the hyaena (*Hyaena crocuta*), the clitoris is very well developed. This extreme development and anatomy make it almost impossible to distinguish a male from a female, especially since the females also has a false scrotum.

TABLE 1.4

Part of Prostate Gound in Female Mammals

Species	*Part of prostate*
Insectivora	
Hemicentetes semispinosus	Dorsal-lateral
Erinaceus europaeus	Dorsal-lateral
Talpa europaea	Ventral
Chiroptera	
Taphozous	Dorsal
Nycteris luteola	Ventral
Coleura afra	Dorsal
Primates	
Macaca mulatta	Dorsal (?)
Cercopithecus aeithiops	Dorsal (?)
Homo sapiens	Dorsal
Lagomorpha	
Oryctolagus cuniculus	Dorsal (?)
Sylvilagus floridanus	Dorsal
Rodentia	
Arvicola sapidus	Ventral (?)
Apodermus sylavaticus	Ventral
Arvicanthis cinerus	Ventral
Rattus norvegicus	Ventral
R. rattus	Ventral
Mastomys erthroleucus	Ventral
Mus musculus	Ventral

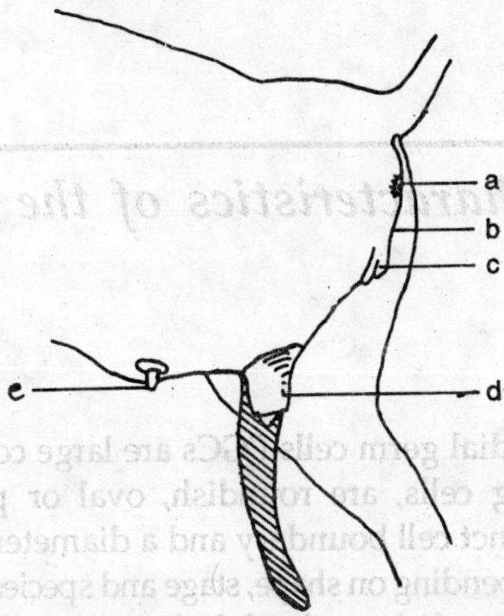

Fig. 1.40. Exterior female genitalia of *Hyaena crocuta*. a = anus, b = perineum, c = false scrotum, d = clitoris not in erection and in erection (hatched), e = teat.

Sexual Skin. Some primates have a so-called sexual skin, which is contiguous with the external genitalia. It consist of a highly vascularized edematous or pigmented dermis. Its colour and vascularization vary during the reproductive cycle.

2

Characteristics of the Primordial Germ Cells

The primordial germ cells PGCs are large compared with the surrounding cells, are roundish, oval or pear-shaped, and have a distinct cell boundary and a diameter varying from 10 to 20 μm depending on shape, stage and species. The voluminous nucleus is roundish, oval or lobular in shape and has a distinct unclear membrane. It is often situated eccentrically; its diameter may vary between 6 and 10 μm. The chromatin is fairly evenly distributed throughout the nucleus contains one or two prominent nucleoli. Depending on the stage of development and the taxonomic group the cytoplasm contains varying amounts of yolk granules, fat droplets and pigment granules. In general the cytoplasm has a dense, granular appearance and contains a roughly hemispherical cytoplasmic body called variously the centre sphere cytoplasmic crescent, mitochondrial crescent or attraction sphere. This consists of one or two centrioles, numerous mitochondria, and small yolk platelets. The unequal distribution of yolk and lipid inclusions and the position of the cytoplasmic body result in the eccentric location of the nucleus and the oval or pear-shaped form of the cell. The PGCs are often also characterised by the presence of lobopodia or filopodia, particularly during their migratory period. The general features of the PGCs have been described in teleost fishes, anuran amphibians, urodele amphibians, coecilians, chelonian reptiles, birds and in mammals.

It is clear that none of the features of PGCs described

above is specific for these cells, so that an unequivocal identification can only be made on the basis of a combination of all or most of them. Moreover, the earlier the stage of development, the more yolk the cells usually contain and the less pronounced are the various cell characteristics. A large amount of yolk also often leads to 'deformation' of both cell and nucleus, and a consequent blurring of several of the above features. The early workers were therefore unable to trace the PGCs back to the earliest stages of development.

The discovery by Bounoure in anurans of a special cytoplasmic structure, the so-called 'germinal plasm', which is originally located in the subcortical layer of the vegetal region of the fertilised egg and later in the PGCs of the tadpole, marked an important step forward in the analysis of the origin of the PGCs in the anuran amphibians. Cytological and cytochemical studies indicated that the germinal plasm contained, besides small yolk platelets and pigment granules, mitochondria and RNA. However, it was the electron microscope which revealed the true nature of the germinal plasm. In recent years numerous investigations have been devoted to the study of this structure that is specific to germ cells. It must however be emphasised that not all chordate or verebrate PGCs are characterised by the presence of a typical germinal plasm and that moreover it is not found during all developmental stages.

Since nothing is known about the ultrastructure of the PGCs in the Cephalochordata and Tunicata the following description will only deal with the ultrastructural and cytochemical characteristics of the PGCs in the various groups of the vertebrates.

The Ultrastructure and Biochemistry of the PGCs in Anamnia

Amphibia

In the amphibians extensive ultrastructural and cytochemical analyses have been carried out on the PGCs of the anurans, few on those of the urodeles, and none on those of the coedlians.

Anura

In the anurans electron-microscopic analysis has revealed, apart from the usual cell organelles found in almost every cell of the embryo, a number of special cytoplasmic and nuclear structures which may be considered as being characteristic of PGCs and their descendants. The first and most striking of these is the 'germinal plasm', which has now been described in different anuran species by many investigators.

The germinal plasm is a well-circumscribe structure occurring in the form of patches 5-12 μm in diameter. These consist of numerous mitochondria, many containing protein crystals, interspersed with so-called 'germinal granules' (electron-dense bodies). These electron-dense bodies are composed of small electron-dense foci which seem to be embedded in a matrix of extremely fine fibrils. It is not clear whether the structure of the germinal granules is essentially granular, fibrillar or fibrillo-granular. The electron-dense bodies are not bounded by a membrane and are frequently found in contact with mitochondrial membranes ('intermitochondrial cement') or with the nuclear membrane ('nuage material'). The germinal plasm also contains pigment granules and some glycogen, as well as a few lipid droplets and a varying amount of small yolk platelets.

Blackler (1958) thought that the islands of germinal plasm found in cleaving eggs arise only after fertilisation, but Czolowska (1969) demonstrated the presence of small patches of germinal plasm in ovarian oocytes. Williams and Smith (1971) found small electron-dense areas devoid of ribosomes within clusters of mitochondria just beneath the vegetal pole cortex of ovarian oocytes. Around the time of germinal vesicle breakdown larger electron-dense areas are seen which are now associated with ribosomes. Since typical germinal granules also appear in progesterone-treated oocytes which are subsequently enucleated, this 'condensation' process does not depend upon germinal vesicle contributions. The characteristic germinal granules therefore seem to coalesce from precursors during oogenesis with a possible synthesis of components *de novo* during maturation.

According to Mahowald and Hennen (1971) the precursors of the germinal plasm in *Rana pipiens*, described as cytoplasmic islands in the unfertilised and fertilised egg and in early cleavage stages, consist essentially of the same components as the germinal plasm in the PGCs of the tadpole. However, a certain structural evolution seems to occur in the germinal plasm during development. In the fertilised egg the germinal plasm is present in the form of small cytoplasmic patches with germinal granules 0.2-0.3 μm in diameter. This is in agreement with Czolowska's (1969) observations on *Xenopus*, where a large number of very small patches was found in the cortical and subcortical cytoplasm of the vegetal, region of ovarian oocytes. In the unfertilised egg the electron-dense foci of the germinal granules measure about 20 nm. At the late blastula stage the germinal granules have acquired a diameter of 0.4-0.5 μm, while the electron-dense foci now have a diameter of up to 50 nm. A condensation process has apparently occurred. In the unfertilised and fertilised egg and at early cleavage stages the germinal plasm seems to contain some RNA in the form of clusters of ribosomes associated with the germinal granules. The number of ribosomes increases upon fertilisation. At the early tadpole stage the germinal granules are in the form of amorphous bodies of a clearly fibrous nature. They are situated in close proximity to the nuclear membrane or are attached to mitochondria, but are no longer associated with clusters of ribosomes. Mahowald and Hennen (1971) ascribe a morphogenetic significance to the temporary association of clusters of ribosomes with the germinal granules, mainly in analogy to similar events observed in the development of the polar granules in the pole cells of certain insects. Protein synthesis has been demonstrated in the metabolically active germinal plasm of the *Xenopus* egg during cleavage.

According to Ikenishi and Kotani (1975) the fibrillo-granular germinal granules found in *Xenopus* at early stages of development first change into irregular string-like bodies at the late neurula to early tail-bud stages, and then into granular material at the feeding tadpole stage.

In *Xenopus* Kalt (1973) distinguishes three different structures which are found in both the male and the female germ line but in no other cells. The first of these is the germinal plasm, which takes the form of large proteinaceous masses of electron-dense material, which he calls 'nuage material'. The second structure, also cytoplasmic in nature, in the electron-dense so-called chromatoid or nucleolus-like body. This is found in germ cells of the indifferent gonad and throughout the female germ line, as well as in the early stages of the male germ line, whether it dissociates into coated vesicles during spermatogenesis. The third structure is a fibrillar nuclear region (fibrillar meshwork) which is free of DNA. A similar structure has been described in *Xenopus*.

Opinions differ on the homology of the various electron-dense structures found in the cytoplasm of PGCs. Some authors, for example Kalt (1973) and Kalt, Pinney and Graves (1975), make a distinction between chromatoid or nucleolus-like bodies on the one hand and germinal granules or nuage material on the other, on the basis of differential sensitivity to various cytochemical treatments. Kalt (1973), Eddy (1974) and Webb (1976) emphasised the great similarity in ultrastructure between germinal granules and nuage material, as well as their resemblance to nucleolar components in the nucleus. Other authors, for example Fawcett (1972), consider the various electron-dense components of the cytoplasm of germ cells as belonging to one and the same family. It must however be emphasised that neither is the homology of the various structures established nor is their functional significance sufficiently understood.

The germinal plasm original present in the fertilised egg

does not change in amount during cleavage. It is distributed unevenly over the exponentially increasing number of blastomeres. Although the number of blastomeres containing germinal plasm initially increased, their total number later remains constant as a consequence of the transfer of the germinal plasm to one daughter cell of a division only (pPGCs). During gastrulation and neurulation a small number of clonal divisions occur, as a result of which the number of PGCs increases about fourfold; these are followed by one more clonal division before stage 35. It is evident that during these clonal divisions the amount of germinal plasm much increase, since otherwise dilution would occur. The frequently observed passage of electron-dense material through nuclear pores may be associated with the formation of new germinal plasm. However, the static electron micrographs do not allow a distinction between a transfer of material from the nucleus to the cytoplasm or vice versa. During the clonal divisions the PGCs also synthesis RNA.

It seems of importance also to mention some negative features of the PGCs in anurans. Gipoulous (1967) was unable to demonstrate either glycogen or alkaline phosphatase in anuran PGCs . The former is at variance with Kessel's (1971) election microscope observations, but the latter is in accordance with Chiquoine and Rothenberg's (1957) studies on PGCs in urodele amphibians.

The distinct cell boundary of PGCs already described by the earlier workers in the various groups of vertebrates turns out to be due to the presence of larger intercellular spaces between the PGCs and the surrounding somatic cells. This emphasises the non-participation of the PGCs in embryonic tissues. The PGCs show pseudopodia and lobopodia which protrude into the intercellular spaces. The demonstration of microtubules and microfilaments in the cytoplasm of spermatogonia constitutes additional evidence for their migratory activity.

Urodela

In the urodele amphibians only very scanty and partially conflicting data on the ultrastructure of PGCs are available.

Recent election-microscopic observations of Ikenishi and Nieuwkoop (1978) on the ultrastructure of the PGCs in *Ambystoma mexicanum* reveal that the PGCs of the urodele larva at stage 46 are characterised by the presence of germinal plasm consisting of clusters of mitochondria and electron-dense bodies of a fibrillo-granular nature. However, the germinal plasm only appears at early larval stages (from stage 40 onwards), at which time the passage of electron-dense material through nuclear pores is also observed. The amount of election-dense material increases rapidly during subsequent developmental stages, until the definitive complement is reached by stage 46.

The late appearance of the germinal plasm in the urodeles seems to be at variance with Williams and Smith's (1971) brief and unconfirmed statement on the presence of germinal granules in the subcortical cytoplasm of the equatorial region of the fertilised, uncleaved egg of *Ambystoma mexicanum.*

Intra-nuclear fibrillar regions, as found by Kalt (1973) in anuran PGCs, have recently also been observed by Ikenishi and Nieuwkoop (1978) in *Ambystoma mexicanum* PGCs but nothing is known about their functional significance.

Fishes

Little is known about the ultrastructure of the PGCs in fishes and the relevant literature is virtually restricted to the Teleostomi. The PGCs in the Agnatha. Chondrichthyes and Osteichthyes have so far only been studied with the light microscope, as described in the first part of this chapter.

Gamo (1961*b*) emphasised the fact that the PGCs in meroblastic embryos contain less yolk than those in holoblastic embryos. This may explain their tendency to associate themselves with nutritional sources, e.g., with the periblast in teleost embryos.

Pala (1970) mentions the presence of small PAS-positive granules in the perinuclear cytoplasm of teleost PGCs, particularly in the attraction sphere Nedelea and Steopoe (1970) were the first to describe the presence in the cytoplasm of PGCs of election-dense bodies, which they call nuage material. Satoh (1974) gives a more detailed description of the electron-dense bodies. They are up to 1 μm in diameter, not bounded by a membrane, and consist of an interwoven meshwork of very fine fibrils. They are often found in close association with clusters of mitochondria, in direct contact with their outer membranes. This description fits that of germinal plasm very well. Nothing is known, however, of the time of appearance of these structures in the development of the teleost PGCs.

The Ultrastructure and Biochemistry of PGCs in Amniota
Aves

So far electron-microscopic studies have only revealed the presence of common cell organelles. No mention has made of structures specific to germ cells, such as germinal plasm or nuage material. However, some cytochemical features can be used to characterise the PGCs in birds.

Although alkaline phosphatase activity had been demonstrated in chick PGCs by Chiquonie and Rothenberg (1975), the activity is too low to characterise them properly, since a similar activity was found in other cells. The most prominent feature is their high content of PAS-positive material. According to McCallion and Wong (1956) it is not an absolute criterion, since PAS-positive material is also found in other cells. We may say, however, that the combination of the various cytological and cytochemical features adequately characterises the PGCs in birds.

The PGCs in birds also go through some form of evolution during development. Whereas in early stages PGCs contain a large amount of yolk and lipids and only a small amount of glycogen, in later stages they contain less yolk and lipids, but abundant glycogen. The amount of glycogen drops again during migration. Clawson and Domm (1963*a*, *b*) and Fujimoto *et al.* (1976*a*) also observed changes in the ultrastructure and regional distribution of the glycogen: it is non-granular and evenly distributed in early stages, but granular and concentrated at one pole of the cell in later stages.

Mammalia

In the mammals McKay *et al.* (1953) were the first to demonstrate high alkaline phosphatase activity in the cytoplasmic rim of PGCs. Clark and Eddy (1975) and Fujimoto *et al.* (1977) mention that the alkaline phosphatase activity along the cell periphery is only faint in early stages but increases markedly in later stages. Asayama (1963) emphasises the fact that high alkaline phosphatase activity is not a feature specific to germ cells, since other cells may also show it. It is moreover stage dependent. During germ cell migration as well as during gametogenesis the germ cells seem to associate themselves with other tissues due to a lack of adequate endogenous energy sources. The alkaline phosphatase activity may thus be connected with an active intercellular exchange of material.

Asayama (1965) observed the presence of rather large amounts of PAS-positive material in mammalian PGCs. This feature is not specific either, since other cells also contain much glycogen. It disappears, moreover, with the onset of meoisis in the development of the female gametes, and at a later stage in spermatogenesis. Fujimoto *et al.* also observed a larger number of lipid droplets in human PGCs.

Eddy (1974) described in detail the presence of electron-dense, granulofibrous material, called 'nuage material', in the cytoplasm of mammalian primordial germ cell, which generally poor in inclusions. It was found either in the form of free, discrete bodies or as intermitochondrial cement, and was sometimes spatially associated with pores in the nuclear membrane. Eddy found these structures in young PGCs but also in the PGCs of the indifferent gonad and in the germ cells of neonatal and adult animals. These observations were corroborated by Motta and Van Blerkom (1974), who found nucleolus-like bodies in the cytoplasm of large oocytes as well as in PGCs of early developmental stages. During spermatogenesis Comings and Okada (1972) had found similar structures, probably formed by the extrusion of nucleolar components into the cytoplasm, which they called chromatoid bodies. Fawcen. Eddy and Phillips (1970), who had seen similar structures, did not assume them to be of nuclear origin. It is not yet clear whether these structures really are a constant feature of mammalian germs cells. While Eddy points out the strong resemblance of the nuage material of the germinal plasm of the anuran egg. Sud (1961) and Motta and Van Blerkom (1974) moreover call attention to its similarity to the chromatoid body found during spermatogenesis. However, in the latter RNA could not be demonstrated. It may finally be mentioned that during their migratory period mammalian PGCs show cytoplasmic lobopodia and filopodia containing numerous microfilaments.

Reptilia

Milaire (1957) mentioned that in the Chelonia both the extra-gonadal and the gonadal PGCs contain large amounts of PAS-positive material. The PGCs of Lacertilia, besides some glycogen, also contain numerous yolk granules and lipid-droplets.

The ultrastructure of the PGCs has been fairly extensively studied only in the Squamata. Hubert (1970*b*) describes the ultrastructural organisation of the PGCs in the Lacertilia. Structures specific to germ cells are not found in the cytoplasm but only in the nucleus. He mentions the presence of a paranucleolar mass as well as of a small, dense fibrillar corpuscle ('ring-shaped' in cross section): both are situated in the vicinity of the large 'ring shaped' nucleolus. When there are two nucleoli both structures are also present in duplicate. The paranuclear mass consists of a network of the fine fibrils and very fine particles. It is probably of a proteinaceous nature, since it is Feulgen-negative but Fast Green-positive and stains green with

Methyl Green Pyronine. A similar structure was described as 'fibrillar meshwork' in PGCs of anuran and urodele amphibians. No general plasm or nuage material has hitherto been demonstrated. In *Lacerta* the paranuclear mass is a constant feature of PGCs, of oogonia and oocytes, and of spermatogonia. Hubert (1968)found the same structure in the snake *Vipera* but not in the slow-worm *Anguis*, so that it may actually not be a constant feature in the squamata. The small fibrillar corpuscle is also found in spermatogonia of *Lacerta*.

During the migratory period the PGCs show micropseudopodia containing bundles of microfibrils, which provide evidence of amoeboid motility.

Conclusions

Summarising, it may be stated that the *general cytological features* of the PGCs are very much alike in the various groups of chordates. Although any of these features in itself is not specific for germ cells, the combination of various features quite satisfactorily serves for the unequivocal identification of PGCs. It must however by emphasised that during certain periods of development, particularly in the very early stages and during rapid multiplication, several of these features are rather indistinct or even unrecognisable. This had led to the suggestion of discontinuity in the germ line and in formation of germ cells *de novo* during certain period of development, particularly during gametogenesis.

Ultrastructural analysis, which unfortunately is still rudimentary or even lacking in certain groups, suggests quite strongly that the electron-dense fibrillo-granular bodies or the intermitochondrial cement found in the germinal plasm are structures specific to germ cells. They have not been found in all groups, however, so that they certainly do not represent a universal feature of germ cells.

It is intriguing that the PGCs, which show almost identical general characteristics in all chordate groups, nevertheless show pronounced differences in ultrastructure or cytochemistry. We have already mentioned the non-universality of the germinal plasm. Glycogen, which is so abundant in avian germ cells, is also found in mammalian and reptilian germ cells, but in smaller amounts, while it is virtually absent from the PGCs of amphibians and fishes. A high alkaline phosphatase activity of the cell periphery seems to characterise the mammalian PGCs. A special nuclear structure in the form of a fibrillar meshwork inside the nucleus, which is free of DNA but RNA-positive, is found in the PGCs of anuran and urodele amphibians as well as in those of certain reptiles.

During the last decades attention has been focussed particularly on the cytoplasmic features of the PGCs, so that specific nuclear structures may have been overlooked in a number of vertebrate groups. In our opinion it is not unlikely that nuclear structures may represent more constant and more specific features of the PGCs than the well-known germinal plasm.

3

The Extra-Gonadal Origin of the Primordial Germ Cells

Although the origin of the primordial germ cells is studied in details, still it has several controversies. Two controversial issues with respect to the origin of the PGCs in vertebrates have dominated the literature for many decades: (*a*) that of the gonadal versus extra-gonadal origin of the PGCs, and (*b*) that of the one-time versus repeated origin of the PGCs during the life span of the organism. We feel that these controversial issues should be dealt with before the place and mode of origin of the PGCs can be satisfactorily discussed.

Gonadal Versus Extra-Gonadal Origin of the PGCs

The early literature on the origin of the PGCs in vertebrates was dominated by the controversial issue of the gonadal versus extra-gonadal origin of the PGCs. Embryologists were sharply divided into opposing camps, i.e., those supporting Waldeyer's (1870) ideas and those advocating Nussbaum's (1880) concepts. *Waldeyer's theory*, which implies an origin of the PGCs from the somatic 'germinal epithelium' of the gonadal anlage, was primarily based on a study of the higher vertebrates, where the PGCs were not clearly distinguishable before they were inside the gonadal anlagen. The concept was later extended to all the vertebrates and even to a number of invertebrate groups, Nussbaum's notion, which implies an early segregation of the PGCs from the somatic cells of the embryo long before the formation of the gonadal anlagen and topographically separated from them was chiefly based on studies in the anuran amphibians and teleosts.

The issue remained controversial until the fourth decade of this century, as a result of the rather unspecific staining methods available. During the nineteen-thirties the balance between the two points of view began to shift towards extra-gonadal origin, particularly in the lower vertebrates. When Bounoure's book appeared in 1939 the advocates of Waldeyer's concept had lost nearly all their ground as regards the lower vertebrates, but not as regards the higher vertebrates, especially the mammals, where the PGCs could only be clearly-distinguished inside the gonadal anlagen.

Brambell (1960) already stated: It is now recognized by a large majority of embryologists that [in the vertebrates] the PGCs arise exceedingly early [in development], long before the rudiments of the gonad are formed and at a distance from the site they will [ultimately] occupy. . . Their state of origin is extra-gonadal in all instances and in the amniotes and some of the fishes it is [moreover extra-embryonic].

This marked shift in the general consensus was mainly due to the development of new histological and histochemical techniques which allowed identification of the PGCs a much earlier stage of development that was possible until then. Recent studies have fully confirmed Brambell's conclusions e.g., in the fishes.

One-time Versus Repeated Origin of the Germ Cells

Towards the end of the nineteenth and the first half of the twentieth century the concept of the origin of the PGCs was obscured by the controversial issue of whether there exist only a single source or multiple sources of germ cells during the life spon of an individual. The concept of the continuity of the germ line as postulated in Weismann's *Keimplasma* theory implies a segregation of the germ cells from the somatic cells-preferably during early development—as well as the persistence of the former throughout fertile life. The adversaries of this concept essentially claimed the existence of several phases of asexual and sexual life.

In his 1945 review Everett distinguished four different categories among those working on the origin of the germ cells in vertebrates: (1) those who deny the early segregation of the PGCs and advocate only a late, somatic origin from the 'germinal epithelium', (2) those who believe in the early segregation of germ cells but postulate the subsequent degeneration of this first generation and the ultimate formation of a second generation of definitive germ cells from the 'germinal epithelium', (3) those who agree to the early segregation of PGCs and the subsequent development into definitive gametes, but in addition postulate a supplementary formation of germ cells from the 'germinal epithelium', and (4) those who defend a one-time, early segregation of the PGCs, which constitute the only source of the definitive gametes. The issue is further complicated by the assumption either of a periodic or of a continuous formation of germ cells from the 'germinal epithelium' throughout the reproductive period.

We have seen in the preceding section that in the vertebrates the PGCs are formed outside the gonadal anlagen, to which they are subsequently transferred. The problem therefore boils down to the question of whether these PGCs represent the sole source of the definitive gametes or whether they are supplemented with or replaced by one or more generations of germ cells formed *de novo* in the gonadal anlagen during later phases. This uncertainty was mainly due to the fact that particularly in the higher vertebrates the PGCs lose most of their cytological characteristics during multiplication in the gonadal anlagen.

The problem can therefore only be definitely solved by experimental analysis. Nevertheless, the evidence for the existence of a single and uninterrupted germ line in the life cycle of vertebrates is both descriptive and experimental. The descriptive evidence consists of quantitative data on germ cell history in normal animals and sterile mutants, while the experimental evidence mainly relates to defect and grafting experiments.

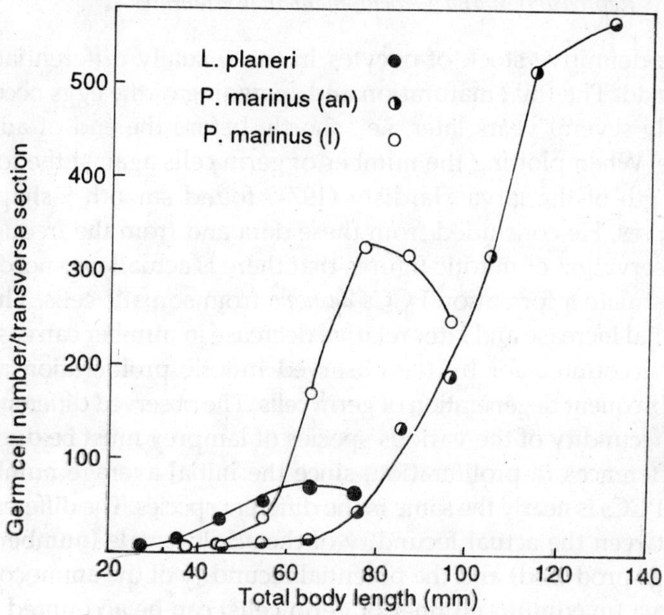

Fig. 3.1. Increase in average germ cell number per transverse body section in relation to total body length, in *Lampetra planeri* and in *Petromyzon marinus* of the anadromous (an.) and landlocked (l.) races.

Quantitative data

In the vertebrates *oogenesis* follows a uniform pattern with two main variants. Either germ cell multiplication continues *uninterruptedly* or *cyclically* throughout the reproductive period (as in most teleosts, all amphibians, the majority of reptiles and possibly some mammals), or it ceases early in life with the onset of meoisis, so that a definitive, finite stock of germ cells if formed for the entire fertile period (as in cyclostomes, elasmobranchs, a few teleosts, perhaps a few reptiles, all birds, the monotremes, and all eutherian mammals, with a few possible exceptions).

Following Okkelberg's (1921) study on the early history of the germ cells in the brook lamprey. Hardisty and Cosh (1966) followed the germ cell history in several Cyclostomata from the appearance of the germ cells to the establishment of the definitive stock of oocytes in the sexually differentiated gonad. The final maturation and spawning of the eggs occurs only several years later, i.e., shortly before the end of adult life. When plotting the number of germ cells against the total length of the larva Hardisty (1971) found smooth S-shaped curves. He concluded from these data and from the frequent observation of mitotic figures that there is actually no need to postulate a formation PGCs *de novo* from somatic cells. Their initial increase and later relative decrease in number can easily be accounted for by the observed mitotic proliferation and subsequent degeneration of germ cells. The observed differences in fecundity of the various species of lamprey must be due to differences in proliferation, since the initial average number of PGCs is nearly the same in the different species. The difference between the actual fecundity of the adult female (number of eggs produced) and the potential fecundity of the ammocoete larva (maximum number of germ cells) can be accounted for by the observed extensive degeneration of germ cells during sexual differentiation.

In the chick Hughes (1963) studied germ cell number in the left ovary from day 9 of incubation till the first day after hatching. The formation of oocytes ceases at about the time of hatching. Oogonial mitoses are responsible for the nearly 25-fold increase in germ cell number between day 9 and day 17, while the subsequent decrease by about 40% between day 17 the first day after hatching is due to the high incidence of degenerating germ cells. Hughes concludes that also in the chick ovary there is no need to postulate a second source of germ cells during oogenesis.

A similar study, on oogenesis in the rat, performed between day 14• *post coitum* (*p.c.*) and day 2 *post partum* (*p.p.*), shows that oogonia are mitotically active until day 17• *p.c.*, after which a sharp decline in mitotic activity occurs at the onset of the leptotene stage of meiosis. The total population increases sixfold between days 14• and 17• reaches a slightly higher maximum value at day 18•, and is reduced by about one-third, between day 18• *p.c.* and day 2 *p.p.* During this period four different waves of germ cell degeneration were distinguished.

In the rabbit ovary Chretien (1966) found two main periods of germ cell multiplication, a first one between days 10 and 12 *p.c.*, leading to a fourfold increase, and a second one between days 16 and 18 *p.c.* resulting in an approximately eightfold increase. The germ cell number reaches a maximum at day 26 *p.c.*, after which a marked decrease takes place by date 28 as a consequence of extensive degeneration. Since in the rabbit PGC migration starts at about day 10 and lasts at least until day 16 to 18, it is evident that PGCs actively multiply during migration. A similar germ cell history was found in the guinea pig by Ioannou (1964), in the monkey by Baker (1966), and in the human foetus by Baker (1963).

The most dramatic increase in germ cell number occurs in the ovary of the human foetus, viz. from 700-1300 during migration to approximately 600000 in the second month of gestation, and further to a maximum of around 7000000 in the fifth month. After several waves of degeneration the number falls to about 2000000 at birth and further to approximately 300000 at the age of seven. Since only about 500 eggs ovulate during the entire fertile period, there is a tremendous overproduction of female germ cells in human development.

Although in the human foetus oogonial multiplication reaches a maximum in the fifth month of gestation. Gailard (1950) observed extensive germ cell formation in explants of ovarian cortex of 14-week to 36-week human foetuses and concluded that the potency for germ cell multiplication is not restricted to the early foetal period but certainly persists until birth. His explants first showed complete degeneration of parenchyma and young oocytes, after which new cord-like structures were formed from the 'germinal epithelium' which surrounds the explant. In these sex cords new oogonia appeared in large numbers. He left open the question of whether these oogonia originated from persisting but unrecognisable germ cells or were formed *de novo* from the somatic cells of the 'germinal epithelium'.

In the mouse Rudkin and Griech (1962) found that [^{3}H] thymidine is incorporated into oocyte nuclei when administered half-way through the gestation period. The label was still found in the oocytes of female offspring at 6 weeks *p.p.*, suggesting that the PGCs present at the time of labelling had actually formed mature oocytes in the adult female. Borum (1966) found up to 100% labelling in the oocytes of female offspring after [^{3}H] thymidine injections into pregnant mice at days 12 to 15 of gestation, but no incorporation into oocytes upon injection at different times after birth. He concluded that all oocytes present in the adult mouse have persisted from foetal life, so that at birth the mouse is furnished with a definitive population of oocytes from which all mature ova will later be derived. Peters and Crone (1967) showed that DNA synthesis occurs in oogonia in premeiotic interphase, the rate of synthesis falling during the foetal period in the mouse but within the neonatal period in the rabbit. Once synthesized the DNA persists in the growing and maturing oocytes. Kennelly and Foote (1966) concluded from [^{3}H]methyl-thymidine injections into female rabbits on the day of birth and at 4 weeks *p.p.* that most, if not all definitive ova have already been formed at birth and that oocytogenesis *de novo* does not occur in the post-pubertal rabbit.

Unfortunately little is known about PGC numbers in forms which show a continuous or cyclic multiplication of the germ cells during the reproductive period, all the above forms belonging to the second variant of germ cell multiplication.

Spermatogenesis is usually characterised by either continuous or cyclic renewal of the germ cell population. In mammals it has been studied in recent years by a number of workers. In these studies emphasis was placed on renewal of the spermatogonial stem cells and on the phenomenon of germ cell degeneration.

Hilscher and Hilscher (1979) state that when female and male gametogenesis in mammals are compared the 'gonia' stage of the female germ cells shows only a single proliferation wave, whereas that of the male germ cells shows a first proliferation wave, which is comparable with that of the oogonia, followed by a second wave after a preparatory interphase. The second proliferation wave is characterised by both stem cell renewal and differentiation into spermatocytes. According to Clermont (1972) the seminiferous epithelium of the mammalian testis is composed of five to six generations of germ cells. The spermatogonial population is renewed by a number of successive mitoses (three in man and up to six in the rat). Two types of spermatogonial stem cells can be distinguished, the type A 'reserve' and type A 'renewing' stem cells. The former type is not involved in the production of spermatocytes, while the latter type renews itself and simultaneously gives rise to differentiating spermatogonia at each mitotic cycle. However, the 'key' division between stem cell and differentiating spermatogonium does not have the character of a differential in the morphological sense. The mechanism which regulates this choice is still unknown.

The general conclusion from the quantitative analysis of both oogenesis and spermatogenesis in the vertebrates is that there actually is no need to postulate any formation of germ cells *de novo* from somatic cells during any part of development or fertile life. It is therefore very likely that all the definitive male and female gametes descend from the initial population of germ cells which has arisen extra-gonadally during early embryonic development.

Experimental evidence

Several experiments have been performed related to the surgical removing of primordial germ cells. Elimination of the PGCs has been achieved surgically by removing that portion of the embryo in which the PGCs are located at the time of operation. It is a well-established fact that complete removal of the gonads leads to permanent sterility, but incomplete removal may lead to regeneration of the gonad and recovery of fertility.

In amphibians removal of the extra-gonadal source of the PGCs prior to their migration to the gonadal anlagen leads to the formation of gonads without PGCs and to permanent sterility. In neurulae of *Ambystoma mexicanum* Nieuwkoop (1947) removed the presumptive lateral plate mesoderm—the source of the PGCs in the urodeles—and obtained larvae which had more or less normal genital ridges but lacked PGCs. Vivien (1964) obtained similar results in *Lebistes* and *Xiphophorus* upon ^{32}P administration. Blackler (1962, 1965a) replaced the fertile region of the endoderm of *Xenopus laevis* neurulae—the source of PGCs in the anurans—by a similar, more anterior region of the endoderm which does not contain PGCs, and obtained completely normal-looking tadpoles with normal gonadal anlagen but without PGCs. When reared to maturity the animals remained sterile.

In the anurans similar results were obtained by the elimination of the so-called 'germinal plasm', which is assumed to act as a germ cell determinant. Pricking of the vegetal pole or UV-irradiation of the vegetal hemisphere performed at the 1-to 4-cell stage can lead to sterility of the gonadal anlagen of the larva and to subsequent permanent sterility.

It may therefore be concluded that destruction of the PGCs or their putative 'determinants' leads to permanent sterility. In other words, no formation of germ cells *de novo* occurs in animals which have no PGCs from an early stage of development. Although this seems a strong argument in favour of the continuity of the germ line, it is essentially negative and therefore inconclusive, for formation of germ cells *de novo* from somatic cells of the 'germinal' epithelium could depend upon a stimulating influence from existing viable germ cells. This rather unlikely postulate can only be disproved by heteroplastic or xenoplastic recombinations of presumptive germ cells and somatic gonadal tissue or by using specific nuclear markers. Such experiments were performed by Blackler and Fischberg (1961), who exchanged the fertile endoderm region between two strains of *Xenopus laevis* of which one contained the Oxford nuclear marker. Blackler (1962) made a similar exchange between two subspecies of *X. laevis* while Blackler and Gecking (1972*a*, *b*) performed the same experiment between *X. laevis* and *X. mulleri*, with subsequent interspecific and interspecific matings. In all these experiments the germ cells showed the characteristics of the donor type that furnished the fertile endoderm region.

Similar experiments were carried out in urodeles by Smith (1964), who exchanged the ventro-lateral marginal zone (presumptive lateral plate mesoderm, representing the source of PGCs in the urodeles) between gastrulae of the white and the black axolotl. The successful cases in which the exchange was complete showed only progeny of the donor type, as tested by mating as well as histological examination. Because in Blackler's experiments the overlying ectoderm and mesoderm were also exchanged, while in Smith's experiments the graft not only furnished the source of the PGCs but also presumptive lateral plate mesoderm, it is possible, though rather unlikely, that the gonadal anlage, which is normally formed from the intermediate mesoderm, partly developed from graft tissue. Disregarding this miner objection, these experiments demonstrate that the presence of PGCs does not lead to a formation of germ cells *de novo* from the somatic components of the ground. Moreover, no recent publications have led to any claim contradictory to the conclusions stated above. The most elegant proof of the continuity of a single germ line was furnished in birds. First the extra-gonadal and extra-embryonic location of the PGCs in the so-called anterior germinal crescent of the blastoderm at early somite stages was demonstrated by surgical extirpation, cauterisation. UV-irradiation, γ-irridiation and X-irridiation.

In birds the PGCs are transported by the blood stream from the extra-embryonic germinal crescent to the gonadal-anlagen. Reynaud (1969) was above to obtain repopulation of the gonadal anlagen after intravenous injection of a PGC suspension made from germinal crescent endoderm into a host embryo which had previously been sterilised by UV-irradiation of its germinal crescent. When he made xenoplastic recombinates of turkey PGCs with chick hosts or vice versa, he found that the gametes formed in the F1 generation were all of donor type. The colonisation of the host gonadal anlagen by foreign germ cells constitutes a crucial experiment, demonstrating that in birds the PGCs formed during early development are the sole precursors of the définitive gametes.

In mammals the evidence for the existence of a single and uninterrupted germ line is not as conclusive as in amphibians and birds, but is nevertheless fairly strong. X-irradiation (more than 168r) of mouse gonadal anlagen after their colonisation with PGCs leads to the development of sterile gonads, which nevertheless who more or less nòrmal proliferation of the gonadal epithelium. Mouse gonadal epithelium — prior to colonisation with PGCs — when grated into the kidney capsule of host embryos remains sterile, whereas similar grafts made after colonisation with PGCs form typical testicular or ovarian tissues containing germ cells. These results plead strongly against any formation of germ cells *de novo*.

Further evidence comes from genetical studies. Several workers have described several alleles of a mutation in mice called *dominant white spotting (W)*, which in homozygous condition leads to sterility at birth. In the mutants the PGCs are formed at 8 days of gestation in the yolk sac endoderm in the normal number, which ranges from 10 to 100. The cells behave normally and migrate towards the gonadal anlagen. However, in normal mice the number of PGCs increases exponentially and reaches a value of approximately 5000 at the end of the migration period, whereas in the mutants the number of PGCs does not increase during migration. They subsequently degenerate, leading to total and permanent sterility of the gonadal anlagen. Twenty-five per cent of the offspring of heterozygous parents of the *W* series show the defect and are sterile at birth. This germ cell deficiency, which begins to manifest itself on the ninth day of embryonic development and is fully expressed at birth, furnishes a strong argument against any formation of germ cells *do novo* from the gonadal epithelium. This conclusion is further supported by culture *in vitro* of sterile mutant half-gonads fused with younger normal fertile half-gonads. In these chimaeric explants epithelial cells of the mutant gonad are in close contact with normal PGCs but no proliferation of mutant germ cells was ever observed. A similar germ cell deficiency seems to exist in the steel mutant (*Sl*) described by Bennett (1956). Radiation-induced damage (400 r) likewise results in depletion of the PGC stock leading to permanent sterility. At lower doses the gonads may become repopulated, but all evidence points towards repopulation by remaining germ cells and not from somatic cells of the gonadal epithelium.

Conclusions

Summarising, it may be concluded that in the vertebrates the PGCs have an extra-gonadal origin during early embryonic development. The concept of a continuous, single germ line originating in early embryonic development has become highly probable for amphibians and birds and likely for mammals, while there is circumstantial evidence for it in agnathan fishes. Moreover, in recent years no compelling evidence against the general validity of this concept has been presented. Although further analysis is highly desirable, particularly in fishes and reptiles, it may be assumed for the present that in the vertebrates germ cell history is characterised by the one-time origin of PGCs during early development and that the resulting PGC population gives rise to all the definitive gametes of the adult.

Waldeyer (1870) introduced the term 'germinal epithelium' for the putative germ-cell producing outer layer of the gonadal anlage. The now convincingly demonstrated extra-gonadal origin of the PGCs makes this term, which is unfortunately currently used in the literature, a very confusing one (see also

Roosen-Runge. 1977). It would be better to speak of 'gonadal epithelium' when describing the proliferative layer of the somatic gonadal anlage, and this is the term we shall use in the following chapters.

In the lower chordales the problem of the origin of the PGCs is still unsolved. Moreover, gonadogenesis differs markedly from that in the vertebrates. Since the further history of the germ cells has not been studied, the conclusions drawn above for the vertebrates cannot yet be extended to the entire phylum of the chordates.

4

Site and Mode of Origin of the Primordial Germ Cells

Before considering the site and mode of Origin of primordial germ cells, a clear cut distinction must be made between descriptive and experimental evidence. In the older literature the place where PGCs were first identified during embryonic development was often taken as synonymous with their actual site of origin. We know now that the PGCs are formed extra-gonadally and have amoeboid motility, migrating though various tissues towards their final location in the gonadal anlagen. Although migration usually occurs during later stages of development (e.g., early larval stages), the possibility cannot be excluded that active displacements also take place during early stages. This may for instance be the reason for the often-observed accumulation of PGCs in the endo-mesodermal interspace in the Anamnia. Descriptive evidence may therefore only be considered as circumstantial.

It is clear that the earlier the stage of development at which the PGCs can be recognised, the more likely it is that the place where they are first identified corresponds to the actual site of origin. However, conclusive evidence for the site of origin of the PGCs requires experimental analysis. We may only learn about the causal factors responsible for germ cell formation (in other words, their mode of origin) from adequately planned experiments. These are virtually lacking in the lower chordates, and are mainly restricted in the Anamnia to the amphibians, and in the Amniota to the birds. The anuran and urodele amphibians will therefore be discussed first, followed by the various groups of fishes and the cephalochordates and tunicates. Among the amniotes the birds will be treated first, followed by the mammals and reptiles.

THE ANAMNIA

Amphibia

Anura: The history of the germinal plasm. Since the discovery of the so-called 'germinal plasm' in the egg of *Rana temporaria* by Bounoure in 1931 and the elucidation of its subsequent history through embryogenesis, the anuran amphibians have served as the classical example of the presence of an uninterrupted germ line in the vertebrates. These observation agree very well with the results of similar studies in several insects and other invertebrates (for further information). The correspondence in germ cell history, as well as the striking similarity in ultrastructure of the germinal plasm in groups of the animal kingdom which otherwise differ so much, has led to the idea that the presence of an uninterrupted germ line from the fertilised egg to the adult organism is a basic mechanism for maintaining organismal identity through successive generations. We must however test whether such a postulate is actually justified, and decide to what extent other mechanisms may also be represented in the chordates.

The fate of the germinal plasm during cleavage, gastrulation, neurulation and further embryogenesis was carefully studied for the first time by Bounoure in the years 1931 to 1934. In the vicinity of the vegetal pole of *Rana temporaria* eggs he found small subcortical patches of a special cytoplasm which he called 'germinal plasm'. During the first two cleavages these are more or less evenly distributed among the four blastomeres and are partially pulled inwards along the cleavage furrows. In subsequent cleavage cycles they are usually transferred to only one of the two daughter cells (differential division), and consequently are distributed among a restricted number of vegetal blastomeres. As a consequence of the pre-gastrulation movements the blastomeres containing germinal plasm are displaced together with other endodermal cells towards the centre of the vegetal yolk mass and sometimes even as far as the floor of the blastocoel. At late blastula or early gastrula stages the germinal plasm, which was hitherto situated at the periphery of the cell, is displaced intracellularly to a juxta-nuclear position, which is its' final location in the PGC. Subsequently the PGCs are displaced along with the vegetal yolk mass by the gastrulation and neurulation movements, and are ultimately found among the vegetal blastomeres in the caudal portion of the embryo.

In the anurans a distinction must be made between true PGCs and their forerunners, the presumptive PGCs (pPGCs). During the first two cleavage divisions and the succeeding differential blastomere divisions one can only speak of pPGCs. One may define true PGCs as cells which are characterised by the presence of germinal plasm and which divide by clonal divisions. When the PGCs are first formed they do not divide very frequently, so that a considerable time may elapse between the last 'unequal' division and the first clonal division. It is interesting that this transition from presumptive to true PGCs coincides with the centripetal displacement of the germinal plasm. 'Unequal' divisions occur as long as the germinal plasm is situated peripherally, while 'equal' divisions take place when the germinal plasm has reached its juxta-nuclear position.

It is evident from the literature that the anurans great significance is attributed in the germinal plasm for the determination and differentiation of the PGCs. However, what is the present evidence for this assumption? To test this question the following approaches have been made (*a*) removal or destruction of the germinal plasm or its transplantation to an egg lacing functional germinal plasm, and (*b*) analysis of the chemical composition and possible function of the germinal plasm.

The significance of the germinal plasm for germ cell formation.

As already described above. PGCs develop from blastomeres which contain germinal plasm. A number of workers have tried to remove the germinal plasm surgically. Nieuwkoop and Suminski (1959) were among the first to try to remove the germinal plasm at the 2-to 4-cell stage of *Xenopus* by superficial pricking of the vegetal pole region without causing a considerable outflow of cytoplasm. In general this did not lead to a marked reduction in the number of PGCs in the developing larvae. Fischiarolo (1960), using *Discoglossue* eggs found normal gonads with germ cells in metamorphosed larvae after exovation of about 40% of the egg contents as a consequence of pricking the vegetal region. Librera (1964) obtained varying results after withdrawal of about one-quarter of the egg contents with a micropipette inserted into the vegetal region of uncleaved eggs, 4-cell, blastula and early gastrula stages of *Discoglossus*. Of the 49 animals which reached metamorphosis, 32 had normal gonads, 5 had gonads with only a few germ cells, and 12 were completely sterile. Buehr and Blackler (1970) found either complete sterility or a marked reduction in the number of PGCs at the tadpole stage after pricking the vegetal pole at the 2-to 4-cell stage in *Xenopus*. They checked that some or neariy all of the germinal plasm was present in the exovate and found a quantitative relationship between the amount of germinal plasm remaining and the number of PGCs in the tadpole. From these experiments they concluded that the germinal plasm is indispensable for germ cell formation and that at least upto the tadpole stage germ cells are not formed from secondary sources. Gipoulous (1971, 1975) confirmed these results in *Discoglossus*, applying numerous superficial punctures to the vegetal pole region at the 2- to 4-cell stage. He ascribed the negative results of Nieuwkoop and Suminski to the experimental technique used, which did not lead to any substantial outflow of cytoplasm.

In 1937 Bounoure described for the first time the effect of UV-irradiation on the germinal plasm in the vegetal hemisphere of the fertilised anuran egg. He observed a dramatic reduction (up to *c.* 90%) in the number of PGCs in the treated animals at metamorphosis. Although some animals had very few germ cells, he never actually obtained complete sterility. The gonads, though reduced in size, showed a normal somatic structure. Aubry (19533*a*, *b*) obtained still better results with UV-irradiation combined with animal-vegetal compression of the eggs. Padoa (1963*b*) studied UV sensitivity at different stages of development and found it to be highest in the period from shortly before the initiation of first cleavage till the beginning of second cleavage. He ascribed this to the accumulation of the germinal plasm in the vicinity of the vegetal pole shortly before first cleavage. The decrease in sensitivity after second cleavage seems to be due to the inward displacement of part of the germinal plasm along the cleavage furrows. This interpretation is supported by the observation that irradiation of the vegetal pole of the blastula no longer has any effect, since nearly all the germinal plasm is by now situated in the interior of the embryo. Smith (1966) obtained complete sterility in nearly 100% of the developing larvae after UV-irradiation with 8000 erg/mm^2 of the vegetal hemisphere of *Rana pipiens* eggs at the beginning of first cleavage, whereas similar irradiation of the animal pole had no effect. By testing the effect of different wavelengths he obtained a UV spectrum for PGC inactivation. The maximally effective wavelength of 254 nm corresponds to the UV sensitivity of nucleic acids. Tanabe and Kotani (1974) also found complete sterility after irradiation of the vegetal hemisphere of the 2-cell stage of *Xenopus* with UV of wavelength 253.7 nm and dose *c.* 6000 erg mm^2. When the stage of irradiation was gradually shifted from the 2- to the 4-cell stage an increasing number of PGCs was found in the larvae. When the dose of UV-irradiation at the 2-cell stage was decreased from 6000 to 750 erg/mm^2 the number of germ cells found in tadpoles increased proportionally. The number of germ cells was also inversely proportional to the irradiated area.

In *Xenopus* Ikenishi, Kotani and Tanabe (1974) observed a swelling and vacuolation of the mitochondria and a fragmentation of the germinal granules as a direct effect of UV-irradiation of the germinal plasm. They concluded that maintenance of the normal structure of the germinal plasm is indispensable for its role as germ cell determinant, however, warn against the conclusion that the effect of UV-irradiation represents a specific inactivation of the germinal plasm. They observed an inhibition of cleavage in the vegetal hemisphere in *Xenopus*, leading to a markedly abnormal segregation of the germinal plasm into particulars blastomeres, and suggest that the sterilising effect of UV-irradiation is more likely to be due to a derangement of the normal cell lineage than to a direct deteriorating effect on the germinal plasm. Grant, Wacaster and Turner (1976) observed a disturbance of the cleavage process in the UV-irradiated vegetal half of the anuran egg and a subsequent delay in the gastrulation process.

The most elegant experiment indicating the great significance of the germinal plasm in PGC determination and differentiation was done by Smith (1966). He found that he could restore the germ line in *Rana pipiens* eggs previously sterilised by UV-irradiation by injecting cytoplasm from the vegetal pole of normal eggs, but that cytoplasm from the animal hemisphere was ineffective. However, whole vegetal cytoplasm was injected, *not* pure germinal plasm, so that this is not absolute proof of the indispensability of the germinal plasm. Nevertheless the restoration of the PGC-forming capacity by injection of certain fractions of a crude homogenate of vegetal pole plasm into UV-irradiated *Rana* eggs suggests that germinal granules rather than mitochondria are indeed involved.

Very recently Dixon (personal communication) found that in contrast to all previous observations in *Xenopus* early PGC formation is probably not affected at all by UV-irradiation, but that the migration of the PGCs toward the genital ridges is markedly retarded, so that the gonadal anlagen become colonised only at a late stage of development. This finding, which requires further confirmation, seems to call for a reinterpretation of several of the results mentioned above.

A positive regulation after UV-irradiation towards the normal germ cell number at metamorphosis was noted by Bounoure *et al.* (1954). This may explain the negative results of Fischiarolo (1960) as well as the varying results obtained by Librera (1964). In Gipouloux's (1972) experiments negative regulation occurred, i.e., a restoration of the normal number of germ cells after an initial excess in a parabiont with one set of gonadal anlagen and two endodermal masses as sources of PGCs. Positive or negative regulation must be due to an increases or decrease in the number of mitoses in the PGCs between the tadpole stage and metamorphosis. Bounoure *et al.* (1954) actually observed a much higher mitotic rate in the PGCs of animals with a markedly reduced number of germ cells. The fact that Bounoure and Librera obtained metamorphosing tadpoles with only a few PGCs suggests that regulation may be impaired when the number of germ cells falls below a certain minimum level. This lack of regulation may, however, also be due to the remaining germ cells being damaged too much by the treatment used.

Composition and possible function of the germinal plasm. Gipouloux (1975) distinguishes three successive phases in germ cell formation. The first phase is characterised by the peripheral position of the germinal plasm during cleavage (pPGCs). In the second phase the germinal plasm occupies a juxta-nuclear position in contact with the nuclear membrane (PGCs). According to Gipouloux germ cells determination occurs in this phase, which is moreover characterised by an arrest of mitosis. The third phase is that of cellular-differentiation of the PGCs in the gonadal anlagen.

Several authors have studied the composition of the germinal plasm in order to analyse its possible function in germ cell determination.

The UV sensitivity of the vegetal hemisphere of the anuran egg seems to correspond to that of nucleic acids (Smith, 1966). However, these nucleic acids are not necessarily in the germinal granules themselves, but may be those in the ribosomes and polysomes that are associated with the germinal granules at the unfertilised egg to early cleavage stages or even the mitochondrial DNA, the germinal plasm being very rich in mitochondria.

Smith (1965) investigated the capacities of germ cell and somatic nuclei for supporting normal development when injected into enucleated eggs of *Rana pipiens*. Eggs with germ cell nuclei showed a much higher incidence of cleavage than eggs with somatic nuclei (43% compared with 18%), while 40% compared with 10% of completely cleaved eggs formed normal embryos. He concludes that the nuclei of PGCs are apparently still totipotent, whereas the somatic nuclei have lost part of their potency. He therefore makes the suggestion that the germinal plasm in effect protects the nucleus from becoming channelled into a somatic pathway of differentiation.

Smith and Ecker (1970), reviewing the relevant literature, came to the conclusion that nucleic acids in the form of stable templates are present in the germinal plasm and act as germ cell determinants. Using autoradiography, Eddy and Ito (1971) failed to demonstrate any RNA synthesis in the electron-dense bodies of ovarian oocytes but found a rapid incorporation and turnover of amino acids. They also failed to detect any RNA cytochemically. They pointed out the great similarity between the electron-dense bodies found in oogonia and oocytes and the germinal plasm found in PGCs. Wallace, Morray and Langridge (1971), studying gene amplification during gametogenesis, inferred the presence of self-replicating ribosomal DNA particles in the nucleolar core of oocyte nuclei. They suggest that during maturation these particles pass into the cytoplasm and end up in the germinal plasm. It must however be stressed that DNA has never been demonstrated in germinal plasm.

Deuchar (1972), who also considers a large store of ribosomal DNA as a characteristic of the germ line, assumes that during embryonic development and early larval life this DNA is transcribed into RNA, which is sequestered in the germinal plasm. Webb (1976) demonstrated ribosomal RNA synthesis associated with nuage material in oogonia. He suggests that the negative results of Eddy and I to (1971) may be due to a protective effect resulting from the predominantly proteinaceous nature of the germinal plasm. Dziadek and Dixon (1977) found three phases of DNA synthesis in PGCs of *Xenopus laevis*, viz. in the stages 10 to 12, 22 to 24, and 37 to 39 no DNA synthesis occurring in the intervening periods. The PGCs distinguish themselves from surrounding endoderm cells as early as the mid-gastrula, since in the endoderm cells DNA synthesis is not temporarily arrested. In the PGCs RNA synthesis begins in the mid-gastrula (stage 11) and ceases at the early tail-bud stage (stage 24). Dziadek and Dixon did not find RNA synthesis in the germinal plasm and conclude that it is probably confined to the nucleus. Zust and Dixon (1977) state that the germ cell lineage consists of a number of successive cell generations, so that the process of germ cell differentiation is a stepwise process. This may also hold for the process of germ cell determination.

Both the descriptive and the experimental evidence point to a crucial role for the germinal plasm in the determination and differentiation of the PGCs in the anuran amphibians. Only those blastomeres which contain a sufficient (?) amount of intact germinal plasm, or which receive it after previous UV treatment, develop into PGCs. Smith (1966) therefore calls the germinal plasm of the anuran egg *a germ cell determinant*. This means that the presence or absence of this plasm — more particularly of us nucleic acid component-determines whether a cell will become a PGC or a somatic cell. The present authors feel, however, that one should be cautious in drawing such a far-reaching conclusion. So long as the germinal plasm or its active component has not been isolated and tested separately, Smith's conclusion must be considered premature. On the basis of the available evidence it can be said only that the vegetal pole region of the anuran egg contains factors which apparently act in germ cell determination.

Urodela

Descriptive evidence. The descriptive evidence concerning the site of origin of the PGCs in the urodeles is somewhat scanty. On the basis of their yolk content Abramowicz (1913) concluded that in *Triturus* the PGCs arise from the endoderm and migrate into the mesoderm at a relatively advanced stage of development. In *Hemidactylium* and other urodeles, however, Humphrey (1925) identified the PGCs at an early tail-bud stage (from the 7-somite stage onwards) in the medial and later in the dorsal region of the lateral plate mesoderm. He stated that there is no evidence for Abramowicz's conclusion that their initial position is in the endoderm. Humphrey's observations were confirmed by Nieuwkoop in *Triturus* embryos at older tail-bud stages and recently in *Ambystoma* by Ikenish and Nieuwkoop (1978), who could not, however, trace the PGCs back to stages earlier than stage 23 (H.).

Experimental evidence. Humphrey (1929) confirmed the localisation of the PGCs in the medial to dorsal lateral plate mesoderm by unilateral extirpation any by transplantation of this portion of the mesoderm to the ventro-lateral side of host embryos at stags 21 to 25. The former experiments led to the complete absence of PGCs on the operated side and the latter to additional germ cell formation. Nieuwkoop (1947) carried out a more systematic analysis of the origin of the PGCs in the urodeles. Neither removal of vegetal cytoplasm at the I-cell stage, nor removal of the central portion of the vegetal blastomeres at the early gastrula stage in *Ambystoma* and *Triturus*, respectively, affected the number of PGCs in the developing larvae. Removal of the entire endoderm from early neurulae led to endoderm-free larvae containing a restricted number of PGCs situated in that portion of the mesoderm which had already invaginated and had apparently come into contact with the caudal endoderm prior to operation. Transplantation of the entire endoderm between neurulae of different *Triturus* species and between *Triturus* and *Ambystoma* showed that in the chimaeric embryos the PGCs always belonged to the species which had furnished the ecto-mesodermal envelope. Removal of the presumptive lateral plate mesoderm at the early neurula stage led either to complete sterility or to a very marked reduction in the number of PGCs, the remaining PGCs being situated in the unaffected, most cranial section of the fertile region. Heteroplastic transplantation of ventro-lateral marginal zone (presumptive lateral plate mesoderm) at an early yolk-plug stage (11-11• H.) gave rise to chimaeric larvae in which the regional distribution of the PGCs of the two species corresponded to the regional composition of the chimaeric lateral plate mesoderm. These experiments demonstrated unequivocally that in the urodeles the PGCs are of mesodermal origin and are evenly distributed throughout the presumptive lateral plate mesoderm.

Contrary to Nieuwkoop, Takamoto (1953) observed complete absence of PGCs after removal of entire endoderm from late neurulae of *Triturus pyrrhogaster*, but in Amanuma's (1958) experiments extirpation of the dorso-caudal endoderm of early neurulae of *T. pyrrhogaster* did not affect the number of PGCs. Amanuma (1957) moreover found that extirpation of presumptive lateral plate mesoderm at mid-gastrula or early neurula stages of *T. pyrrhogaster* and *Hynobius nebulosus* yielded larvae with a varying number of PGCs. He assumed that regulation occurred and concluded that the determination of the PGCs is not yet completed at the neurula stage. Orthotopic and heterotopic transplants of lateral plate mesoderm at tail-bud stages of *T. pyrrhogaster* by Asayama (1950, 1961) led to similar conclusions.

Lacroix and Capuron (1966) concluded from grafts of intermediate mesoderm of a tail-bud stage (stage 26) between Iberian and Moroccan races of *Pleurodeles* (which are characterised by a different lampbrush chromosome structure) that at this stage the PGCs are actually localised in the intermediate mesoderm. Maufroid and Capuron (1972) used defect experiments to determine the exact location of the PGCs in the lateral plate mesoderm at neurula and tail-bud stages of

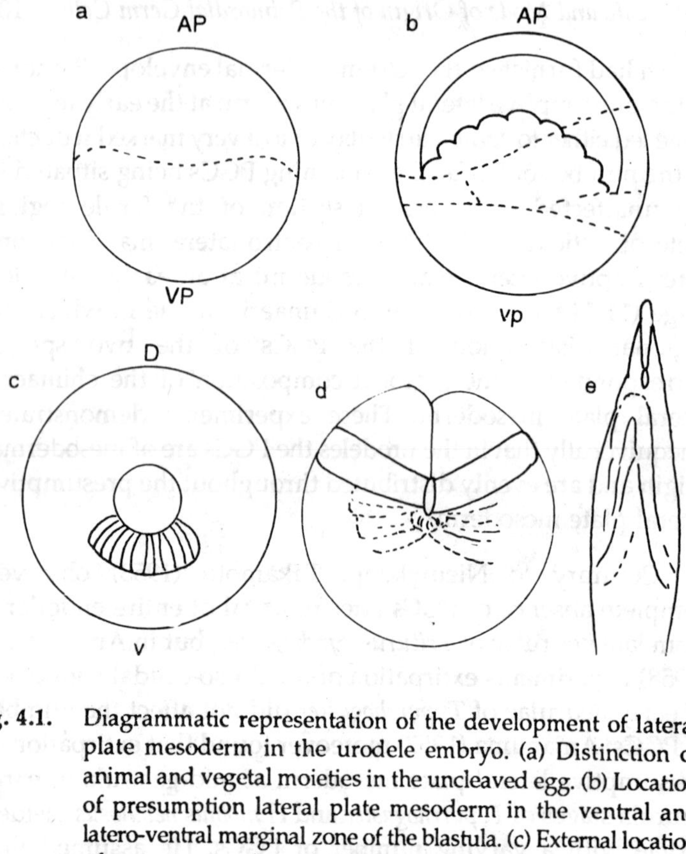

Fig. 4.1. Diagrammatic representation of the development of lateral plate mesoderm in the urodele embryo. (a) Distinction of animal and vegetal moieties in the uncleaved egg. (b) Location of presumption lateral plate mesoderm in the ventral and latero-ventral marginal zone of the blastula. (c) External location of presumptive lateral plate mesoderm at yolk-plug stage. (d) Formation of left and right wings of lateral plate mesoderm (solid lines uninvaginated, broken lines invaginated). (e) Cranial and ventral extension of lateral plate mesoderm at tail-bud stage; ventral view. A.P., animal pole; V.P., vegetal pole.

Pleurodeles, and found that in the early neurula the PGCs are located in the neighbourhood of the blastopore, from where they are displaced anteriorly (along with the lateral plate mesoderm) during mesodermal mantle formation. At tail-bud stages (stages 21 to 23, G and D) the PGCs move inside the lateral plate mesoderm from a ventral and lateral position towards the vicinity of the Wolffian ducts. Ablation of lateral plate mesoderm containing PGCs did not lead to any formation *de novo* from adjacent mesoderm. The seemingly contradictory results of Asayama and Amanuma can be adequately explained on the basis of the exact location of the PGCs at successive stages of development. Maufroid and Capuron (1973) transplanted presumptive ventral and lateral marginal zone of early gastrulae of *Pleurodeles* into the blastocoel of host embryos of the same age, and observed the formation of extra ventral mesoderm with PGCs. They concluded that the PGCs are of mesodermal origin and are distributed throughout the entire ventral and ventro-lateral marginal zone at the early gastrula stage.

Asayama and Amanuma (1957) grafted an extra dorsal blastoporal lip into the ventral marginal zone of a host embryo, and concluded from the distribution of PGCs in the induced secondary anlage that PGCs are induced *de novo* in the secondary embryo. The conclusion seems to be at variance with the observation by Capuron (1968) that in *duplicitas* embryos the total number of PGCs never exceeded that of a normal embryo and was often considerably lower. This would indicate that the PGCs of the secondary embryo come from the same stock as those of the primary embryo and are not formed *de novo* from mesodermal tissues. However, the lateral plate mesoderm formation was not investigated in these *duplicitas* embryos.

Isolates of the ventral equatorial region of urodele blastulae, used as controls for mesoderm induction experiments, in a small number of cases formed some ventral mesodermal structures, which frequently contained PGCs. Maufroid and Capuron (1978) concluded from isolation and recombination experiments involving ventro-lateral marginal zone and ventral yolk endoderm that in *Pleurodeles* during gastrulation an additional inductive action by the endoderm plays a role in PGC formation. Similar experiments with ventral marginal zone by Nieuwkoop and Sutasurya (unpublished) suggest, however, that in *Ambystoma* PGC formation does not require an additional inductive action by the endoderm. Here PGC determination seems to occur very early in development, the

PGCs constitution one of the self-differentiation pathways of the ventral mesoderm. These contradictory results show that in the urodeles the time and mode of determination of the PGCs are still an open question.

UV-irradiation of the vegetal hemisphere of the *Pleurodeles* egg has no effect on PGC formation when performed at cleavage, blastula or early gastrula stags, but leads to a dramatic reduction of the number of PGCs or even to complete sterility when applied at a late yolk-plug stage. At that stage the presumptive ventro-caudal mesoderm — the source of the PGCs — is localised in the ventral blastoporal lip and is therefore accessible to the irradiation. In our opinion this result indicates that the PGCs pass through a sensitive period at the late gastrula stage, which possibly represents a (first) decisive step in their determination.

Attention must now be drawn to a different line of approach. Kotani (1957) replaced presumptive lateral plate mesoderm of early gastrulae of *T. pyrrhogaster* by presumptive epidermis and found PGC formation in the grafts. He concluded that the presumptive epidermis of early gastrulae possesses germ-cell-forming potencies. Smith (1966) replaced ventro-lateral marginal zone of stage 11 gastrulae by presumptive epidermis of early gastrulae, using white and black axolotls. When testing the progeny of the operated embryos, he did not find PGC formation of graft origin. Kotani's results were later confirmed by Kocher-Becker and Tiedemann (1971), who found PGC formation in partially mesodermised and endodermised 'ectodermal' explants from *Triturus* early gastrulae treated with a vegetalising factor isolated from chick embryos.

In the course of the analysis of mesoderm induction in the urodeles many workers found that PGCs are a normal constituent of induced ventral mesoderm. When studying the mesoderm inducing capacities of the dorsal, lateral and ventral portions of the yolk mass found that both the lateral and ventral portions induce predominantly ventral mesoderm, while the dorsal portion induces massive dorsal, axial mesodermal structures. When comparing the inductive capacities at various stages in *Ambystoma* they found that by a very early gastrula stage (stage 10H.) the mesoderm-inducing capacity of the dorsal endoderm has virtually vanished, that of the lateral endoderm is already markedly reduced, and that of the ventral endoderm is slightly diminished. This renders it very likely that the entire endoderm will have lost its mesoderm-inducing capacity by stage 11: this may explain the contradictory results of Kotani and Smith, who for their grafting experiments used early and middle gastrula stages, respectively. In *Pleurodeles*, however, the germ-cell-inducing capacity of the ventral endoderm seems to persist during gastrulation. Kotani's conclusion is corroborated by the study of Sutasurya and Nieuwkoop (1974), who found that under the inductive influence of ventral endoderm PGCs can be formed from any part of the totipotent animal-moiety of the urodele blastula, the competence being highest in the 'ectoderm' near the equator and lowest at the animal pole. In normal development PGC formation is restricted to the ventral and ventro-lateral marginal zone by an inhibiting influence from the presumptive notochord spreading dorso-ventrally through the mesodermal mantle. From these experiments Sutasurya and Nieuwkoop conclude that in the urodeles PGCs *do not develop from predetermined elements, but arise strictly epigenetically from common, totipotent cells of the animal moiety of the blastula* as part of the regional induction of the mesoderm by the vegetal yolk endoderm.

When comparing PGC formation in the urodeles with that in the anurans, one is unavoidably led to the conclusion that not only do the PGCs originate from two different sites in the two groups, but that there are moreover two fundamentally different inechanisms at work. (This conclusion is clearly in flat contradiction to a statement by Blackler (1968) to the effect that the different site of origin of the PGCs in anurans and urodeles is only a minor variation and has no development significance.) In the anurans all the PGCs originate from the endodermal moiety of the egg in the vicinity of the vegetal pole, whereas in the urodeles they arise from the animal 'ectodermal' moiety more particularly the presumptive lateral plate mesoderm in the ventral to ventro-lateral equatorial region. In the anurans all the descriptive and experimental evidence pleads in favour of the *predetermined* nature of the pPGCs, based on the presence of a germ-cell-specific cytoplasmic component, the 'germinal plasm', which is present in the embryo from the very beginning of development. In contrast, in the urodeles the PGCs develop strictly epigenetically from common, totipotent cells of the animal moiety under the inductive influence of the ventral yolk endoderm.

Anuran and urodele PGCs seem to have one characteristic in common, i.e., their irreplaceability. Removal of the region from which they originate leads to sterility of the developing larvae in both groups. Regulatory formation of new PGCs from adjacent regions apparently does not occur. This could lead to the conclusion that the PGCs are already determined or at least partially determined at the stage of operation.

The most striking difference in the mode of origin of the PGCs between anuran and urodele amphibians seems to be the role of the germinal plasm in germ cell determination and differentiation. It was shown that the presence of germinal plasm is also a characteristic of urodele PGCs. Although ultrastructurally it looks identical to that of the anuran PGCs, it appears only at a late stage of development. i.e., when the cellular differentiation of the PGCs has already begun and thus long after their actual determination must have taken place. In the urodeles the germinal plasm cannot therefore play the role of germ cell determinant. It is very difficult to conceive that in one animal group a certain organelle acts as a determinant of a special cell type whereas in another group it constitutes only an attendant structure, being a differentiation product of that same cell type. This incongruity rises the question of whether in the anurans the characteristic structures of the germinal plasm actually do act as germ cell-determinant, or whether another, non-structural component of the cell is responsible for germ cell determination.

Coecilia

There is only one reference, Marcus (1938), on the possible site of origin of the PGCs in the Coecilia, the most primitive group of recent amphibians. In histological sections of rather late embryonic stags of *Hypogeophis* Marcus identified the PGCs in the endoderm in the vicinity of the blastopore. At later stages he found them in the dorsal mesentery and concluded that they had arrived there either direct from the dorsal endoderm or by passing through the lateral plate mesoderm. They are finally located in the genital ridges ventral to the Wolffian ducts. Marcus states that the PGCs have an 'endodermal' appearance. The evidence is inconclusive, however, the more so since it is known that the PGCs may migrate from one tissue into another at relatively early stages of development and may engulf yolk platelets from cytolysing cells.

The fishes

As regards the fishes a clear distinction must be made between the development of the four classes: the Agnatha or cyclostomes, the Chondrichthyes or elasmobranchs, the Osteichthyes, and the Teleostomi or teleosts.

The early development of the eggs of the Agnatha and Osteichthyes, which both represent ancient groups of fishes, is rather alike and shows close parallels with the early development of the amphibians. The development of the yolk-rich egg of the Chondrichthyes is rather different from that of the former groups, while the Teleostomi constitute a highly specialised group with a very aberrant development.

Agnatha

The only data available at present on the origin of the PGCs in the Agnatha are those of Okkelberg (1921) on the brook lamprey. He found PGCs first at an early tail-bud stage, in the most caudal portion of the trunk where they are scattered in the lateral plate mesoderm. Later they accumulate in the vicinity of the Wolffian ducts. This descriptive evidence suggests a mesodermal origin of the PGCs but is inconclusive.

Osteichthyes

On the origin of the PGCs in the Osteichthyes only descriptive data from a few species are available. In *Amia* some PGCs are found in the peripheral region of the endoderm at an early stage, but the majority the present in adjacent lateral plate mesoderm. In *Lepisosteus* Allen (1911) found them in the ventral and lateral portions of the single-layered gut endoderm. Reinvestigating the original material of *Polypterus* collected by Kerr. De Smet (1970) found PGCs were first visible in an early larval stage among the yolk-laden cells of the roof of the gut, at the boundary between the gut endoderm and the lateral plate mesoderm and in the lateral plate mesoderm itself. Finally, in *Acipenser* Maschkowzeff (1934) was first able to detect PGCs in the intestinal endoderm. On the basis of the yolk content of the PGCs the various authors concluded that in all the four species studied they are of endodermal origin. They are supposed to migrate from their endodermal site of origin towards the lateral plate mesoderm and then to the dorsal midline, where they accumulate near the nephrotomes. It must however be said that the available descriptive evidence suggesting an endodermal origin of the PGCs in the Osteichthyes is very scanty and inconclusive.

Chondrichthyes

Data on the origin of the PGCs in the elasmobranchs are similarly scanty and inconclusive. In *Raja batis* Beard (1900,1902*a*, *b*) found them for the first time at an open neural plate stage. The majority were located in the peripheral dorso-lateral endoderm, some in the adjacent lateral plate mesoderm, and even a single one in the overlying ectoderm. In older embryos they were found in the splanchnic mesoderm or between this and the intestinal epithelium. Beard (1920*a*) found a similar location of the PGCs in *Raja radiata*, *Pristiurus melanostomus*,

Scyllion vulgaris, Torpedo ocellata and *Acanthias vulgaris.* This descriptive evidence points towards a relatively early and endodermal origin for the PGCs in the Chondrichthyes, but is again far from conclusive.

Teleostomi

Eigenmann (1891) concluded from the size of the PGCs in *Micrometrus* (= *Cymetogaster*) that they must have segregated from the somatic cells at an early cleavage stage. In *Findulus* Richards and Thompson (1921) were first able to identify the PGCs in the peripheral endoderm in 46 to 53-hour embryos (early tail-bud stages). They stressed the close association of the PGCs with the syncytial periblast, an observation which was confirmed by Hann (1927) and Wolf (1931). In *Scardinius* Reinhard (1924) described their possible segregation from the syncytial periblast. In the salmonid *Micropterus* Johnston (1951) identified the PGCs first in an excrescence of the periblast protruding over the caudal edge of the blastodisc (erroneously called the dorsal blastoporal lip). At a slightly later stage they were found in the caudal (erroneously called ventral) periblast, from where they moved into the caudal extension of the yolk sac. From there they wedged their way between the splanchnic mesoderm and the gut endoderm towards the mid-dorsal region, and subsequently towards the genital ridges. In *Oryzias* PGCs were first found in the unsegregated mesendoderm, occasionally also in the ectoderm, at an 'early gastrula stage'. After the segregation of mesoderm and endoderm they were found in all three germ layers. Since their number increased in the absence of detectable mitotic divisions. Gamo assumed that they segregated during the entire early development. In early stages they were closely associated with the syncytial periblast. After myotome formation they were found predominantly in the peripheral endoderm, where they seemed closely associated with yolk protrusions. Gamo also noticed the capacity of the PGCs to absorb yolk. Only those PGCs which are incorporated into the gonadal anlagen become definitive germ cells, while the ectopic germ cells degenerate.

According to Pala (1970), in *Gambusia* PGCs are first discernible among the deeper cells of the embryonic node in the caudal half of the blastodisc at an 'early gastrula stage'. During the further segregation of the embryonic anlage they are found in the mesodermal portion of the mesendoderm. Nedelea and Steopoe (1970) localised the PGCs in the mesodermal layer of the embryonic shield of *Cyprinus* shortly before the completion of the overgrowth of the yolk.

The only experimental evidence available seems to argue for a mesodermal origin of the PGCs. Randomly reassociated fragments of posterior thirds of embryonic shields of 'middle gastrulae' (about half way in the overgrowth of the yolk), grafted into the extra-embryonic area of a host embryo, formed gonads with PGCs in the absence of any endodermal structures.

Summarising, it must be said that there is no concordant evidence for a particular site of origin of the PGCs in the teleosts. The often observed close association of the PGCs with the syncytial periblast may well be of a secondary nature in view of the tendency of the PGCs to associate themselves with a nutritional source. Nothing can at present be said about the mode of origin of the PGCs.

The Lower Chordates

Tunicata

Except for the Larvacea, which are sexually mature tadpole-like organisms, all the other tunicates undergo metamorphosis, during which the axial structures (notochord and somites) so characteristic of the chordates are resorbed. As far as is known PGCs are only found in metamorphosed, sessile individuals.

Berrill (1975) states that among the tunicates various forms of asexual reproduction are found, during which a small fragment of the parental body is set apart for the formation of a new zooid. In several forms sexual and asexual generations alternate, and in these forms no germ line is therefore traceable. The germ cells in the hermaphroditic gonads — the ovotestes

- seem to arise from small masses of undifferentiated mesodermal cells. These cells are called haemoblasts, suggesting an origin from the vascular system.

In summary it may be said that in the tunicates the PGCs probably originate in the mesoderm, although their mode of origin is unknown.

Cephalochordata

According to Boveri (1892) the PGCs in *Branchiostoma* larvae are found in clusters at the ventro-cranial extremities of the eleventh to thirty-sixth of the 61 somites. The *'Grenzzelle'* which Hatschek (1888) found at the ventral extremity of each somite in a younger larva may represent the segmental stem cell of the PGCs. The strictly segmental origin of the PGCs pleads in favour of their differentiation *in situ* from the intermediate mesoderm.

The Amniota

Aves

Swift (1914) was the first to locate the PGCs at early somite stages, in an anterior, extra-embryonic crescent-shaped area at the boundary between the are pellucide and the area opaca, which he called the 'germinal crescent'. In the next decade this observation was both confirmed and contradicted, but it gradually found wider approval. Willier (1937) found PGCs in chorioallantoic grafts of portions of the chick blastoderm containing part of the germinal crescent, whereas grafts without this region were sterile. Simon (1960) observed complete sterility or a very marked reduction in the number of PGCs after removal of the germinal crescent. Reynaud (1969) destroyed all the PGCs by UV-irradiation of the germinal crescent, and was able subsequently to restore a normal PGC population in the irradiated embryo by intravenous injection of a cell suspension made from the germinal crescent of another embryo.

The spatial extension of the germinal crescent apparently varies markedly in different species and breeds. In the chick

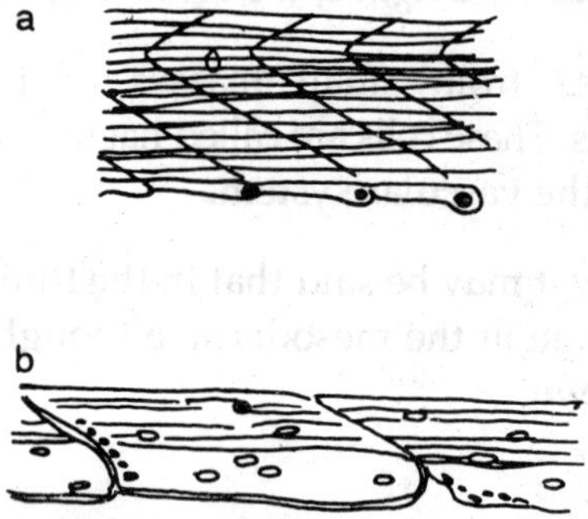

Fig. 4.2. Lateral view of part of a row of somites (s) in larva of *Branchiostoma lanceolatum.* (a) Segmental location of presumed stem cells of PGCs (black) in early larva (Hatschek's 'Grenzzelle', 1888). (b) Similar arrangement of clusters of PGCs in slightly older larva.

and turkey it usually surrounds only the anterior half of the embryonic anlage, but in the quail it extends around the anterior two-thirds, while in a chick hydrid and in the duck its extension is even more variable and may reach the posterior side of the embryonic anlage.

The PGCs are initially located in the endodermal hypoblast layer of the germinal crescent. They then segregate from the hypoblast and accumulate at the junction of the ectoderm and endoderm. The begin to enter the extra-embryonic blood islands as soon as the invaginated mesoderm reaches the germinal crescent area.

Although the extra-embryonic location of the PGCs at early somite stages is by now well established, little is known about the *primary origin* of the PGCs and the actual time of their determination.

Quite a few experiments have been done at the unincubated blastoderm stage. Dubois (1967) concluded from in-vitro culture of parts of the unincubated blastoderm that the PGCs originate from the primary hypoblast in the posterior region of the blastoderm and are transported anteriorly by its morphogenetic movements. Dubois later found the PGCs at the time of oviposition to be distributed in a transverse band in the posterior half of the embryonic anlage. This seems to be at variance with the findings of Fargeix (1969), who obtained a 'normal' number of PGCs (i.e., equal to that of a normal embryo) from anterior halves of unincubated blastoderms, while posterior halves formed either a 'normal' or a deficient number of PGCs. Lateral halves always formed a 'normal' number of PGCs. The number of PGCs decreased in cultured posterior halves as they were isolated at later stages. Rogulska (1968) found that duck blastoderms transected *in vivo* formed double the 'normal' number of PGCs. In those cases where the embryonic anlagen were oriented parallel to each other, the PGCs were predominantly located in the zone between them. Working with freshly laid chicken 'winter' eggs which did not yet show hypoblast formation. Eyal-Giladi, Kochav and Menashi (1976) found that both anterior and posterior halves formed the 'normal' number of PGCs.

Fargeix (1975) showed that X-irradiation of the entire blastoderm can lead ultimately to the sterility of the embryo. In such embryos the number of PGCs is unaffected at an early somite stage but decreases later, whereas in unirradiated embryos the number of PGCs increases steadily. Fargeix found that in the unincubated blastoderm irradiation of either the anterior or the posterior half led to the same reduction in PGC number. However, irradiation of the posterior half of the blastoderm incubated for 10 hours led to only a slight reduction in the number of PGCs, whereas a marked reduction was obtained after the same treatment of the anterior half. From these experiments he concluded that the pPGCs are evenly distributed throughout the unincubated blastoderm, but accumulate in the anterior half during the first hours of incubation. This is in accordance with the segregation *in situ* of the primary hypoblast from the epiblast and its subsequent anterior displacement as a result of the expansion of the secondary hypoblast from Koller's sickle.

The formation of a 'normal' number of PGCs from anterior and posterior halves of unincubated blastoderms seems to be at variance with the assumption the predetermined elements are present in the primary hypoblast — unless a mechanism exists which leads to a doubling of the number of PGCs after transaction. In this connection two contradictory observations are of interest. Fargeix (1969) states the transected blastoderms are always retarded in development, so that there would seem to be enough time for an extra clonal division of the PGCs. In contrast Swartz and Domm (1972), who studied mitoses in PGCs by colchicine treatment, observed that the clonal divisions of the PGCs start late, i.e., not before the migration period.

A possible alternative explanation for the 'normal' number of PGCs in half embryos, first suggested by Fargeix (1969) and again by Eyal-Giladi *et al.* (1976), is that predetermined elements in the sense of the anuran pPGCs do not exist in the early development of birds, but that (in both half and normal embryos) the PGCs are induced (in 'normal' number) at a later stage of development.

If an early segregation of pPGCs does occur in birds, it nevertheless seems to differ from that in the anuran amphibians, since no germinal plasm has hitherto been demonstrated in birds. The prime characteristic of avian PGCs, i.e., their high glycogen content, cannot be regarded as a germ cell determinant since it is not specific to PGCs and appears relatively late in development. It probably only represents a condition for their migratory activity. In birds the role of germ cells determinant cannot at present be attributed to any particular ultrastructural or physiological characteristic of the PGCs.

Reynaud (1968) observed accelerated multiplication of the germ cell population re-established in previously sterilised chick hosts by injection of a small number of turkey PGCs. Lutz and Lutz-Ostertag (1972) called this phenomenon auto-regulation. Reynaud (1971*a*) studied 'normal' germ cell multiplication in chick, quail and turkey. Auto-regulation was also observed by Fargeix (1976) X-irradiated anterior and posterior halves of unincubated duck blastoderms.

Summarising, it may be said that the early segregation of the avian PGCs from the primary hypoblast shows certain similarities to the history of the germ cells in the anuran amphibians. However, the avian PGCs are not characterised by the presence of germinal plasm or any other germ-cell-specific structure. The suggestion by Fargeix (1969) and by Eyal-Giladi *et al.* (1976) that avian PGCs may be formed epigenetically during embryogenesis (as they are in the urodeles) cannot be excluded, but does not seem to be easily reconciled with the endodermal origin of the PGCs and with the results of X-irradiation at early blastoderm stages. It must therefore be concluded that the actual mode of origin of the PGCs in birds is still unknown.

Mammalia

The data relevant to the site and mode of origin of the PGCs in the mammals come predominantly from the analysis of the early development of the mouse embryo, to which we will therefore mainly refer.

Brambell (1927) was the first to recognise the extra-gonadal origin of the PGCs when he detected them in the extra-embryonic yolk sac endoderm of the 8-day mouse embryo. During gut formation they are displaced towards the hindgut endoderm. Leaving the endoderm, they migrate through the dorsal mesentery towards the genital ridges. This observation was confirmed by Everett (1943), Chiquoine (1954), on the other hand, stated that in the 8-day mouse embryo the PGCs originate from the yolk sac splanchnic mesoderm and perhaps even from the caudal portion of the primitive streak, thus suggesting a mesodermal origin. In 1960 Brambell again inferred an endodermal origin for the PGCs. Mintz (1960*a*) observed them first in the yolk sac endoderm in the 8-day presomite mouse embryo. Blandau, White and Rumery (1963) found the PGCs at the base of the allantoic evagination. The endodermal origin of the PGCs was again questioned by Ozdzenski (1967), who found them in the caudal region of the primitive streak as well as in the anlage of the allantois of presomite embryos. He concluded that they migrate from the primitive streak mesoderm into the endodermal layer. Bruel-Beaudenon and Hubert (1968) speak of a 'posterior germinal crescent' in the mammals. Hamilton and Mossman (1972) again have the PGCs originating from the extra-embryonic yolk sac endoderm. Merchant and Zamboni (1973) identified the PGCs in the gut epithelium of the 9-day mouse embryo. They came to the conclusion that the PGCs associate themselves with other tissues because as the result of insufficient yolk material they need an exogenous energy source, the alkaline phosphatase activity at the cell periphery denoting an active intercellular exchange of material. The location of the PGCs in the yolk sac endoderm of the 8- to 9-day mouse embryo was confirmed by Spiegelman and Bennett (1973). Clark and Eddy (1975) agree with Merchant and Zamboni (1973) that the PGCs reside only temporarily in various somatic tissues during their life history, and depend upon the somatic tissues for their nutrition. Although they found the PGCs for the first time in 8- to 8•-day embryos in that portion of the yolk sac endoderm that later forms the hindgut, the PGCs do not resemble the surrounding endodermal cells in their ultrastructure but look much more similar to the adjacent mesodermal cells. They therefore propose that the PGCs do not in fact originate from the yolk sac endoderm but from the mesoderm.

In other mammalian species the PGCs are likewise first detected in the extra-embryonic yolk sac endoderm and the allantoic endoderm and mesoderm of presomite embryos in the guinea pig; and in the human. At slightly older stages they are found in the hindgut endoderm in the guinea pig; Chretien (1966) in the rabbit; Sawano (1959) in the human.

The only disparate observation concerns a study by Vanneman (1917) in the marsupial *Armadillo*. In this polyembryonic form the PGCs were identified for the first time between the extra-embryonic ectodermal and endodermal layers, outside the primary embryonic buds. This more or less resembles the situation in birds, but the finding must be viewed with caution since it has never been confirmed.

The only experimental evidence comes from Gardner and Rossant (1976), who produced climaeras for coat colour genes in the mouse by injecting embryonic ectoderm cells into host blastocysts, and found that PGCs were formed from the injected cells. The concluded that the PGCs originate from (totipotent) precursor cells of the definitive embryo rather than from the extra-embryonic yolk sac endoderm. The totipotent embryonic 'endoderm' would yield both PGCs and somatic cells, the segregation of which would occur only later. This situation rather resembles that in the urodele amphibians where the lateral plate mesoderm, from which the PGCs originate, arises in the process of mesoderm formation from the totipotent animal 'ectodermal' moiety of the blastula. Recently Falconer and Avery (1978) concluded that in the mouse the variability observed in chimaeras produced by aggregation of cleavage stages and in mosaics resulting from X chromosome inactivation provides additional evidence that the PGCs originate in the primary ectoderm and not in the yolk sac endoderm.

Reptilia

The origin of the PGCs in the reptiles has been traced back no further than the early somite stages. In all reptiles studied the PGCs were found in the extra-embryonic area of the blastoderm. As to the more precise location of the PGCs the reptiles fall into two categories: (*a*) those with an anterior germinal crescent, which may extend around the entire embryonic anlage, and (*b*) those with a posterior germinal crescent. The former localisation more or less resembles the situation in birds, the latter that in mammals. A strictly anterior germinal crescent was found in the snake *Vipera* by Hubert (1969), while an anterior localisation with extensions to the lateral and posterior sides seems to exist in the Lacertilia *Mabuya* and *Chamaeleo* and the slow-worm *Anguis*. Both anterior and posterior germinal crescents were described in *Sphenodon* by Tribe and Brambell (1932). A strictly posterior germinal crescent was found in the chelonians as well as in several *Lacerta* species. It is evident that the two types of localisation do not coincide with particular taxonomic groups, since the Lacertilia show both types. According to Hubert (1976), in the forms having an anterior germinal crescent (e.g. *Mabuya, Chamaeleo*) the PGCs are displaced anteriorly together with the primary hypoblast by the gastrulation movements, whereas in those having a posterior germinal crescent (e.g., the chelonians and *Lacerta*) the localisation of the PGCs is not affected by the gastrulation process, which occurs more anteriorly.

At the stages examined the PGCs were found in the endodermal layer before the mesoderm has spread between the ectoderm and endoderm. According to Hubert (1976) the PGCs originate either in the primary hypoblast or in the definitive hypoblast formed during gastrulation.

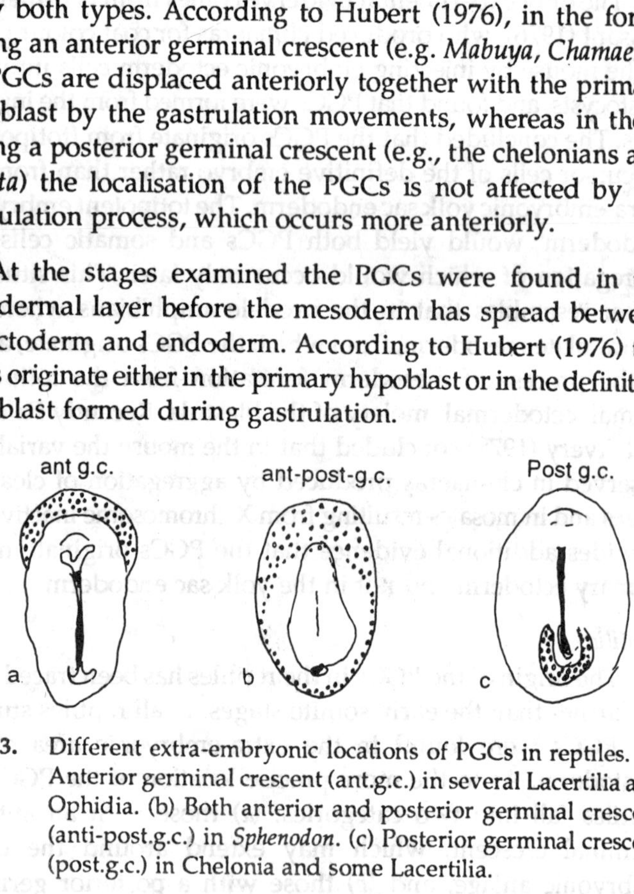

Fig. 4.3. Different extra-embryonic locations of PGCs in reptiles. (a) Anterior germinal crescent (ant.g.c.) in several Lacertilia and Ophidia. (b) Both anterior and posterior germinal crescent (anti-post.g.c.) in *Sphenodon*. (c) Posterior germinal crescent (post.g.c.) in Chelonia and some Lacertilia.

The only experimental evidence available, which was obtained by the extirpation of the posterior germinal crescent in *Lacerta vivipara* either before or after the migration of the PGCs towards the genital ridges, confirms the descriptive data.

Summarising, it must be concluded that little is known about the actual site of origin of the extra-embryonic PGCs in the reptiles. The observed localisation in the anterior or posterior germinal crescents probably represents a secondary situation. About the mode of origin of the PGCs in the reptiles we are entirely in the dark.

General Conclusions

Surveying the entire phylum of the Chordata with respect to the site and mode of origin of the PGCs, it is evident that at least two essentially different mechanisms are involved in individual groups, viz. (*a*) a very early segregation of the germ line from the somatic cell lines, and its determination by a special cytoplasmic component, and (*b*) a strictly epigenetic development of the germ cells from totipotent embryonic cells under an inductive influence. These two alternative mechanisms correspond to an endodermal and a mesodermal origin of the PGCs, respectively. The existence of the two mechanisms has been deduced from the experimental analysis performed in anuran and urodele amphibians. The late appearance of the germinal plasm in the urodele PGCs moreover throws doubt on the determinative role of the germinal plasm generally.

When surveying the chordates we must unfortunately state that it is at present only in the two groups of amphibians where the situation is sufficiently clear. Our knowledge of the development of the urochordates, in which as far as we know the PGCs are only detectable after metamorphosis, does not allow any conclusion. In the cephalochordates the PGCs also seem to arise rather late in development, i.e., during the final differentiation of the mesoderm. Their clearly segmental localisation ventral to the somite segments pleads in favour of differentiation *in situ* within the epigenetically formed mesoderm, suggesting a 'urodele' type of germ cell formation. In the Agnatha the PGCs are first found scattered in the lateral plate mesoderm at an early tail-bud stage, a situation which again resembles the 'urodele' type of germ cell formation. In the Osteichthyes there is some purely descriptive evidence for an endodermal localisation of the PGCs, but this does not say anything about their actual site and mode of origin. The same uncertainty holds for the situation in the Chondrichthyes. The purely descriptive evidence regarding the Teleostomi is far from concordant and does not allow a decision one way or the other. No experimental evidence is available for the Coecilia.

When considering the amniotes, the Reptilia are almost unknown terrain, apart from some evidence that the situation in the Chelonia differs from that in the Squamata. In contrast much experimental evidence is available on the birds. The PGCs seem to originate in the extra-embryonic primary hypoblast during early development. The lack of germinal plasm in avian PGCs possibly means their actual mode of origin is different from that in the anuran amphibians. It also calls into question once more the determinative role of the germinal plasm. In the mammals the stili scanty experimental evidence suggests an epigenetic mode of origin of the PGCs, in line with the 'urodele' type of germ cell formation.

5

The Migration of the Primordial Germ Cells

In all the vertebrate groups a displacement of the PGCs occurs from their extra-gonadal (in the amniotes extra-embryonic) site of initial appearance to their ultimate destination in the gonadal anlagen: Cephalochordates are the exceptions where the primordial germ cells arise *in situ* in their definite location and no migration is therefore required.

The migration of the PGCs from their site origin towards the gonadal anlagen has been most thoroughly studied in birds, and reasonably well studied in the anuran amphibians. Since the migration route is quite different in the two groups, and since as far as we know these groups essentially typify the two main routes along which PGCs may reach the gonadal anlagen, we will first focus our attention on the anuran amphibians, deal only briefly with the urodeles and fishes, subsequently discuss extensively the situation in the birds, and finish with some additional data on the mammals and reptiles. It seems desirable to distinguish between two different kinds of movement of the PGCs during their migration, i.e. (*a*) 'passive' displacement by morphogenetic movements, and (*b*) 'active' displacement by their own amoeboid motility. During their active displacement the PGCs may make use of the circulatory system of the embryo. This is often called 'passive transport' but should be clearly distinguished from the passive displacement by morphogenetic movements, during which the PGCs do not show any motility. The term 'vascular transfer' therefore seems more appropriate.

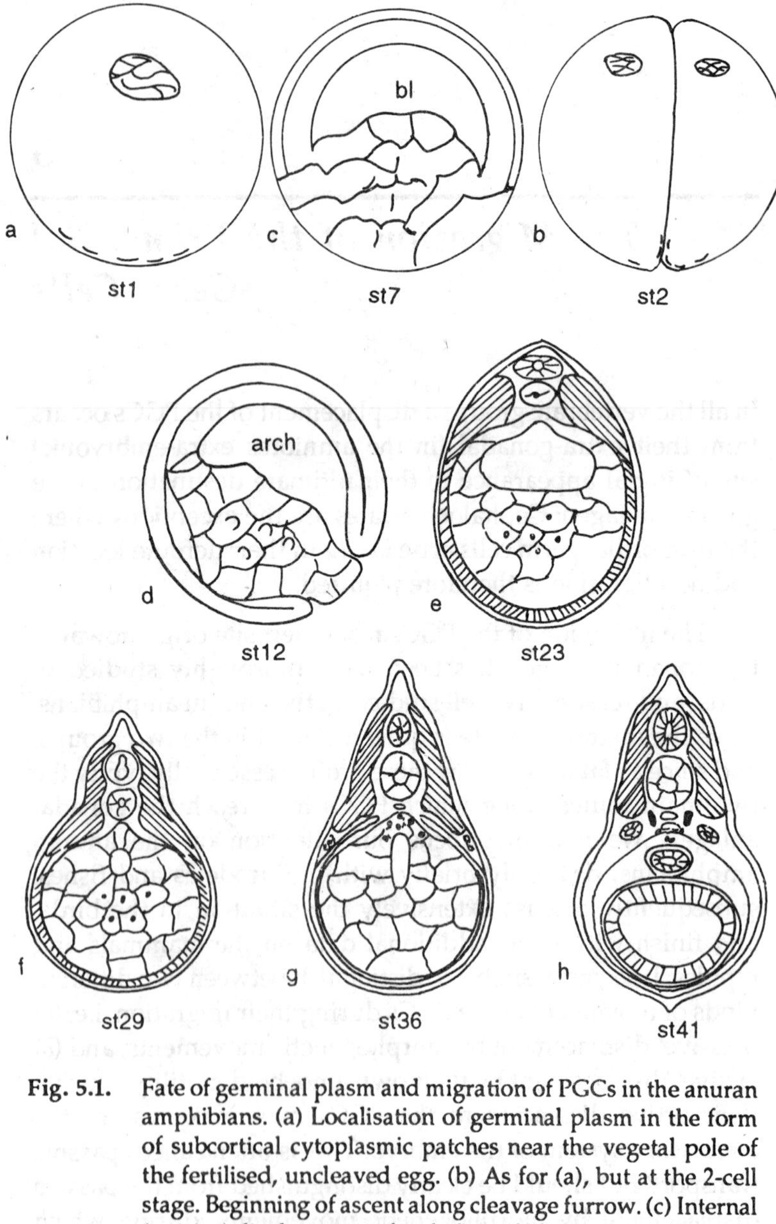

Fig. 5.1. Fate of germinal plasm and migration of PGCs in the anuran amphibians. (a) Localisation of germinal plasm in the form of subcortical cytoplasmic patches near the vegetal pole of the fertilised, uncleaved egg. (b) As for (a), but at the 2-cell stage. Beginning of ascent along cleavage furrow. (c) Internal (passive) displacement of endodermal blastomeres containing germinal plasm by pregastrulation movements at early blastula

stage. (d) Further (passive) displacement of endodermal blastomeres by gastrulation movements, and intra-cellular displacement of germinal plasm from a peripheral towards a juxta-nuclear position. (e) and (f) Nearly unchanged localisation of PGCs at tail-bud stages. (g) Active dorsal migration of PGCs inside endoderm at early larval stage and accumulation in dorsal endodermal ridge. (h) Further active migration of PGCs through dorsal mesentery towards genital ridges, arch., archenteron; bl., blastocoel.

THE ANAMNIA

Amphibia

Anura

The displacements of the presumptive and true PGCs in the anuran amphibians have already been described in the proceeding chapters.

It is now generally agreed that the displacement of the germinal plasm from the vegetal surface of the egg towards a more internal position during cleavage is due to the morphogenetic movements involved in the cleavage process. The subsequent displacements of blastomorers containing germinal plasm (pPGCs) towards the centre of the yolk mass and even as far as the floor of the blastocoel can likewise be satisfactorily explained by the morphogenetic pregastrulation movements, which were first described by Schechtmann (1934) in *Triturus* and analysed further by Gipouloux (1962) in *Discoglossus*. The displacements of the PGCs during gastrulation and neurulation also seem to be due to morphogenetic movements affecting the endodermal yolk mass. Since up to this stage of development no active movements of the PGCs are required, these displacements of the PGCs and their forerunners may be designated as *passive displacement*.

At the end of this first period of development the PGCs are located in the centre of the yolk endoderm in the caudal half of the embryo. Kamimura *et al.* (1976) found that in *Xenopus* the PGCs subsequently move to a peripheral position in the yolk mass (stage 31, N. & F.). They then move to the dorsal endoderm during stages 33 to 36, accumulate in the form of a dorsal endodermal crest or ridge around stage 40, and separate from the endoderm at stage 41. Subsequently they move to the dorsal root of the mesentery, and finally laterally towards the genital ridges. Wylie *et al.* (1976) observed that in *Xenopus* the PGCs actually accumulate underneath the gonadal epithelium of the genital ridges.

Vannini and Giorgi (1969) and Giorgi (1974) hold the opinion that the displacements of the PGCs from the dorsal endodermal crest towards the dorsal root of the mesentery and subsequently towards the genital ridges are mainly caused by growth processes in the surrounding tissues (passive displacement). There is, however, rather convincing evidence that the displacements of the PGCs during this phase of development are for the greater part to their active movements.

The PGCs show a distinct cell boundary in the light microscope. When studied with the electron microscope this is found to be due to wide intercellular spaces between the PGCs and the surrounding cells. Cambar *et al.* (1970) described the ultrastructure of the PGCs in *Rana* during their migration and after they have settled in the genital ridges. The PGCs show cytoplasmic processes extending into the intercellular spaces. With time-lapse cinematography Wylie and Roos (1976) actually observed active amoeboid movements *in vitro* of PGCs isolated during the migration period.

Gipouloux (1967) carried out an experimental analysis of the displacements of the PGCs in *Discoglossus* and other anurans. After dorso-ventral transposition of the endodermal mass the PGCs migrate through the endoderm in a direction opposite to normal, though in reduced numbers. In non-rotated endoderm there was normal dorsal migration (PGCs in dorsal root of mesentery) when either notochord plus somites plus Wolffian ducts, notochord plus somites, or somites plus Wolffian ducts were present. Partial migration (PGCs partially in dorsal endoderm and partially in mesentery) was observed when somites alone were present, and still more restricted migration (PGCs in dorsal endoderm only) in the presence of part of the somites or of the Wolffian ducts only. Implanted notochord or Wolffian duct exerted a weak attraction on only some of the host PGCs. The effect was stronger with implanted somites, stronger still with implanted notochord plus Wolffian ducts or with notochord plus somites, and strongest with implanted notochord plus somites plus Wolffian ducts. Gipouloux concludes from these experiments that the dorsal mesodermal organs exert an attracting influence upon the PGCs, leading to the establishment of a dorso-ventral gradient of a chemotactic agent inside the embryo. This primary attraction of PGCs is a property of the caudal, but not of the cephalic mesoderm. In embryos brought into ventral parabiosis, so that two axial systems are situated in opposite sides of a single endodermal mass, the PGCs are partially immobilised by the opposite attracting influences, leading to a markedly reduced migration.

That the attractant is a chemical agent is rendered likely by the observation that PGCs are also attracted by an implanted piece of agar previously immerse in an extract of dorsal mesodermal structures (Gipouloux, 1967). However, Gipouloux's study of the diffusibility of the attractant through millipore filters and of its heat and enzymatic degradation is in our opinion too preliminary to allow any definite conclusion regarding the possible chemical nature of the attractant.

Gipouloux (1967) also found that suppression of genital ridge formation by removal of the axial structures or by mercuric chloride treatment leads to a 'permanent' accumulation of the PGCs in the dorsal root of the mesentery. Nieuwkoop (1947) had observed and attraction of host PGCs by an extra genital ridge formed in urodele larvae after implantation of either notochord or Wolffian duct anlagen. This was confirmed by Gipouloux (1967) for the anuran embryo. Moreover, there is the interesting observation by Martin (1959) that in *Alytes obstetricans*, which has L and R genital ridges of unequal lengths, the PGCs are also distributed unequally between the two gonadal anlagen. These data plead strongly in favour of a chemotactic action exerted by the genital ridges.

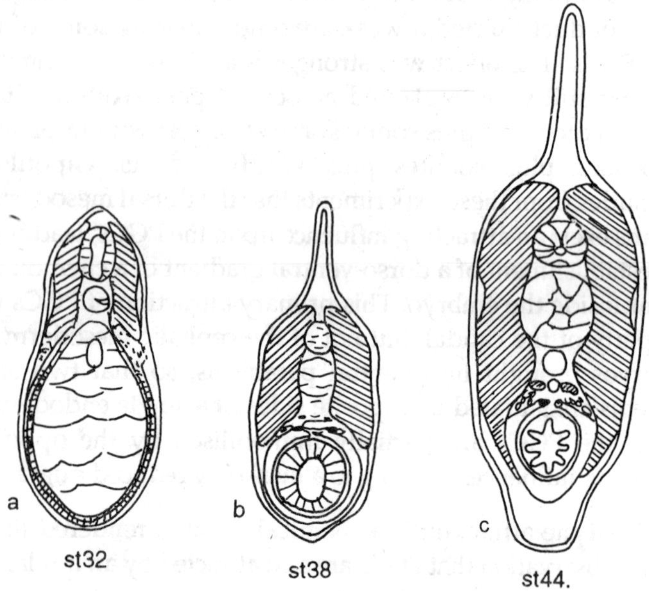

Fig. 5.2. Migration of PGCs (black) in the urodelen amphibians. (a) Successive steps in the migration of PGCs along the mesoendodermal interspace towards the dorsal midline, and (b and c) their subsequent migration towards the genital ridges.

Urodela

Nieuwkoop (1947) concluded for his xenoplastic transportations of ventro-lateral marginal zone that in the urodeles the PGCs are initially more or less evenly distributed throughout the presumptive lateral plate mesoderm. This was confirmed by Maufroid and Capuron (1973) by means of grafts of ventral and ventro-lateral marginal zone into the blastocoelic cavity of host embryos. Maufroid and Capuron (1972) studied the dorsal displacements of the PGCs in *Pleurodeles* during stages 16 to 27 by means of unilateral ablation of various dorso-ventral regions of the lateral plate mesoderm. Ikenishi and Nieuwkoop (1978) found cilia on some of the PGCs in *Ambystoma* during the migration period. However, nothing is known about the mechanisms upon which the displacement of the PGCs through the lateral plate mesoderm and their ultimate accumulation in the genital ridges are based.

Coecilia

The migration of the PGCs in the Coecilia has not been studied.

Fishes

The only evidence available on the migration of the PGCs in the fishes is of a descriptive nature.

In the lamprey (Agnatha), where according to Okkelberg (1921) the PGCs seem to originate in the lateral plate mesoderm, the migration route resembles that in the urodele amphibians.

Among the Osteichthyes Allen (1911) described the migration route in *Amia* and *Lepisosteus*. In *Amia* the PGCs migrate from the peripheral endoderm into the lateral plate mesoderm, then accumulate in its mediodorsal region, and migrate from there to the genital ridges. This situation seems more or less intermediate between that in the anuran and that in the urodele amphibians. In *Lepisosteus* the PGCs first migrate towards the dorsal portion of the gut and subsequently through the dorsal mesentery to the genital ridges, a route which resembles that in the anuran amphibians.

No relevant data are available on the migration of the PGCs in the Chondrichthyes.

In the Telestomi the available data are scanty and controversial. The authors who assume an endodermal origin for the PGCs take them to migrate from the yolk sac endoderm— or even earlier from the periblast — through the splanchnic mesoderm towards the dorsal midline, and subsequently through the dorsal mesentery towards the genital ridges in

Fundulus; Micropterus; Channa; and in *Oryzias*. The authors who defend a mesodermal origin of the PGCs assume them to arise in the unsegregated mesendoderm. After segregation of the two layers they are found in the mesodermal layer or distributed throughout all three germ layers, with a preference for the mesodermal layer. Although the PGCs may show a temporal association with the peripheral endoderm, they subsequently migrate through the splanchnic towards the somatic mesoderm and further towards the genital ridges (*Gumbusia;* and in *Cyprinus*).

Several authors have observed pseudopod formation on the surface of the PGCs, which they took to be an indication of active migratory movements, but other authors were more inclined to ascribe the movements of the PGCs entirely (Richards and Thompson, 1921, in *Fundulus*) or at least partially to passive displacement by the surrounding tissues. Stolk (1958), who observed many highly ectopic PGCs in *Abramites* and *Cirrhina*, even assumed vascular transfer to be involved, but this has never been confirmed.

The Amniota

Aves

The presumptive PGCs are very probably passively displaced by the morphogenetic movements occurring during the formation of the primary and secondary hypoblast and the subsequent formation of the primitive streak and the invagination of the tertiary hypoblast or embryonic endoderm. The subsequent, active displacements of the PGCs from the germinal crescent towards the genital ridges may be divided into four phases. The first phase is their segregation from the endodermal layer and their accumulation between the endoderm and ectoderm during stages 4 to 8 (H. & H.). In the second phase the PGCs being to penetrate the vascular network. This occurs after the invagination of the extra-embyronic mesoderm, its subsequent extension between the ectoderm and endoderm, and the start of the differentiation of the blood islands (stage 10, H. & H.). In the next phase the PGCs are found successively

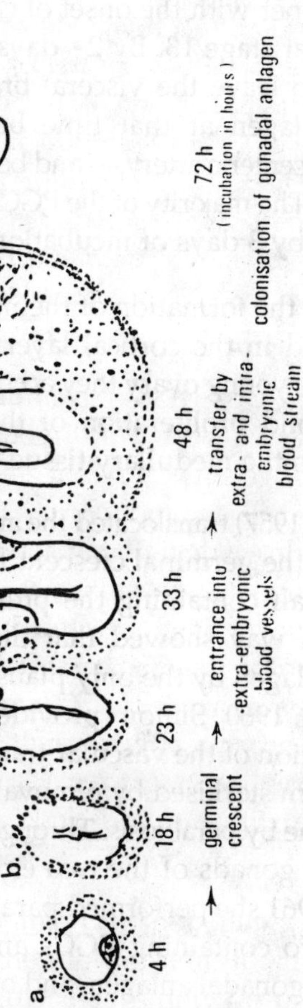

Fig. 5.3. Migration of PGCs in the avian embryo. (a) Absence of identifiable germ cells prior to primitive streak formation (b) and (c) (Passive) accumulation of PGCs in the anterior germinal crescent at head-process and head-fold stages, respectively. (d) (Active) penetration of PGCs into mesodermal blood islands and beginning of transfer by blood circulation. (e) Circulation of PGCs through entire vascular system and start of their egress from the vascular system in the neighbourhood of the gonadal anlagen. (f) Continuing colonisation of gonadal anlagen by actively migrating PGCs.

in the extra-embryonic blood vessels at stage 12, and in the embryo proper with the onset of cardiac pulsation and blood circulation at stage 13. By 2• days of incubation (last phase) they start to leave the visceral branches of the aorta — the gonadal anlagen at that time being situated close to the omphalomesenteric artery — and begin to penetrate the gonadal epithelium. The majority of the PGCs have settled in the gonadal epithelium by 3 days of incubation.

During the formation of the indifferent gonads the PGCs are scattered in the cortical layer but also in the medullary tissue. In the young ovary they accumulate in the cortex formed by the second proliferation of the gonadal epithelium, but migrate into the medullary tissue in the young tetis.

Simon (1957) translocated the anterior half of the blastoderm containing the germinal crescent laterally with respect to the posterior half containing the presumptive gonadal anlagen, and in this way showed that the PGCs actually reach the gonadal anlagen by the only plausible route, i.e., the vascular system. In 1960 Simon provided an even more elegant demonstration of the vascular transfer of the PGCs by joining a blastoderm sterilised by removal of the germinal crescent to a normal one by parabiosis. Through the anastomosing vascular system the gonads of the two embryos were colonised with PGCs. In 1961 she performed parabiosis between and anterior half embryo containing PGCs and a posterior half embryo containing gonadal anlagen, and obtained normal colonisation of the gonadal anlagen with PGCs through the connecting vascular system.

Dantschakoff's (1936) notion that the PGCs would leave the vascular system by being mechanically trapped in the fine capillaries of the gonadal region — an idea also adhered to by Blocker (1933) — was disposed of by Van Limborgh, Van Deth and Tacoma (1960), who found that the capillaries of the head and the yolk sacare in fact finer and this more suitable for trapping PGCs than those of the gonadal region; nevertheless the PGCs only leave the vascular system in the vicinity of the gonadal anlagen. Van Limborgh *et al.* proposed instead that a slowing-down of the blood stream in the capillary network promotes the egress of the PGCs from the vascular system. Simon (1960) came to the same conclusion and suggested that in addition a chemotactic process is responsible for the directive migration of the PGCs (see below). The former idea was confirmed by Dubois (1968), who observed that an experimental reduction of the blood flow causes a greater number of PGCs to leave the large blood vessels. The discovery of PGCs in blood smears of 2-day embryos, using both phase contrast microscopy and PAS staining, demonstrated unequivocally that the PGCs do indeed use the vascular route to reach the gonadal anlagen. Further support for this route is given by Reynaud (1969), who was able to repopulate the gonads of embryos previously sterilised by UV-irradiation of the germinal crescent by intravenous injection of a cell suspension made from germinal crescent endoderm of another embryo.

The insertion of a piece of shell membrane between the hypoblast and the presumptive vascular mesoderm does not prevent the PGCs from leaving the hypoblast (although it of course prevents their entering the vascular system), and an isolated hypoblast cultured *in vitro* shows a spontaneous but random migration of the PGCs. Thus, the first step in the migration of the PGCs, i.e., their movement from the endodermal layer of the germinal crescent into the vascular system, is an active process. PGCs not only have amoeboid motility but actively invade various tissues under in-vitro conditions. The entry of the PGCs into the vascular system apparently does not require a specific attraction by the vascular mesoderm.

As already suggested by Simon (1960) the PGCs leave the vascular system under the influence of a chemotactic attraction exerted by the gonadal epithelium. *In vitro* the PGCs may leave the germinal crescent when it is associated with the gonadal region of another embryo, and may colonise the gonadal epithelium in the absence of a vascular system. The attraction exerted by young gonadal epithelium can also act upon the PGCs in a colonised but still undifferentiated gonad. It turned out that even spermatogonia still possess migratory capacity, sensitivity to the attractive influence, and the ability to colonise gonadal epithelium at least up to the 12th day of incubation. The majority of the oogonia lose their migratory capacity on the 8th day. Since the attraction can be exerted across a permeable barrier such as the shell membrane, the most satisfactory explanation is that of positive, selective chemotaxis.

By delayed injection of a PGC suspension made from germinal crescent endoderm, Reynaud (1969) was able to show that a sterile host gonad obtained by UV-irradiation of the germinal crescent can still trap PGCs at 5 days of incubation. Thus the competence of the PGCs to react to the attracting influence lasts much longer than the normal period of PGC immigration and only disappears with the onset of gametogenesis, while the attracting capacity of the gonad disappears during the formation of the primary sex cords. Rogulska (1969) demonstrated a direct transfer of PGCs from an intracoelomic graft of germinal crescent to an adjacent host genital ridge at $3.^{3}/_{4}$ days of incubation, i.e., long after normal migration is completed. By means of intracoelomic grafts of mouse hindgut to the gonadal region of a chick embryo, Rogulska, Ozdzenski and Komar (1971) found that chick gonadal epithelium also attracts mouse PGCs, demonstrating that the attractant is not species or class specific.

Let us now consider the properties of the PGCs and the gonadal epithelium more closely in order to understand better the chemotactic process. The PGCs in birds undergo a measure of cytogenesis. Clawson and Domm (1963*a*, *b*) described the PGCs of the germinal crescent as containing much yolk and only little glycogen. The migrating PGCs, however, contain far less yolk but much more glycogen. The glycogen content diminishes during migration, so that after arrival in the gonadal ridges the PGCs again contain little glycogen. This was confirmed by Fujimoto *et al.* (1975). Dubois and Cuminge

(1968) described more or less parallel changes in the amount of osmophilic lipid droplets, an observation confirmed by Ukeshima and Fujimoto (1975). In addition, Fujimoto *et al.* (1976*a*) described a structural change in the glycogen and its unipolar intracellular localisation during the settling of the PGCs in the gonadal epithelium. The glycogen and the lipids are considered to play an important role as energy sources during the migration of the PGCs.

The PGCs show pseudopodial extensions during their migration. Microfilaments and microtubules, however, have not been demonstrated. According to Cuminge and Dubois (1971) the PGCs show an intense lytic activity and actively penetrate the gonadal epithelium. There they take up soluble substances and cellular debris by pinocytosis and phagocytosis, respectively.

The gonadal epithelium has a syncytial and spongy character and shows large intercellular spaces. The same features are found in sterile gonadal epithelia, so they are not due to the lytic activity of the PGCs. Autoradiographic electron-microscopic analysis has shown that the gonadal epithelium synthesis exportable proteins or protein complexes, which are set free by a merocrine excretory process. Although in the nucleus only a single slow type of protein replacement is observed, two classes of proteins are synthesised in the cytoplasm, i.e., slowly replaced structural proteins and rapidly replaced exportable glycoproteins. However, it has not been conclusively shown that the attractive factor produced by the gonadal epithelium is in fact a protein. Swartz (1975) obtained some evidence that the chemical attractant may be steroid in nature, since injection of androgens or oestrogens interferes with PGC migration. In the synthesis of the exportable proteins the Golgi complex plays an important role. The fact that concanavalin A, which binds to glycoproteins, inhibits the migration of PGCs placed under the attracting influence of young gonadal epithelium, suggests that glycoproteins play a decisive role in the chemotactic migratory capacity of the PGCs.

In contrast to those of the gonadal epithelium, the glycoproteins synthesised by the PGCs show only a slow turnover and are not excreted. They are found preferentially in the lysosomal apparatus and on the plasma membrane, where they are concentrated on microvilli, pseudopodia and at sites of pinocytosis and phagocytosis.

Very little is known about possible attractive influences that play a role in the ultimate distribution of the PGCs between the cortical and medullary tissues of the differentiating gonad.

Summarising, it may be said that the initial step in the migration of the PGCs, i.e., their movement from the hypoblast of the germinal crescent into the vascular system, is an active but non-directional process. Impressive evidence argues for the involvement of a chemotactic mechanism in the migration of the PGCs from the vascular system into the gonadal epithelium. Very little is known about the mechanisms involved in the ultimate displacements of the germ cells inside the differentiating gonad.

Mammalia

Leaving aside their possible primary site of origin, the PGCs are clearly recognisable for the first time in presomite and early somite stages in the yolk sac endoderm and adjacent splanchnic mesoderm near the allantoic evagination, in the marsupial *Tatusia;* man; mouse, the guinea pig; and in the rabbit. Slightly later they are found in the endoderm and adjacent mesoderm of the hindgut. They then leave the gut and migrate through the splanchnic mesoderm towards the dorsal mesentery, and from there towards the genital ridges *Tatusia,* The guinea pig; the mouse; rabbit; and in man. It is still unclear whether the PGCs migrate through the endodermal epithelium of the yolk sac and hindgut, or through the adjacent splanchnic mesoderm, associating themselves temporarily with the endodermal cells for nutritive purposes only.

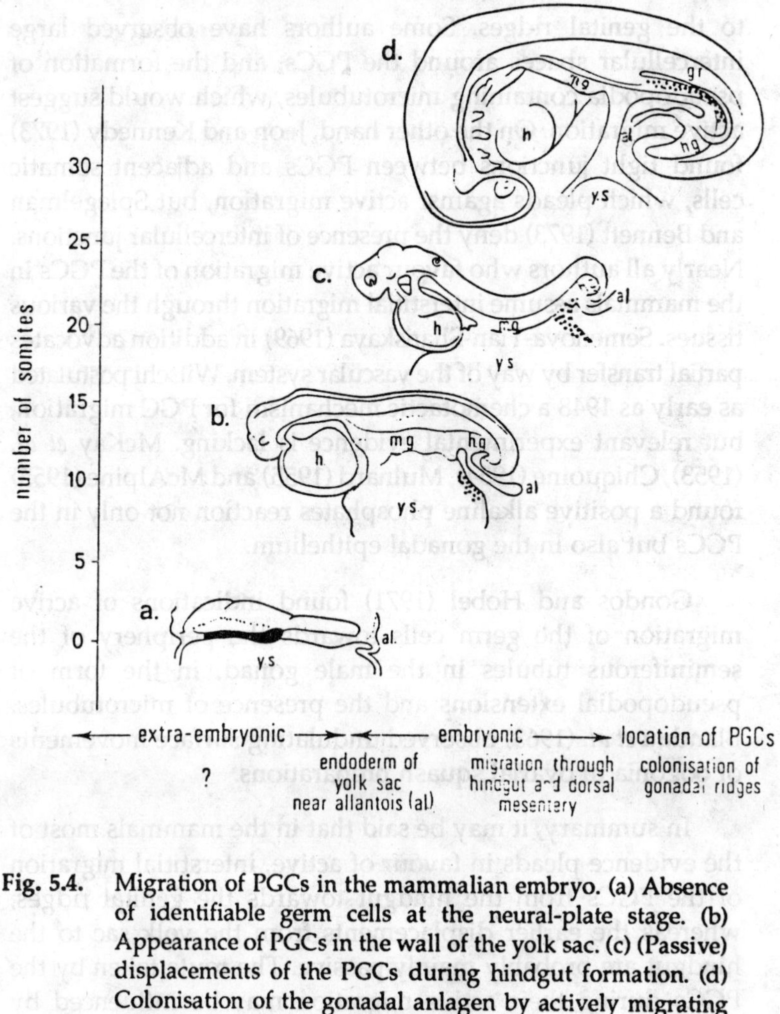

Fig. 5.4. Migration of PGCs in the mammalian embryo. (a) Absence of identifiable germ cells at the neural-plate stage. (b) Appearance of PGCs in the wall of the yolk sac. (c) (Passive) displacements of the PGCs during hindgut formation. (d) Colonisation of the gonadal anlagen by actively migrating PGCs, al., allantois: g.r., gonadal ridge; h., heart; h.g., hindgut; m.g., midgut; y.s., yolk sac.

The great majority of authors the opinion that the initial displacements of the PGCs during gut formation are mainly or entirely passive, i.e., they result from the morphogenetic movements associated with gut formation. Opinions differ as to the mode of their subsequent translocation from the hindgut to the genital ridges. Some authors have observed large intercellular spaces around the PGCs, and the formation of pseudopodia containing microtubules, which would suggest active migration. On the other hand, Jeon and Kennedy (1973) found tight junctions between PGCs and adjacent somatic cells, which pleads against active migration, but Spiegelman and Bennett (1973) deny the presence of intercellular junctions. Nearly all authors who favour active migration of the PGCs in the mammals assume interstitial migration through the various tissues. Semenova-Tian-Shanskaya (1969) in addition advocates partial transfer by way of the vascular system. Witschi postulated as early as 1948 a chemotactic mechanism for PGC migration, but relevant experimental evidence is lacking. McKay *et al.* (1953), Chiquoine (1954), Mulnard (1955) and McAlpine (1955) found a positive alkaline phosphates reaction not only in the PGCs but also in the gonadal epithelium.

Gondos and Hobel (1971) found indications of active migration of the germ cells towards the periphery of the seminiferous tubules in the male gonad, in the form of pseudopodial extensions and the presence of microtubules. Blandau *et al.* (1963) observed undulating surface movements of oogonia in ovarial squash preparations.

In summary, it may be said that in the mammals most of the evidence pleads in favour of active, interstitial migration of the PGCs from the hindgut towards the genital ridges, whereas the earlier displacements from the yolk sac to the hindgut are probably mainly passive. The route taken by the PGCs during their active migration may be influenced by their tendency to associate with somatic cells for nutritional purposes.

Reptilia

The reptiles are an interesting group exhibiting different types of PGC migration. Risley (1933) found the PGCs to be localised in a posterior crescent in the turtle *Sternotherus*. He described what he called active interstitial migration of the

PGCs from the site towards the gonadal anlagen through the splanchnic mesoderm and the dorsal mesentery, in much the same way as described above for the mammals. A similar situation was found by Hubert (1976) in various *Lacerta* species. Hubert (1970*b*) described the formation of pseudopodia containing microfilaments on the surface of the PGCs. In *Sphenodon* Tribe and Brambell (1932) described an anterior as well as a posterior germinal crescent. They postulated that the PGCs of the posterior crescent would reach the gonadal anlagen by interstitial migration through the splanchnic mesoderm and dorsal mesentery, but those of the anterior crescent would enter the vitelline veins and reach the gonadal ridges by way of the vascular system — i.e., passing through the vitelline veins, the heart, and aorta and the omphalo-mesenteric artery. The presence of an anterior germinal crescent and the subsequent vascular transfer of the PGCs was also described in *Mabuya* and *Chamaeleo* by Pasteels (1953), Hubert (1969) found a strictly anterior germinal *crescent* in *Vipera*, but in *Anguis* the PGCs were found around the entire embryonic enlarge, though markedly predominating anteriorly. He observed vascular transfer of the PGCs towards the gonadal anlagen from both the anterior and the posterior extra-embryonic regions.

Summarising, it may be said that the Chelonia the migration of the PGCs from the posterior germinal crescent towards the gonadal anlagen seems to be exclusive interstitial. Interstitia migration as well as vascular transfer is found in the Squamata. PGCs from a posterior site generally take the interstitual migration route, whereas those from an anterior site over the much longer instance to the gonadal ridges by means of vascular transfer.

Conclusions

Surveying the displacements of the PGCs in the chordates, it is evident that in the lower chordates, where the gonadal anlagen-arise in the vicinity of the PGCs, no transport mechanism is required. In the vertebrates, where the PGCs have an extra-gonadal site of origin, they are both passively and actively displaced towards the gonadal anlagen. The early displacements occurring during embryogenesis are mainly passive and result from morphogenetic movements of the surrounding tissues. Although morphogenetic processes may still play role in germ cell displacement during later phases of development, and even during gonadogenesis, the considerable distance between the location of the germ cells in the early embryo and the gonadal anlagen is bruged in the main by active migration, in which chemotactic processes play a leading role. The chemotactic attraction originates first from the dorsal mesodermal organs, leading to an accumulation of the PGCs near the dorsal root of the mesentery, and later from the epithelium of the genital ridges. The former phase has been extensively studied in the amphibians, the latter in the birds. Depending upon the absolute distance to be covered, the PGCs may make use of the blood circulation to reach the neighbourhood of the genital ridges. Vascular transfer occurs mainly in forms with meroblastic eggs, but is also related to the time of onset of blood circulation — early in the Amniota and late in the Anamnia. The use of the vascular route seems to be correlated with the nature of the passive displacements of the PGCs during early embryogenesis, i.e., whether they lead to an anterior or posterior localisation of the PGCs. Posterior localisation is usually associated with interstitial migration, whereas vascular transfer occurs in the case of anterior localisation. A careful analysis of the displacements of the PGCs during early embryogenesis seems of crucial importance. The reptiles, which show example of both types of migration, are a very interesting group in this respect and should be studied much more extensively.

6

The Testis

The testis is essentially a mass of small seminiferous tubules, along the walls of which the primordial germ cells are formed. These germ cells undergo spermatogenesis forming sperms.

As stated already that in vertebrates the testis are never more than two (one pair) in an adult organism. Testis are formed from the embryonic mesoderm. Instead of performing the spermatogenesic activity, the testis functions as an endocrine gland and severity *testosterone.*

Spermatogenetic Function

Spermatogenesis is the process forming sperms. Spermatogenesis involves meiotic divisions as well as mitotic divisions, so that the resulting spermatozoa have a number of chromosomes. The simplest basic scheme of spermatogenesis is as follows:

Gametes	*Number of chromosomes per somatic cell*	*DNA content of the cell*
Spermatogonia **mitotic multiplication**	**2n**	**2C**
↓		
Spermatogonia **maturation**	**2n**	**2C**
↓		

cont.....

Primary spermatocytes meiosis I	2n*	4C
↓		
Secondary spermatocytes meiosis II	n	2C
↓		
Spermatids spermiogenesis	1n	
↓		
Spermatozoa	n	1C

* At this stage there are 2n chromosomes, but each chromosome consists of two strands. This explains why the number of chromosomes in 2n, but the DNA content per cell is 4C.

There are two major types of spermatogenesis in vertebrates. In type 1, found in fishes and amphibians, spermatogenesis consists of the development of spermatogenic clones within cysts, which are located in tubular or lobular compartments. In type 2, found in amniotes (reptiles, birds and mammals), spermatogenesis takes place in the seminiferous tubules, in several layers of epithelium, and there are no subcompartments. In type 1 spermatogenesis, the cells within a cyst are all in the same stage of development. The primary spermatogonia lie against the basal membrane and multiply several times in succession by mitotic divisions, as do the follicle cells, so that one follicle may contain several secondary spermatogonia. The secondary spermatogonia within a follicle divide synchronously, and after about 8 mitotic divisions, the follicle contains more than 200 secondary spermatogonia, when maturation begins. The maturation process consists mainly of enlargement of the nucleus. The resulting cell is now called the primary spermatocyte. At this stage of development, the follicle cells become glandular in appearance.

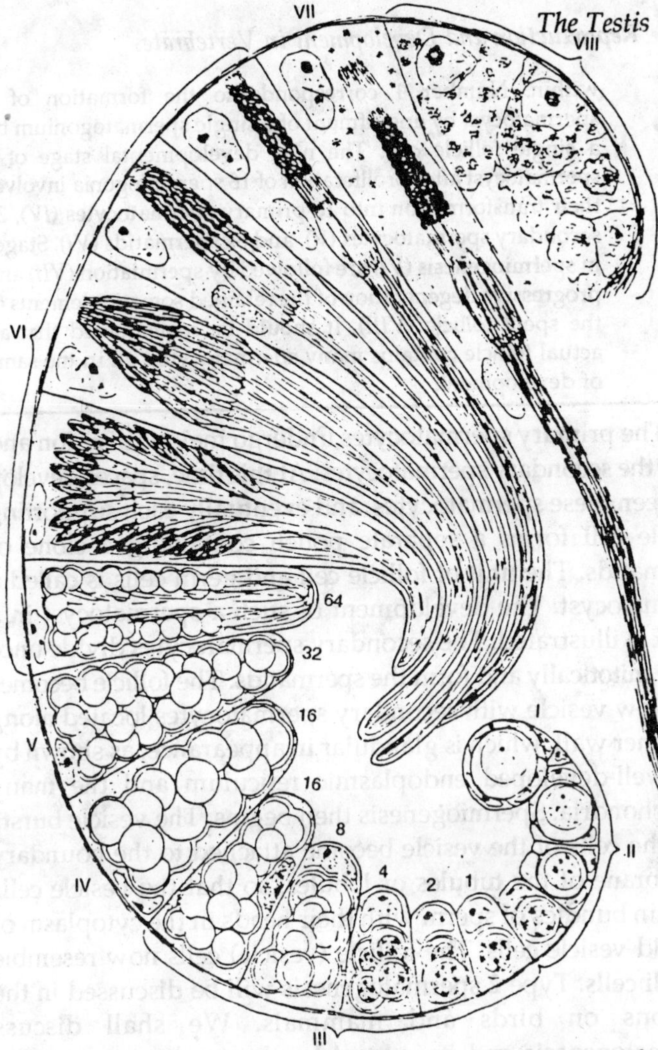

Fig. 6.1. Diagram indicating developmental history of single spermatocyst [in a shark]. Development proceeds from lower right (I) around to the upper right (VIII). *I* indicates a newly formed seminiferous follicle. The segment marked *II* indicates the period of mitotic proliferation if spermatogonia and of follicle cells as separate individuals *III* marks the period in which spermatogonial mitoses continue after engulfment by a follicle cell. The arabic numeral at the inner end of each spermatocyst indicates the number of germ cells contained *cont...*

within. Number 1 corresponds to the formation of a spermatocyst by engulfment of a single spermatogonium by a single follicle cell. The next developmental stage of a spermatocyst after proliferation of 16 spermatogonia involves their transformation into 16 primary spermatocytes (*IV*), 32 secondary spermatocytes (*V*), and 64 spermatids (*VI*). Stages in spermiogenesis (*VI*) are followed by spermiation (*VII*) and progressive degeneration of the retained somatic elements of the spent follicle (*VIII*). It should be remembered that an actual follicle contains many spermatocytes, all in the same of development.

The primary spermatocytes undergo meiotic division and form the secondary spermatocytes. At this time, spaces develop between these spermatocytes, and eventually each supporting follicle cell forms a separate pouch containing a clone of spermatids. The unit of follicle cell and germ cells is called a spermatocyst. The development of such a spermatocyst in a shark is illustrated. The secondary spermatocytes divide once more mitotically and form the spermatids. The follicle becomes a hollow vesicle with secondary spermatocytes located along the inner wall, which is glandular in appearance, as shown by the well-developed endoplasmic reticulum and the many mitochondria. Spermiogenesis then begins. The vesicle bursts and the cells of the vesicle become attached to the boundary membrane of the tubules or lobules, so that the vesicle cells contain bundles of sperm with their heads in the cytoplasm of the old vesicle cells. The follicle (vesicle) cells now resemble Sertoli cells. Type 2 spermatogenesis will be discussed in the sections on birds and mammals. We shall discuss spermatogenesis and its control by the pituitary in various classes of vertebrates that have type 1 spermatogenesis.

Cyclostomes

Spermatogenesis takes place in follicle. In the primitive hag fish *(Eptatretus stouti)* of the Pacific Ocean, hypophysectomy has little effect on spermatogenesis, although it results in an increased incidence of abnormal testicular follicles. In the lampreys *(Lampetra* spp.), the anterior pituitary (AP) is not required for spermatogenesis to proceed beyond the stage of the primary spermatocyte formation. The rate of spermatogenesis after is slower than in controls, but the process is otherwise normal.

Elasmobranchs

Spermatogenesis takes place in follicles. The testis of the dogfish (*Squalus acanthias*) has a definite rostral caudal zonation, which is the result of the migration of the ampullae, a migration closely related to the supporting cells of the ampullae. In the developing ampulla, the lumen is lined with prismatic epithelial cells: at this stage, the germ cells are secondary spermatogonia. After several cell divisions, the germ cells form eight layers, and the epithelial cells move between the spermatogonia to the periphery. The processes of the epithelial cells envelop some but not all spermatogonia. At this stage of development, the former prismatic epithelial cells are transformed into nurse cells, which may be analagous to the mammalian Sertoli cells. By the time the nurse cells are formed, the cytoplasmic processes contain 64 spermatids. The number of nurse cells equals the number of bundles of spermatozoa, about 424 per ampullae. There are probably 10 spermatogonial divisions between the primordial cells and the formation of the spermatocyst.

Removal of the ventral lobe of the AP of the dogfish (*Scyliorhinus canicula*) results in resorption of those spermatogonia that are to undergo the last four series of mitotic divisions, although this resorption occurs only if the water temperature is above 13°C. Injection of homogenates of ventral lobes prevents testicular degeneration, which normally follows ventral lobectomy, but hCG and PMSG are not effective. The main effect of the gonadotrophic hormone is exerted at the mitotic division that converts 8 spermatogonia into 16 primary spermatocytes. The breakdown zone that appears after ventral lobectomy consists of late spermatogonial ampullae, in which the degenerating germ cells are phagocytosed by the Setoli cells.

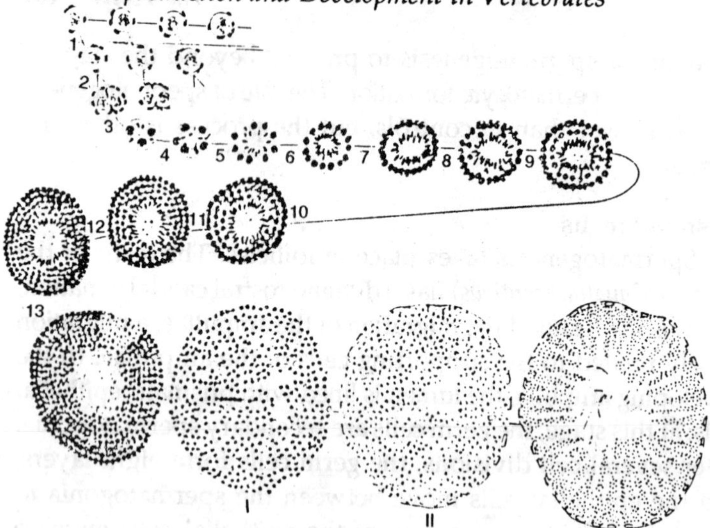

Fig. 6.2. Schematic diagram of spermatogenesis in the dogfish (*Squalus acanthias*) from spermatogonia to spermatid formation. The last three ampullae contain primary spermatocytes (I), secondary spermatocytes (II), and spermatids.

Administration of testosterone does not reverse the effects of ventral lobectomy. This observation and the fact that blood testosterone concentrations do not decrease after ventral lobectomy suggest that testosterone cannot maintain spermatogenesis. This, of course, does not imply that testosterone does not play some role in normal spermatogenesis. It is possible that testosterone is required for one of the stages following spermatocyte formation. This hypothesis could be tested by the use of anti-androgens or antisera against androgens. Complete leads to a statistically significant ($P < 0.05$) decrease in plasma testosterone concentration (12.10 ± 0.67 ng/ml in controls, 9.95 ± 0.98 ng/ml in hypophysectomized dogfish). These data indicate that the AP affects testosterone secretion to some extent. In higher vertebrates, usually results in nearly undetectable concentrations of plasma testosterone.

Teleosts

Spermatogenesis occurs in spermatocysts (also called follicles). The development of germ cells and the spermatocysts, with their germ cells developing synchronously within a spermatocyst, has been reviewed extensively by Roosen-Runge (1977). In general, type 1 spermatogenesis is found in teleosts, except in the sea perch (*Cymatogaster* sp.), in which the spermatogonia divide asynchronously, and in which spermatocysts are formed not during the spermatocysts are formed not during the spermatogonial stage but during the spermatocyte stage. Grier (1975) found that in *Poecilia latipinna*, the spermatocysts develop when the primary spermatocytes become surrounded by a single layer of Sertoli cells at the periphery of the testis. As spermatogenesis proceeds, they move centrally in the tests toward the ductuli efferents.

After AP, the tests regress and spermatogenesis does not continue beyond formation of spermatogonia in the species investigated, i.e., the goldfish (*Carassius auratus*); the guppy (*Poecilia reticulata*); the lake club (*Couesius plumbeus*); the killifish (*Fundulus heteroclitus*); the Indian catfish (*Heteropneustes fossilis*); and the longjaw goby (*Gillichthys mirabilis*).

The study of the control of spermatogenesis in teleosts has been difficult because of the specificity of the gonadotrophins; that is, spermatogenesis may require the gonadotrophin secreted by the species being investigated or the gonadotrophin of a related species. It is difficult, however, to obtain enough pituitaries and to purify a sufficient amount of gonadotrophin for injection purposes, except for some commercial fishes.

In the species, that have been investigated, gonadotrophin from the same or from related species — but usually not those from unrelated species — are capable of restoring testicular weight and spermatogenesis. Mammalian gonadotrophin preparations have (except for FSH, which has no effect on either testicular weight or on spermatogenesis) varying effects, according to the species investigated. For example, human chorionic gonadotrophin (hCG) can restore testicular weight and spermatogenesis in AP Indian catfish, but not in AP goldfish, guppies, or the longjaw goby. Similarly, varying results have been obtained with mammalian LH preparations. Androgen administration can restore spermatogenesis in AP Indian catfish and killifish.

Amphibians

Spermatogenesis takes place in spermatocysts, which are located in ampullae in the urodeles and in tubules in the anurans. In the anurans (frogs and toads), three types of spermatogenesis can be distinguished as follows:

1. *Continuous spermatogenesis.* Spermatogonia mature into spermatocytes during the entire year.

2. *Discontinuous spermatogenesis.* Primary spermatogonia lose their capability to divide mitotically during part of the year, and administration of gonadoptrophins does not restore this capability.

3. *Potentially continuous spermatogenesis.*Primary spermatogonia continue to divide slowly, and the spermatocysts continue to develop; however , development during part of the year does not continue beyond the primary spermatocyte stage, and these spermatocytsnormally degenerate. Nevertheless, spermatogenesis can be stimulated during the season when spermatogenesis normally ceases, either by injection of gonadotrophins or by exposure of the animals to high temperatures, such as they experience during the normal period of spermatogenesis of the species. Table 6.1 lists the distribution of these three types of spermatogenesis among various species.

Spermatogenesis takes place within spermatocysts and, when the flagella of the spermatids start to grow, the spermatocysts rupture, opening into the lumen of the tubule. At the peripheral pole of the spermatocyst, the spermatocyst cells are transformed into so-called supporting cells of Sertoli, to which bundles of spermatids (60-150) become attached. These supporting cells are the targets of gonadotrophic hormones, which cause the release of the sperm cells (spermiation) by increasing the fluid intake of the supporting cells. Both ovine FSH (oFSH) and ovine LH (oLH) can induce spermiation. In the grass frog (*Rana pipiens*), oLH was more potent than oFSH, but in the Pacific tree frog (*Hyla regilla*) and in the coqui (*Eleutherodactylus coqui*), oFSH was about twice as potent as oLH. Neither the α not the β subunits of LH nor the α unit of FSH induced spermiation, but the β unit of FSH retained about 25 per cent of the activity of the native oFSH. Crude extract of AP of the bullfrog (*Rana catesbeiana*) had a higher potency than the most active bovine preparations. Both purified bullfrog FSH and bullfrog LH caused spermiation in AP bullfrogs.

TABLE 6.1

Continuous, Potentially Continuous, and Discontinuous Spermatogenetic Cycles in Some Anura

Continuous cycles

Bufo paracnemis
B. arenarum
B. melanostictus
B. granulosus
Leptodactylus prognathus
L. ocellatus reticulatus
L. laticeps
Physalaemus fascomaculatus
Presudis paradoxa
P. mantidactyla
Rana hexaxactyla
R. esculenta (southern range)
Discoglossus pictus (southern range)

Discontinuous cycles

Rana temporaria
R. arvalis
R. dalmatina

Potentially continuous cycles

Rana esculenta (northern range)
R. tigrina
Discoglossus pictus (northern range)
Leptodactylus bufonius
L. ocellatus tyica (two breaks in spermatogenesis: winter and summer)
Leptobatrachus asper

Our understanding of the normal control of spermatogenesis is incomplete; because in many experiments, mammalian gonadotrophic hormones have been use. It is now known that mammalian and amphibian gonadotrophins differ biochemically and physiologically. The following information is relevant: First, AP results in degeneration of the tests and of the germinal epithelium. According to Rastogi *et al.* (1976), the primary spermatocytes are the most sensitive elements of the germinal epithelium of the green frog (*Rana esculenta*), whereas the spermatids are not affected. In the toad (*Bufo bufo bufo*), AP reduces the rate of multiplication of primary spermatogonia a few days after AP; however, the mitotic divisions of secondary spermatogonia are not affected for about a week and then diminish. The differentiation of the secondary spermatogonia fails to occur, and the spermatocysts with secondary spermatogonia degenerate. The primary spermatocytes and the subsequent stages of spermatogenetic development are not affected byAP. In general, one can state that AP affects premeiotic but not postmeiotic stages of spermatogenesis in this species.

Second, after AP of green frogs, spermatogenesis can be restored either by injection of homogenates of amphibian pituitaries or by homiotropic, but not by ectopic pars distalis transplants. In AP bullfrogs, either bull frog FSH or bullfrog LH stimulates spermatogenesis; however, androgen production is stimulated by bullfrog LH but not by bullfrog FSH. Third, PMSG or mammalian FSH can stimulate spermatogenesis up to the secondary spermatocyte state in AP larval salamanders *(Ambystoma tigrinum)*, but hCG and mammalian LH are not effective.

Finally, The role of testosterone in the spermatogenesis of the Anura appears to be a complicated one. In the leopard frog *(Rana pipiens)*, testosterone not only fails to restore spermatogenesis in AP frogs but inhibits the restoration induced by exogenous FSH plus LH. According to Rastogi *et al.* (1976), testosterone stimulates spermatid formation, but it does not affect other stages of spermatogenesis of green frogs *(Rana esculenta)* and thus cannot restore spermatogenesis in AP frogs. Experiments by Moore (1975) with AP larval *Ambystoma tigrinum* further illustrate the complex interaction between testosterone and FSH. After AP, treatment for 40 days with oFSH resulted in animals with large germinal cysts, many with secondary spermatocytes. Testosterone enhanced the effect of oFSH when the A I[3] larvae were treated for 20 or 40 days with oFSH plus testosterone. Comparisons made with 40 days of oFSH treatment showed that testosterone *inhibited* the action of oFSH during the last 20 days (of the 40-day treatment) if the animals had *not* received testosterone during the first 20 days, but *enhanced* spermatogenesis if testosterone had been given during the first 20 days. Rastogi *et al.* (1976) have proposed that the interaction between the environment, the AP, and the gonadal hormones in *Rana esculenta* can be explained as shown in Figure 6.3. The sequence of events and hormone concentrations found in *Rana esculenta* around Naples, Italy.

Besides the control of spermatogenesis by the AP, there is an important influence on spermatogenesis by the fat bodies. In green frogs, removal of fat bodies in February leads to degeneration of the testes. This degeneration cannot be prevented by the administration of exogenous gonadotrophins but can be prevented by implanting fat bodies, obtained in March or October, or by injecting extracts of such fat bodies. Unilateral removal of fat bodies leads to degeneration of the ipsilateral testis, suggesting that there is a direct transfer of lipid materials from fat body to testis. The mobilization of such materials appears to be under the control of the pars distalis. There is also a reciprocal effect of the testes on the fat bodies, because castration leads to a decrease in the weight of the fat bodies.

Urodeles. The urodeles have lobules filled with follicles, and all follicles within a lobule are in the same stage of spermatogenesis. The testes are divided into different zones, with spermatogenesis being most advanced in the posterior zone and least advanced in the anterior zone. In the newt,

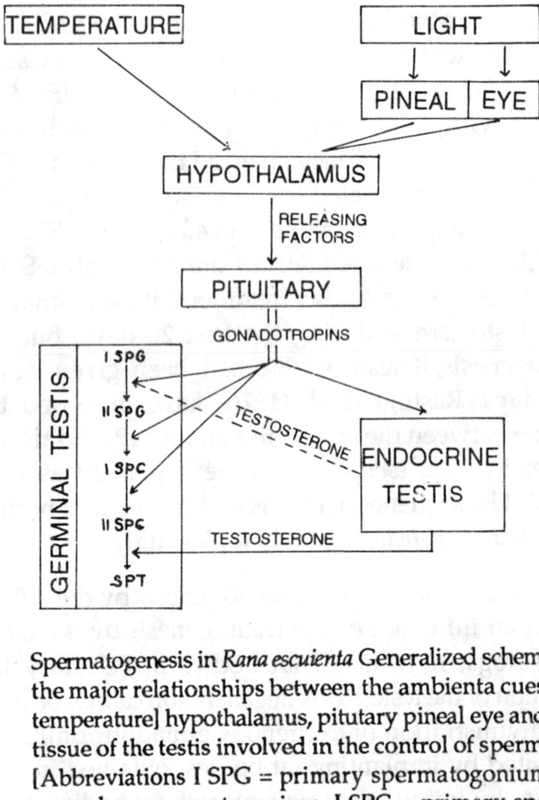

Fig. 6.3 Spermatogenesis in *Rana escuienta* Generalized scheme depicting the major relationships between the ambienta cues [light and temperature] hypothalamus, pitutary pineal eye and endocrine tissue of the testis involved in the control of spermatogenesis [Abbreviations I SPG = primary spermatogonium. II SPG = secondary spermatogonium, I SPC = primary spermatocyte II SPC = secondary spermatocyte, SPT = spermatid].

Trituroides hongkongensis, a yellow zone is formed by the hypertrophied walls of the lobules and the Sertoli cells. These two structures together form an interstitial gland that is eventually resorbed. The yellow zone is most prominent from February through April, when testicular weights are at their minimum. When spermatogenetic activity starts, the yellow zone begins to be resorbed, and the resorption is completed by May. Spermatid formation starts in June, and spermiogenesis begins in July, followed by sperm evacuation in August and September.

The ductuli efferentes conduct the sperm cells into the sexual segment of the kidney, after which the sperm are discharged via the ductus deferens into the cloaca. In all urodeles with internal fertilization (*Cryptobranchidae* and *Hynobiidae* have external fertilization), the sperm are packaged into spermatophores.

Reptiles

In reptiles, birds, and mammals, type 2 spermatogenesis is found. Spermatogenesis takes place in tubules and several layers of germ epithelium are formed. The spermatogenetic process can be divided into three main periods : (1) the muultiplication of spermatogonia, which after differentiation

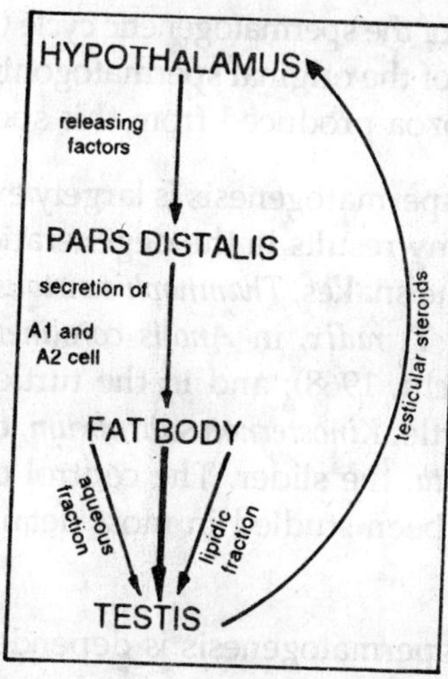

Fig. 6.4. Schermatic atic diagram showing possible position of frog that bodies in the hypothalamo-hypophyseal-gonadal axis. [A_1 and B_2 cells are gonadotrophin cells].

into spermatocytes move inward to form a new layer, the germinal epithelium (2) spermatocyte division, which results in spermatid formation; and (3) spermiogenesis, the period in which spermatozoa are formed from the spermatids. In ducks and several mammalian species, different stages of spermiogenesis have been correlated with different cell associations of the germ cells, as will be discussed in the sections on birds and mammals.

The testis consists of seminiferous tubules, with Leydig cells among the tubules. Teiid lizards have a circumstesticular layer of Leydig cells. Before the breeding season, the tubules are enveloped by a peritubular membrane, which consists largely of myoid cells. At the end of the breeding season, these cells are replaced by fibroblasts.

Little appears to be known about the quantitative aspects of spermatogenesis in reptiles, in contrast to the information available on some birds and especially mammals; e.g., Roosen-Runge's excellent review (1977) cites no data on any reptiles for the duration of the spermatogenetic cycle (the period between the appearance of the original spermatogonium and the release of the spermatozoa produced from this spermatogonium).

Control of spermatogenesis is largely exercised by the AP. Hypophysectomy results in the degeneration of the germinal epithelium in the snakes, *Thamnophis sirtalis,* the grater snake, *T. sauritus* and *T. radix,* in *Anolis corolinensis,* the American chameleon, (Licht, 1968); and in the turtles, *Chrysemys picta,* The painted turtle, *Kinosternon subrubrum,* the mud turtle, and *Pseudemys scripta,* the slider. The control of spermatogenesis by the AP has been studied in most detail for the American chameleon.

Although spermatogenesis is dependent upon the AP, it is also affected by the environmental temperature. At 31° C, AP resulted in complete gonadal regression in about 2 weeks, whereas at 20C, even three weeks after AP, little effect could be detected. The effect of temperature on testicular degeneration after AP is thus quite similar to that observed in *Scyliorhinus canicula.* After injection of oLH and oFSH, spermatogenesis was restored in AP *Anolis carolinensis* at 31 ° C, but the amount of LH required was several hundred times greater than the amount of FSH. When LH and FSH were given together, the effect on spermatogenesis was less than that obtained by FSH alone. Testosterone retarded post-AP regression of the testes, but it did not restore spermatogenesis. Subsequent studies with *Anolis carolinensis* have shown that :

1. All the gonadotrophic activity for spermatogenesis is in the β subunit of oFSH.

2. After removal of residual LH by LH antiserum. oFSH retains its capacity to stimulate and maintain spermatogenesis of animals.

3. Ovine LH retains its capability to stimulate spermatogenesis after removal of FSH activity.

4. Testicular growth and spermatogenesis can be maintained in animals with all species of amphibian, reptilian, and avian FSH and LH examined, although the LH of certain species may be less effective than the FSH. The greater effectiveness of the FSH can be explained by the lower clearance rate of FSH in comparison with LH.

Investigation on the relationship between AP and tests of reptiles have so far shown that:

1. Snakes and lizards lack LH, as tested by induction of ovulation in ovaries of *Hyla regilla* and *Xenopus laevis*, but LH is present in the AP of turtles and several Crocodilia.

2. After, AP the testes of snakes and lizards respond to oFSH or to very high doses of oLH by restoration of spermatogenesis. In neither *Anolis carolinensis* nor in *Thamnophis sirtalis* does testosterone restore spermatogenesis in the AP animal.

3. In snakes and lizards, the testis shows little specificity for FSH and LH, which may be related to a lack of specificity in gonadotrophin-binding sites for mammalian gonadotrophins.

4. Ovine FSH is more potent than oLH in restoring spermatogenesis in *Pseudemys scripta, Kinosternon subrubrum* and *Chrysemys picta.*

As we shall discuss later in this chapter, exposure of the testes of scrotal mammals to higher than their normal temperature interferes with spermatogenesis. Exposure of the lizard *Urosaaurus ornatus* for ten days to temperatures 1-2°C above its preferred environmental temperature leads to degeneration of the germinal epithelium. Thus, in reptiles and mammals, high temperature has a profound effect on reproduction.

In most species, the Leydig cells are located among the testicular tubules, but in teiid lizards, the Leydig cells form a circumtesticular capsule. In *Cnemidophorus tigris,* the capsule binds radioiodinated human FSH intensely. In the garter snake, the slider, and the American chameleon, the binding occurs in the interstitial cells and probably in the spermatogonia and Sertoli cells.

Birds

So far we have discussed spermatogenesis in poikilotherms, but birds and mammals are homeotherms. As we shall discuss in the section Mammals, exposure of the testes to body temperature, by interfering with spermatogenesis, causes sterility in many scrotal mammals. Not only are birds homeothermic, but they also generally have high body temperature; e.g., the temperature of the domestic hen fluctuates daily between 39°C and 42°C. In the scrotum, the testes are exposed to temperatures lower than body temperature. In all birds, the testes are internal and the question has been asked whether exposure to temperatures either lower than or higher than body temperature would affect spermatogenesis. Exposure of testes of sexually immature roosters to temperatures lower than body temperatures, by creating an artificial "scrotum," caused precocious spermatogenesis, and exposure of testes to temperature higher than body temperature (i.e., in vitro exposure to warm Ringer's solution followed by regrafting of the testes) interfered with spermatogenesis. The air sacs, which might act as cooling organs for the testes, probably do not function in this way, because the testicular temperature is the same as that in the rest of the body cavity, and destruction of the air sacs has been found not to affect spermatogenesis in roosters.

During spermiogenesis, 10 different stages of spermatid development can be identified accurately in the mallard drake (*Anas platyrhynchos*). These 10 stages are correlated with 8 different types of germ cell associations in the drake. The study of such cell associations makes it possible to determine which spermatogonial cells are renewed.

Control of spermatogenesis is exercised by the hypothalamo-anterior pituitary system. Lesions made in the mammillary nuclei and the posterior part of the ventromedial nucleus resulted in testicular atrophy and regressed combs (20 per cent of controls). Lesions in the posterior part of the mammillary nuclei, impinging on the arcuate nucleus, caused degeneration of the germinal epithelium, but the combs regressed to 40 per cent of the size of combs in controls.

As expected, the testes of all birds investigated (Japanese quail, chickens, drakes, pigeons) regress after; although mammalian FSH can maintain spermatogenesis in roosters.

Brown *et al.* (1975), using chicken FSH and LH, showed that in Japanese quail, FSH stimulated testicular growth, the formation of spermatocytes to the pachytene stage, and the Sertoli cells; whereas LH administration resulted in differentiation of the interstitium, some spermatogonial differentiation, and partial differentiation of the Sertoli cells. These last two effects may have been the result of testosterone secretion stimulated by LH.

The effects of testosterone on the testes of birds seem to vary with the species. Testosterone maintained and also restored spermatogenesis in pigeons, however, in Japanese

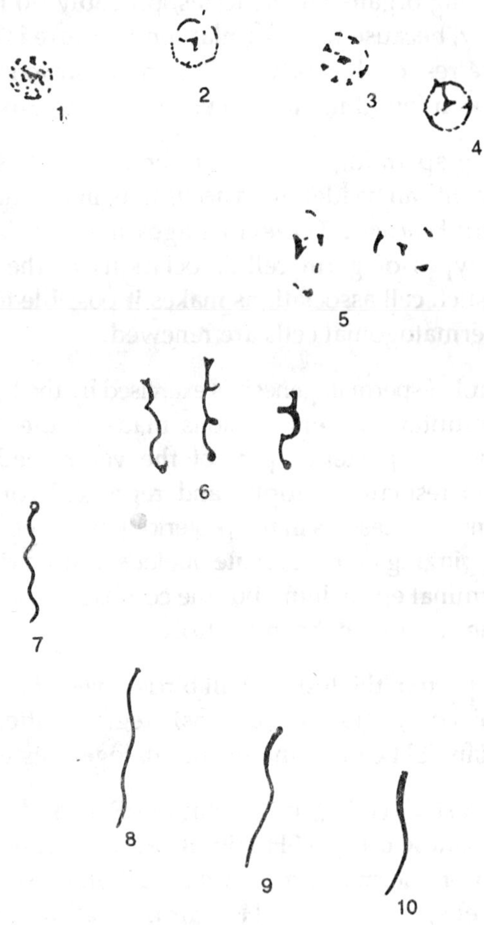

Fig. 6.5. Spermiogenesis in the drake as observed in sections stained with PAS-hematoxylin, showing the series of changes which the spermatid undergoes. 1, young spermatid; 10, mature sperm. The proacrosomic granule (in the cytoplasm in light gray) is shown as a black dot (2), which is attached to the nucleus (3) and becomes a triangular acrosome (4-6). The acrosome becomes more lightly stained (7) but persists as a fine point at the tip of the nucleus (8-10).

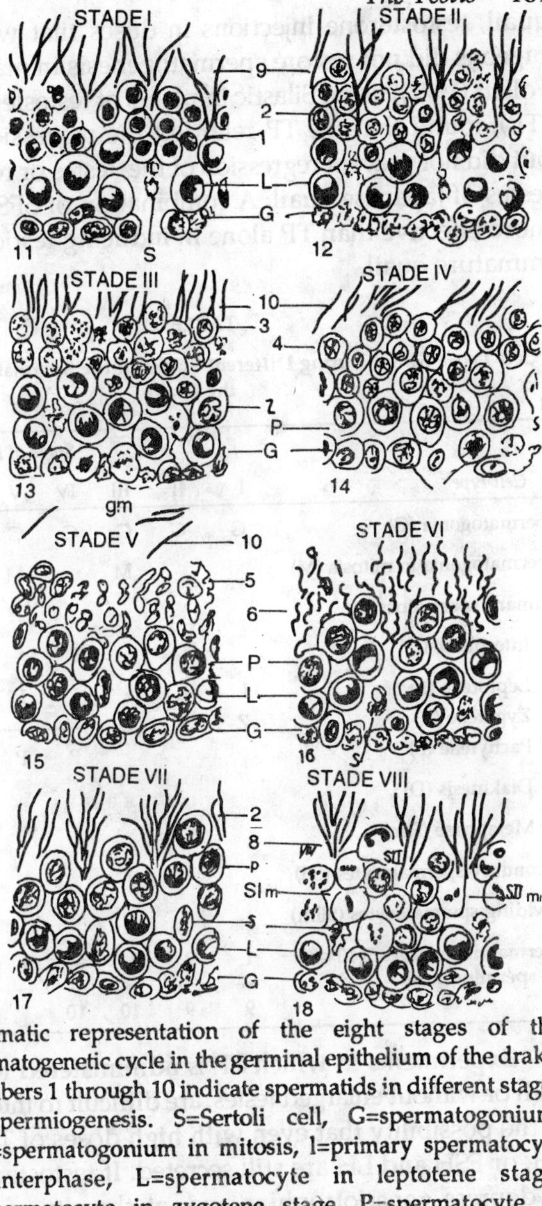

Fig. 6.6. Schematic representation of the eight stages of the spermatogenetic cycle in the germinal epithelium of the drake. Numbers 1 through 10 indicate spermatids in different stages of spermiogenesis. S=Sertoli cell, G=spermatogonium, GM=spermatogonium in mitosis, l=primary spermatocyte in interphase, L=spermatocyte in leptotene stage, Z=spermatocyte in zygotene stage, P=spermatocyte in pachytene stage. SIm=primary spermatocyte in metaphase, SII=secondary spermatocyte, SIIm=secondary spermatocyte in metaphase.

quail, testosterone injections in doses that were effective in pigeons did not restore spermatogenesis in males. Brown and Follett (1977), using Silastic implants of testosterone propionate (TP), found that the TP retarded the regression of the testes but did not prevent regression of the testes or restore regressed testes of Japanese quail. A combination of oFSH plus TP was more effective than TP alone in inducing testicular growth in immature quail.

TABLE 6.2

Cell Types Found During Different Stages of the Spermatogenetic Cycle of the Drake

	Stage of the cycle							
Cell type	I	II	III	IV	V	VI	VII	VIII
Spermatogonia (G)	G	G	G	G	G	G	G	G
Spermatogonia in mitosis (M)			M		M			M
Primary spermatocytes								
Interphase (I)				I				
Leptotene (L)	L	L			L	L	L	L
Zygotene (Z)	Z	Z	Z					
Pachytene (P)				P	P	P	P	
Diakinesis (D)							D	D
Metaphase (Slm)								Slm
Secondary spermatocytes (Sll)								Sll
Dividing spermatocytes (Sllm)								Sllm
Spermatids in stages 1-10 of spermiogenesis	1 — 9	2 — 9	3 — 10	4 — 10	5 —	6 —	7 —	8 —

Experiments in which TP is administered to intact birds, with or without enlarged testes, are difficult to interpret because of the possibility that even with high doses of TP, some FSH, LH, or FSH and LH are still secreted. If testosterone acts both to depress gonadotrophins and, at the site of the testes, to stimulate a step in spermatogenesis after FSH, LH, or FSH plus LH, then one is dealing with too many uncertainties to give a clear interpretation [see, however, Brown and Follett (1977) and Desjardins and Turek (1977)].

Testosterone injections restored spermatogenesis in the regressed testes of photorefractory African weaver finches (*Quelea quelea*) and in sexually immature house sparrows (*Passer domesticus*). They maintained spermatogenesis in slate-coloured juncos (*Junco hyemalis*) and white-throated sparrows (*Zonotrichia albicollis*) after transfer of the birds from long to short days, a transfer that normally causes involution of the testes. In roosters, androgen administration did not restore spermatogenesis in animals bearing hypothalamic lesions that cause testicular regression. So far, the only bird in which testosterone has been shown to restore spermatogenesis in the animal is the pigeon.

The inherent disadvantage of using animals which are deficient not only in FSH and LH, but also in ACTH, TSH, GH, and prolactin, could be overcome by selective "GTH hypophysectomy," i.e., by antibodies to FSH or LH obtained either by passive or active immunization. However, this requires the availability of pure avian LH and FSH in quantities that so far have not been available.

Mammals

In mammals, as in birds, the occurrence of each stage of spermiogenesis is correlated with certain cell associations in the germinal epithelium. The occurrence of these cell associations and the labelling of germ cells by [^{3}H]-thymidine can be used to study several aspects of spermatogenesis quantitatively, such as: (1) the manner in which spermatogonial stem cells are renewed, (2) the duration of the seminiferous epithelial cycle, i.e., the time that elapse between two successive appearances of the same cellular associations; and (3) the duration of the spermatogenetic cycle, i.e., the interval between the appearance of the original spermatogonium and the appearance of the sperm produced from it. The number of spermatogonial mitoses for different species is given in Table 6.3; the duration of the spermatogenetic cycle is given in Table

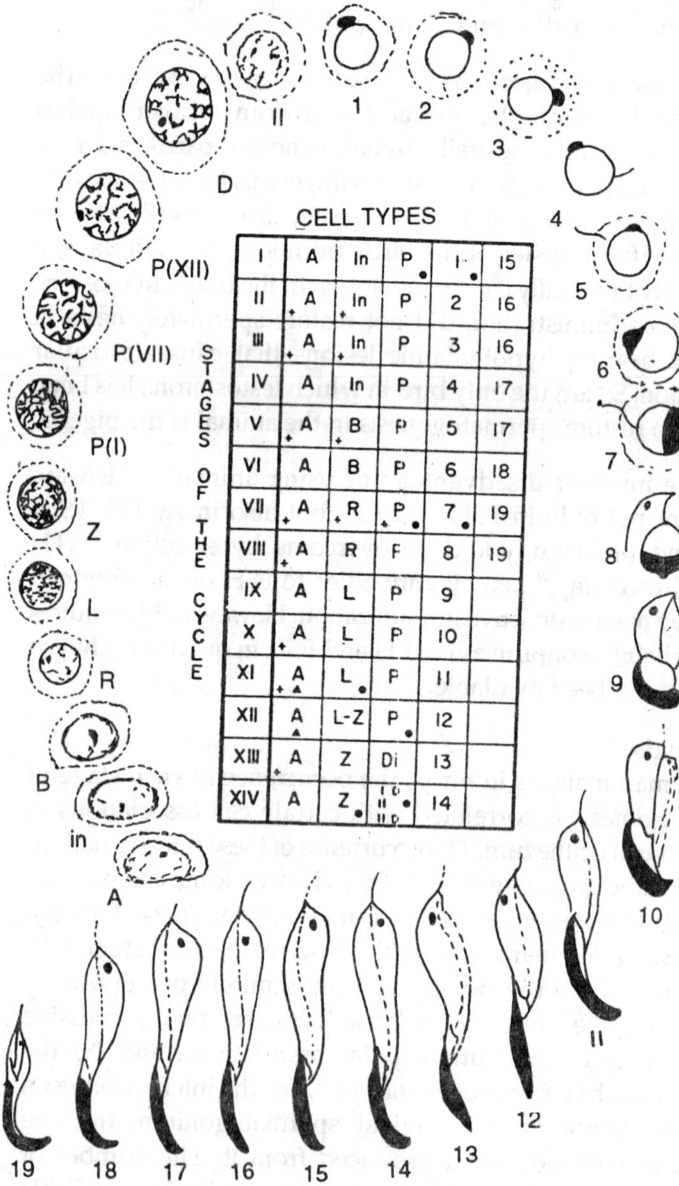

CELL TYPES

I	A	In	P	1	15
II	A	In	P	2	16
III	A	In	P	3	16
IV	A	In	P	4	17
V	A	B	P	5	17
VI	A	B	P	6	18
VII	A	R	P	7	19
VIII	A	R	P	8	19
IX	A	L	P	9	
X	A	L	P	10	
XI	A	L	P	11	
XII	A	L-Z	P	12	
XIII	A	Z	Di	13	
XIV	A	Z	II	14	

Fig. 6.7. (*Opposite page*) Starting from the lower left and encircling the table drawings illustrate the steps of spermatogenesis in the rat. A=type A spermatogonium; In=intermediate type spermatogonium; B=type B spermatogonium; R=resting primary spermatocyte; Z=zygotene spermatocyte; $P_{(i)}$, $P_{(vii)}$, $P_{(xii)}$=early mid, and late pachytene spermatocytes, respectively; roman numerals in brackets indicate the stages of the cycle at which they are found Di=diplotene; II=secondary spermatocyte, 1-19=steps in spermiogenesis. (From Leblond and Clermont, 1952).

The table in the center of the figure gives the cellular composition of the stages of the cycle of the seminiferous epithelium (roman numerals, I-XIV). I^D and II^D=first and second maturation divisions of spermatocytes (+) (●), (Δ) = cells counted in the first, second, or third series of counts respectively.

6.4. In mammals, the spermatogenetic cycle lasts about four times as long as the epithelial cycle.

What we know about the control of spermatogenesis is based on the results obtained from in vitro and in vivo studies, largely of rat testes. Hypophysectomy of rats results in an increased incidence of degeneration of midpachytene spermatocytes and step 7 and 19 spermatids. Administration of FSH does not affect the incidence of such degenerating cells, but LH injections in physiological doses reduce the incidence of degenerating germ cells in stages VII and VIII of the cycle. Administration of FSH plus LH reduces the incidence of degenerating cells to that of the intact rat.

Raj and Dym (1976), using carefully characterized antisera, found that administration of LH antiserum to 20-day-old male rats for 14 days resulted in a reduction of the testicular weight to about one-third of that of the controls; a reduction in the weights of the epididymides, seminal vesicles, and ventral prostates; and a reduction in plasma testosterone concentration to about 20 per cent of that of the controls. The histological appearance of the testes was similar to that seen in rats, i.e., small-diameter tubules, few spermatids, pachytene secondary spermatocytes reduced in number to about 50 per cent of that of controls, but normal spermatogonia. Antiserum to FSH reduced testicular weights to 55 per cent of normal, but did not affect the weights of the epididymides, seminal vesicles, and ventral prostates or the concentrations of plasma testosterone. Histological examinations of the testes revealed a reduction in the number of spermatids to 83 per cent of that of the controls, and of the number of pachytene spermatocytes to 65 percent of that of the controls. These investigations thus showed that lack of LH, probably because it reduced testosterone secretion, has a more severe effect on spermatogenesis in the prepubertal rat than a lack of FSH. The effect of FSH is probably mediated through its effect on the secretion of androgen-binding protein (ABP) by the Sertoli cells, to be discussed below:

TABLE 6.3

Number of Spermatogonial Mitosis in Different Vertebrates

Species	*Number of mitosis*
Elasmobranchs	
Torpedo marmorata	4
Squalus caniculus	12
Teleosts	
Umbra limi	5-6
Phoxinus laevis	4-5
Perca flavescens	4-5
Brachydanio rerio	4-5
Sebastodes paucispinus	6?
Aequidens portalegrensis	5?
Gambusia holbrookii	9-11
Poecilia reticulata	13
Amphibia	
Ambystoma sp.	6
Proteus anguinus	6 and 7
Triton sp.	6
Mammals	
Bull	5

Ram	4
Gerbil	4
Mouse	5
Rat	5
Rabbit	4
Rhesus monkey (Macaca mulatta)	4
Long-tailed monkey (Cercopithecus aethiops)	4 or 5
Humans	3?

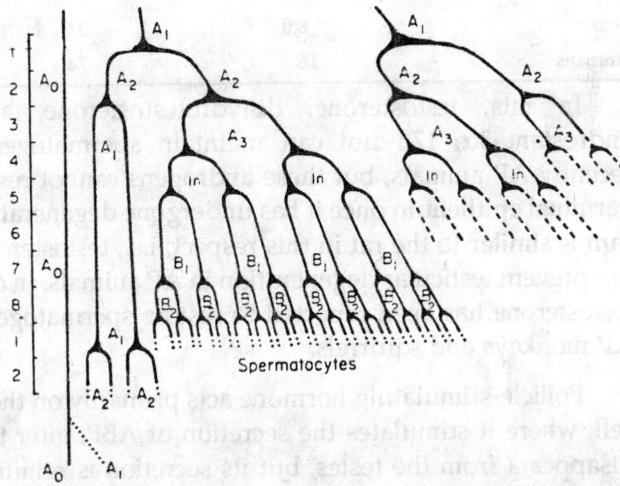

Fig. 6.8. Model of spermatogonial renewal in bull recently proposed by Hochereau-de Reviers. According to this scheme there are three generations of type A spermatogonia (A_1, A_2, A_3), one of intermediate type spermatogonia (In) and two of type B spermatogonia (B_1, B_2). Nodal division, equivalent in nature, is taking place among type A_2 spermatogonia in stages 2-3 of cycle. Degeneration affecting mainly type A_3 and intermediate-type spermatogonia is indicated by the broken line. Some type A_o spermatogonia are present (on left of diagram) and considered to divide slowly and sporadically (dotted line) to give rise to new type A_o and type A_1 spermatogonia. [The numbers on the line at the left side of the diagram indicate the stages of the spermatogenetic cycle.]

TABLE 6.4

Duration of Seminiferous Epithelial Cycle and Spermatogenetic Cycle in Different Species of Mammals

Species	*Epithelial cycle (days)*	*Spermatogenetic cycle (days)*
Rat	12.3	49
Mouse	8.6	34
Rabbit	10.9	44
Bull	14.0	56
Ram	10.0	40-49
Boar	8.0	34
Humans	16	74

In rats, testosterone, dihydrotestosterone, and 5α-androstane-3α, 17β-diol can maintain spermatogenesis in recently AP animals, but these androgens cannot restore the germinal epithelium once it has undergone degeneration. The ram is similar to the rat in this respect; i.e., testosterone does not prevent testicular degeneration in AP animals. In contrast, testosterone has been reported to restore spermatogenesis in AP monkeys and squirrels.

Follicle-stimulating hormone acts primarily on the Sertoli cell, where it stimulates the secretion of ABP, after the ABP disappears from the testes, but its secretion is reinitiated by FSH administration in rats. The secretion of ABP is initiated by FSH, but not by testoster-one, although testosterone stimulates the accumulation of ABP in the testes in AP rats; in vitro, FSH and testosterone each stimulate ABP production. Cyproterone acetate, an anti-androgen, does not influence the FSH-induced initiation of ABP section in vitro, but it inhibits the testosterone-induced ABP production.

In 15-day-old rats, FSH administration results in an increased ABP concentration in the testes and the epididymides. This effect of FSH on the testes is enhanced by concurrent administration of either testosterone or LH, but these two hormones each block the effect of FSH on the secretion of ABP into the epididymis. This blocking effect of testosterone has been found in the immature rat but not in the adult rat. The significance of this finding is obscure.

Steinberger and Steinberger (1973) have proposed the tentative interpretation of the hormonal control of spermatogenesis in the rat. Testosterone, which according to this interpretation is required in the formation of spermatid step 1, is secreted by the Leydig cells under the influence of LH. According to the explanation, testosterone enters the tubule, is bound by ABP, and is transported to the germinal epithelial cells, where, after it is released from the ABP, it acts on the appropriate germ cells. Summarizes the possible molecular effects of gonadotrophins on spermatogenesis. Rommerts *et al.* (1976) have cast doubt on the biological significance of the ABP in spermatogenesis. Presumably, only free testosterone can act on the germ cells; therefore, binding of the free testosterone can act on the germ cells; therefore, binding of the free testosterone by ABP as it enters the tubules would tend to impede the action of testosterone on the germ cells.

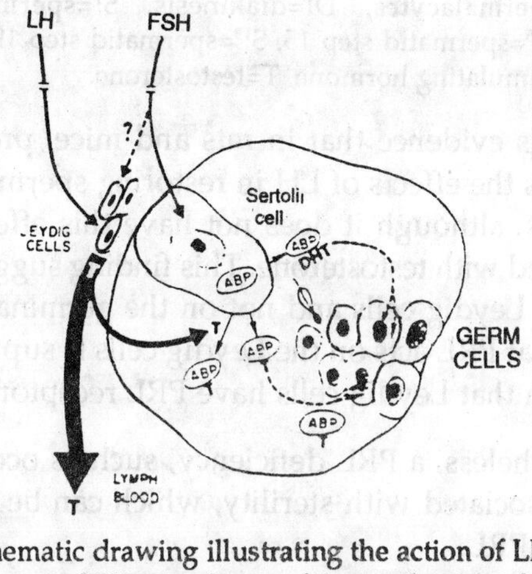

Fig. 6.9. Schematic drawing illustrating the action of LH and FSH in the testis. FSH stimulation of the production of the androgen-binding (ABP) causes an accumulation of androgen (T=testosterone, DHT=5α-dihydrotestosterone) in close proximity to the androgen-dependent cells within the germinal epithelium.

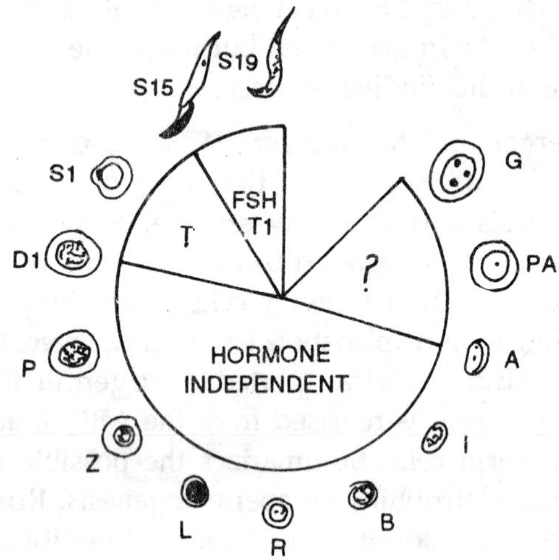

Fig. 6.10. Proposed scheme of the hormonal control of spermatogenesis [in rats]. G=gonocytes, PA=primitive type A spermatogonia, A=type A spermatogonia, I=intermediate spermatogonia, B=type B spermatogonia, R=resting spermatocytes, L=leptotene spermatocytes, Z=zygotene spermatocytes, P=pachytene spermalacytes, DI=diakinesis, S^1=spermatid step 1, S^{15}=spermatid step 15, S^{19}=spermatid step 19, FSH=follicle-stimulating hormone, T=testosterone.

There is evidence that in rats and mice, prolactin (PRL) protentiates the effects of LH in restoring spermatogenesis in AP animals, although it does not have this effect when it is administered with testosterone. This finding suggests that PRL acts on the Leydig cells and not on the germinal epithelium. The idea that PRL acts on the Leydig cells is supported by the observation that Leydig cells have PRL receptors.

Nevertheless, a PRL deficiency, such as occurs in dwarf mice, is associated with sterility, which can be overcome by exogenous PRL.

Several types of evidence suggest that there is negative feedback between the testicular seminiferous tubules and the AP, as will be discussed in the next section.

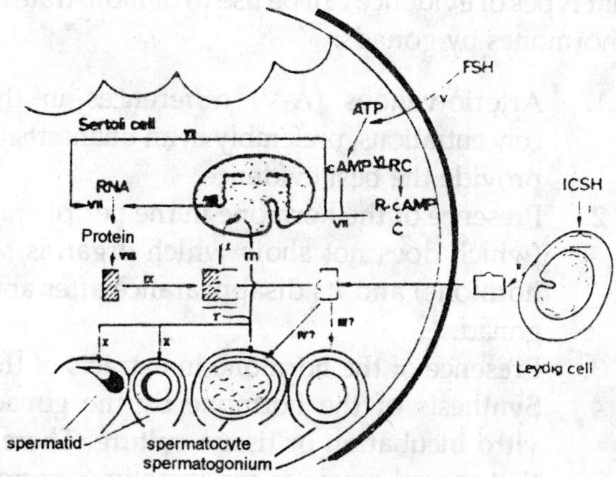

Fig. 6.11. Possible molecular effects of the gonadotrophins and testosterone in relation to control of spermatogenesis. I=stimulation of testosterone production in the Leydig cell by ICSH=LH, II=transport of testosterone into the seminiferous tubules and into Sertoli cells — the "active" androgen from here on is given as testosterone, built may well be a metapolite of testosterone; III=possible direct interaction of testosterone with germinal cells to initiate a wave of spermatogenesis' IV=possible direct action of testosterone on spermatocytes to promote meiotic division; V=production of cyclic AMP (cAMP) in the seminiferous tubules as a result of FSH binding and stimulation of adenylate cyclase; VI = activation of cAMP cAMP-dependent protein kinase (RC) by cAMP VII=increased RNA and protein synthesis effected by cAMP a protein kinase component, or another FSH "messenger" by a variety of mechanisms; VIII=increased androgen receptor protein as a result of increased synthetic activity; IX=complex formation between testosterone and receptor protein, X=direct interaction of the hormone-receptor complex with germinal cells, resulting in the progression of spermatogenesis, or indirect interaction through a "message" (m), generated as a result of events in the Sertoli cell controlled by this complex.

Endocrine Function: Androgens

The testes of most vertebrates have an endocrine function as well as a spermatogenetic one. It is useful to consider briefly what types of evidence can be use to demonstrate the secretion of hormones by gonads:

1. Arterio-venous (A-V) differences in the hormone concentrations, preferably in an unanesthetized animal, provide the best evidence.
2. Presence of the hormone in the peripheral circulation (which does not show which organ is secreting the hormone) and its disappearance after ablation of the gonad.
3. Presence of the hormone in extracts of the gonad.
4. Synthesis of the hormone by the gonad during in vitro incubation or tissue culture. There is evidence that some hormones are sometimes secreted in vitro, although they cannot be demonstrated by the A-V difference method.
5. Regression of secondary sex organs or secondary sex characters after gonadectomy.
6. Correlations between activity of either interstitial cells or Sertoli cells, or both, and the condition of the secondary sex organs or sex characters.
7. Presence of smooth endoplasmic reticulum and/or enzymes involved in steroidogenesis, e.g., 3b-hydroxysteroid dehydrogenase (3b-HSD).

The pathways of androgen and estrogen synthesis make it possible to understand the endocrine function of the testes. Some of the major pathways and the major enzymes. It shows that testosterone can be formed by two pathways. One pathway via steroids that have a double bond between C_4 and C_5 (for example, progesterone) is the Δ^5 pathway; the other pathway via dehydroepiandrosterone (DHA) is the DHA pathway. Testosterone can either be converted to 5α-dihydrotestosterone (and in some species to 5β-dihydrotestosterone), which cannot be aromatized to form estrogens, or testosterone can be aromatized to form estrogens. In discussing the evidence for the control of testicular androgen secretion by the AP for different classes of vertebrates, we shall see that during phylogenetic development, androgen secretion becomes more and more dependent upon the gonadotrophic secretions by the AP.

Cyclostomes

Although hagfishes lack secondary sex characters and show no evidence of the presence of Leydig cells, testosterone and estradiol are present in the blood of *Eptatretus stouti.* Neither AP nor implants of hagfish pituitarles affect the concentration of these steroids, which range from 2.8 to 7.7 ng/ml, in the blood. It appears, therefore, that both the spermatogenetic and endocrine function of the gonads of this species are independent of the pituitary.

Lampreys have interstitial cells showing ultra-structural and histochemical evidence of steroid suretion, and testosterone has been identified in testicular extracts. Indirect evidence (regression of secondary sex characters after AP), however, suggests that the pituitary may control the secretion of androgens.

Elasmobranchs

The presence of testosterone in testicular extracts and in the plasma of some elasmobranchs is a good indication of testosterone secretion by the testes. During the year, the concentration of testosterone in *Scyliorhinus canicula* varies from 2 ng/ml in February to 6 ng/ml in August. These values are considerably lower than those reported for skates. The differences between the testosterone concentration in February and August may be the result of a higher rate of metabolism of testosterone at the higher temperature. It is almost certainly not mediated via the AP because, as discussed earlier in this chapter, removal of the ventral lobe of the AP does not affect plasma testosterone concentration.

Teleosts

As can be seen from Tables 6.5 and 6.6, there is evidence

that teleosts secrete androgens. The major androgen is 11-ketotestosterone, which seems to be limited to this class. The control of steroid production by the pituitary is suggested by the fact that the fish pituitary homogenates injected into fish stimulate steroidogenesis. As is the case for the control of spermatogenesis, the nature and mode of action of fish gonadotrophins have not been elucidated.

Amphibians

Evidence that the gonads secrete androgens is documented in Tables 6.5 and 6.6. After, there is loss of interstitial cell cholesterol in *Rana temporaria*, a loss that is prevented by mammalian LH injections. In intact frogs (*R. esculenta, R. pipiens*), mammalian LH increases 3β-HSD activity. In salamanders (*Pleurodeles waltlii*), mammalian LH is also more effective than FSH in stimulating testicular 3β-HSD, whereas FSH stimulates spermatogenesis. In bullfrogs, bullfrog LH injections but not bullfrog FSH injections increase plasma testosterone and dihydrotestosterone (DHT) concentrations. Neither bullfrog FSH nor ovine PRL affects this response to bullfrog LH.

In vitro, the secretion of DHT and testosterone by bullfrog testes was stimulated by LH from bullfrogs, tiger salamanders, snapping turtles (*Chelydra serpentina*), alligators (*Alligato mississippiensis*), and sheep, but not by bullfrog or oFSH. Moreover DHT was present in the medium in greater amounts than testosterone. This evidence all supports the concept of control of steroid secretion by the interstitial cells through LH and the control of spermatogenesis by FSH, while spermiation may be induced by either of these gonadotrophins.

Reptiles

The excellent review by Licht et al. (1977) is the main basis for the discussion on reptiles that follows. In AP *Anolis carolinensis* males, plasma testosterone can be substantially increased (from less than 0.6 ng/ml to more than 150 ng/ml) by oFSH (treated with LH antiserum). In vitro, oFSH and oLH stimulated androgen production by the testes of *Anolis* and snakes. After acute injection, oFSH and oLH were equally potent in increasing plasma testosterone, but after chronic injection oFSH was more potent.

When the FSH and LH of birds, alligators, sea turtles, snakes and bullfrogs were tested in lizards, both hormones increased the plasma testosterone concentration. Similarly, the plasma testosterone concentrations in turtles *(Kinosternon spp.)* were increased either by FSH or LH obtained from snapping turtles (*Chelydra*).

In vitro, the testes of all three orders of reptiles respond to FSH or LH from any of the three orders of reptiles tested by an increase in plasma testosterone concentration. The relative potency of a particular hormone, however, may vary among groups, with snake hormones, for instance, being more potent in snakes than in lizards and turtles. The ratio of FSH: LH potency varies with the source of the hormones and the species used for incubation of the testes.

Clever use of the circumtesticular capsule of Leydig cells in the lizard *Cnemidophorus*, has shown that it is specifically the Leydig cells that produce the androgens, with oFSH and oLH having the same potency in stimulating androgen production in vitro.

Birds

The secretion of androgens by the testes of several species of birds is documented in Tables 6.5 and 6.6. Some interesting phenomena have been noted in different species. Nalbandov *et al.* (1951) noted that after AP, the combs of roosters decreased in size and vascularity. Mammalian LH injections maintained the comb only for a short period or could restore the comb for a short time, after it had atrophied, whereas crude AP extracts from roosters could maintain or restore the comb indefinitely. On the basis of these and other experiments, Nalbandov *et al.* (1951) proposed the existence of a specific avian LH, with properties different from mammalian LH, i.e., the existence of a third gonadotrophic hormone. This suggestion does not seem to have been tested since purified chicken LH has become available.

Lofts and Marshall (1959) found that induced the secretion of progesterone (as measured by the Hooker-Forbes bioassay) by pigeon testes. Progesterone induces incubation behaviour in pigeons (Lehrman, 1961), so that the secretion of progesterone by males need not be surprising. However, Silver *et al.* (1974) found no fluctuations of the plasma progesterone concentration in male ring doves (*Streptopelia risoria*) during the reproductive cycle.

It would be of some importance to repeat Loft and Marshall's experiments, using RIA and competitive protein-binding assays for the progesterone determination in AP pigeon plasma and testes. The crucial experiments to date are those of Brown *et al.* (1975), in which purified avian FSH and LH were used in Japanese quail. In these experiments, LH stimulated the Leydig cells, but unfortunately no testosterone determinations were made.

Dispersed cells of rooster testes, when incubated in vitro produced androgens. The following hormones have been found to stimulate androgen production; toad (*Bufo* sp.) AP extract, turtle (*Chrysemys* sp) AP extract, avian LH and FSH and human FSH and LH. Snake (*Natrix* sp.) AP extracts, however, failed to stimulate steroidogenesis under these conditions (Callard *et al.*, 1978).

Mammals

As stated already that many mammals have testes that are either permanently or seasonally in the scrotum and that in still other species the scrotal testes are pulled into the inguinal canal, apparently under the control of the animals' central nervous system. For several scrotal mammals, it has been demonstrated that exposure of the testes to body temperature—either by making the animal cryptorchid or by insulating the scrotum — leads to damage of the germinal epithelium, but leaves the Sertoli cells and the Leydig cells relatively undamaged. More research is required, however, to assess possible damage to Sertoli and Leydig cells, since these cells clearly may make important contributions to normal spermatogenesis.

The concentration of plasma testosterone in different species varies over a wide range, as can be seen in Table 6.6. In the Chiroptera, the plasma testosterone concentration in higher than in any other mammalian order, whereas in Insectivora the concentration has the lowest range, although such a generalization may be premature, since only one species seems to have been investigated. The source of testosterone and androstenedione is the Leydig cells. The secretory activity of the Leydig cells is controlled by the AP and specifically by LH. The following evidence can be presented for this control: (1) specific LH receptors have been identified on Leydig cells (Hansson *et al.*, 1976); (2) in isolated Leydig cells, LH stimulates androgen secretion in vitro; (3) after treatment of male rats with LH antiserum, plasma testosterone concentrations drop sharply from 616 pg/ml to 113 pg/ml, whereas FSH antiserum has no significant effect on plasma testosterone secretion (Raj and Dym, 1976); (4) in AP animals, LH injections stimulate testosterone production. This evidence, of course, does not exclude the possibility that other hormones may have an effect on testicular androgen production either alone or by interaction with LH.

There is evidence from various experiments that FSH may play a role in testicular steroid production; e.g., (1) plasma testosterone concentrations in humans are better correlated with plasma FSH than with plasma LH concentrations; (2) the testosterone response to hCG is better correlated with FSH than with LH concentrations; (3) in developing male rabbits, plasma testosterone concentration is correlated with plasma FSH, not with plasma LH concentration; and (4) in rats, FSH prevents the loss of testicular responsiveness to LH. Moreover, FSH may induce LH receptor formation in the testes and thus may modulate the response of the testes to LH. In order to have these effects, however, FSH would need to bind to Leydig cells, and there is no evidence that it does.

Another AP hormone that may play a role in the control of Leydig cell steroidogenesis is PRL. Bartke *et al.* (1978) cite the following kinds of evidence:

1. PRL potentiates the effect of LH in restoring spermatogenesis and in stimulating testosterone production in mice.

2. Testes of dwarf mice, which are deficient in PRL, show greater testosterone production in response to PRL plus hCG than in response to hCG alone.

3. PRL increases the binding of LH by Leydig cells of dwarf mice, of AP rats, and of the regressed testes of hamsters on short photoperiods.

4. Treatment of immature male rats, with bromoergocryptine, which inhibits release of PRL, reduces the binding of LH by the testes.

5. Iodinated PRL becomes localized in the interstitial cells.

6. The concentration of esterified cholesterol in the testis increases after PRL administration. Esterified cholesterol is a precursor of cholesterol, and thus of testosterone.`It is thus possible that the potentiating effect of PRL on testosterone production in LH-treated AP rats is the result of a greater availability of the precursor. Ewing and Brown (1977) dismiss this as an unlikely mechanism without presenting data to support their position.

7. PRL administration increases plasma FSH concentrations in immature female rats and in dwarf mice, and thus, through its effect on FSH secretion, promotes steroidogenesis. Rats with the testicular feminization syndrome (Tfm) provide evidence that androgens may affect the Leydig cells. In such rats, the proportion of Leydig cells to other

testicular cells is greater than in normal rats. Such rats also show the paradox of low plasma testosterone concentrations in spermatic vein blood, but elevated systemic plasma LH concentrations. Such a discrepancy could be the result of a defective androgen biosynthesis, probably as the result of a sharp reduction of LH receptors on the Leydig cells of Tfm rats. Hansson *et al.* (1976) have presented evidence that androgens are required for differentiation of the Leydig cells at puberty and may be required for the induction of LH receptors on the Leydig cell membrane. In vitro incubation of testes with pregnenolone results in production of testosterone by normal testes, but of androstenedione by Tfm testes, thus suggesting 17β HSD deficiency. Furthermore, adult Tfm rats have a higher 5α-reductase activity than normal adults. Thus the low testosterone concentrations in Tfm rats may result from a synthesis of testosterone lower than normal and a metabolism higher than normal.

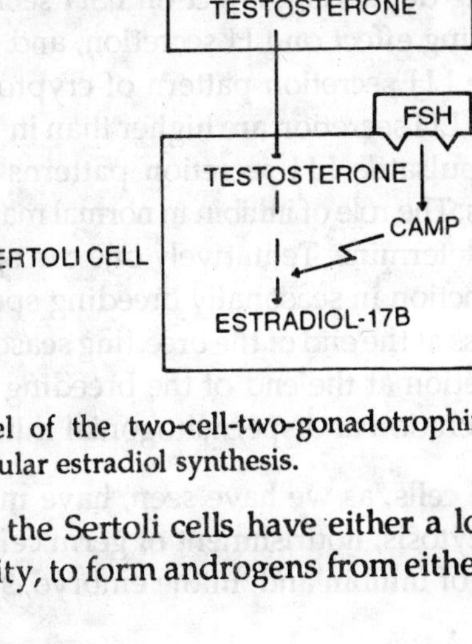

Fig. 6.12. Model of the two-cell-two-gonadotrophin hypothesis for testicular estradiol synthesis.

Although the Sertoli cells have either a low capacity, or lack the capacity, to form androgens from either cholesterol or pregnenolone, they can convert testosterone to estradiol. This estradiol synthesis by the testes may be a two-cell compartment process, with the estrogen formed under the influence of FSH. It has been found that the in vitro formation of estradiol by Sertoli cells in 16-20 d old rats in considerably higher than in 30 and 40 d old rats. The physiological significance of estradiol in the male is still not clear.

Endocrine Function: Inhibin

After destruction of the germinal epithelium by either X-rays, cryptorchidism, or deficiency of vitamin A or vitamin E, so-called castration or signet ring cells appear in the AP, and FSH secretion increases. Further investigation has revealed that the Sertoli cells secrete what is probably a glycoprotein with a molecular weight of 15,000-20,000, with an isoelectric point between pH 5.0 and 6.0. In the rat, this protein, called *inhibin,* decreased the endogenous LH-RH concentration of isolated rat hypothalamic tissue incubated in vitro, decreased FSH secretion, decreased LH-RH-induced FSH secretion, and in higher doses decreased LH secretion. In addition to these endocrine feed-back effects, inhibin inhibits DNA synthesis of dividing spermatogonia of rats. In rams, injections of rete testis fluid, which has a high concentration of inhibin, results in a long-lasting depressing effect on FSH secretion, a short-lasting depressing effect on LH secretion, and disappearance of the pulsatile LH secretion pattern of cryptorchid rams, in which FSH and LH secretion are higher than in intact controls, but in which pulsatile LH secretion patterns are similar to those in controls. The role of inhibin in normal male reproduction remains to be determine. Tentatively one can speculate that it may have a function in seasonally breeding species, in which the testes regress at the end of the breeding season. An increase in inhibin secretion at the end of the breeding season would inhibit GTH secretion and spermatogonial mitosis.

The Sertoli cells, as we have seen, have many functions, such as phagocytosis, nourishment of germ cells, secretion of ABP, secretion of inhibin and, in the embryo, secretion of the

Müllerian duct-inhibiting factor. Inhibin secretion is reduced after AP, but can be restored by either FSH, LH, or testosterone treatment, the effects of FSH and testosterone being direct and the effect of LH probably being indirect by stimulating testosterone secretion. Exposure of the testes to high temperatures by experimental cryptorchidy reduces the ability of Sertoli cells to secrete inhibin in vitro cultures. The decreased inhibin secretion during cryptorchidy explains the changes in the AP and the increased secretion of FSH, but one would expect an increase in dividing spermatogonia; instead, the germinal epithelium degenerates. Steinberger (1981) has speculated that impairment of other Sertoli functions may be the cause of the degeneration of the germinal epithelium. One of these impaired functions, is the secretion of ABP, which may partly explain the damage to the germinal epithelium.

Function of Pampiniform Plexus and Scrotum

In scrotal mammals, the testes are at a temperature that is lower than that of the body cavity. Two principal mechanisms are involved in the process of maintaining this lower-than-body temperature. First, in most species, the tunica dartos, a smooth muscle under the scrotal skin contracts at low temperatures, thus bringing the testes closer to the body, and relaxes at high temperatures, thus allowing greater heat loss through the scrotal skin. Interestingly, however, in Afrikaander cattle — living in very high environmental temperatures (above body temperature) — the tunica dartos muscle contracts and brings the testes closer to the (cooler) body. The cremaster muscle is a striated muscle that is not capable of sustained contraction, and its role in temperature regulation is not certain. Second, there is an efficient heat exchange mechanism between the cooler testicular vein blood and the warmer internal spermatic arterial blood. This heat exchange is possible because of the extensive coiling of the artery. This coiled artery and associated vein is called the *pampiniform plexus*.

The significance of maintaining the testes at lower-than-body temperature is shown by the following effects of body

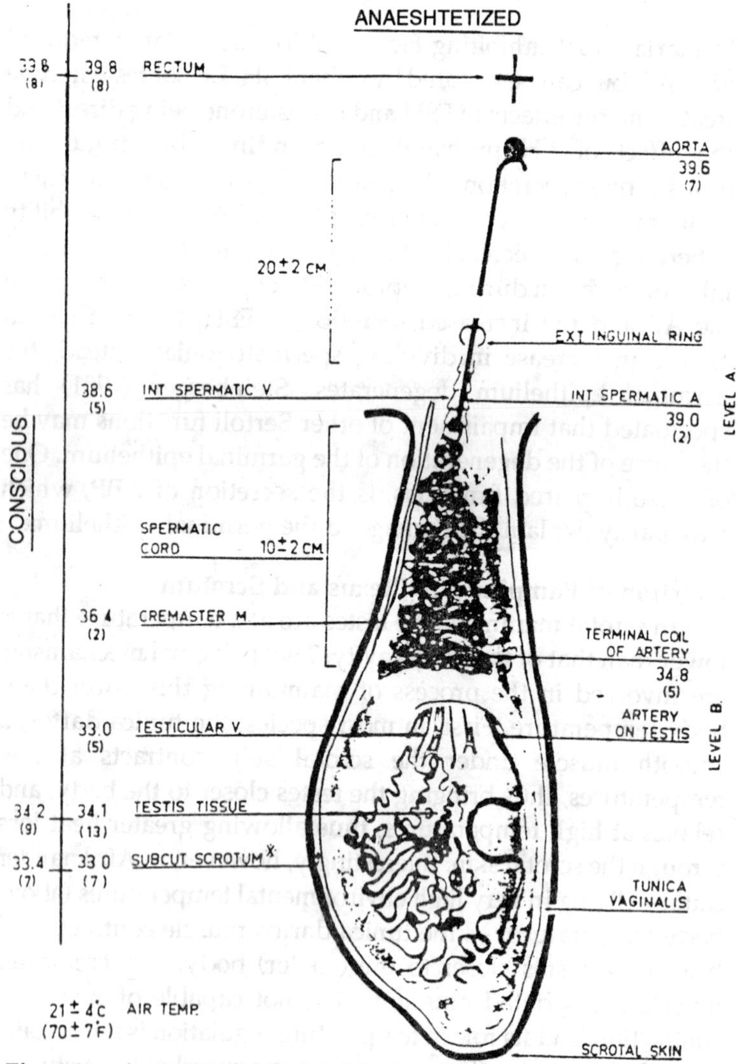

Fig. 6.13. Anotomical relationships and experimental temperature readings at various sites in the scrotum of conscious and anesthetized rams. The internal spermatic artery had been filled with neoprene, and the cast had been exposed by removal of the pampiniform plexus and tunica albuginea over the artery on the testis. Figures in parentheses give the number of temperature readings from which the average was obtained.

temperature: Exposure of the testes of scrotal mammals either for a short time (e.g., 46-47^{0}C for 15-30 minutes in the guinea pig) or for a longer duration (1 to 2 weeks, depending on the species) to body temperature, or for 2• months (in sheep) by insulation of the scrotum, results in damage to the germinal epithelium, as does high environmental temperature or fever. Indeed, exposure to body temperature may lead to irreversible damage and permanent sterility.

In a species that normally has scrotal testes, some animals have testes that remain in the body cavity or in the inguinal canal; this is called *cryptorchidism*. In some species, e.g., the rat, the inguinal canal remains open after testicular descent, and one can make rats cryptorchid experimentally by pushing the testes into the inguinal canal and ligating the canal. Studies with cryptorchic bulls have revealed that: (1) plasma LH concentrations are higher than in controls; (2) episodic releases of LH have greater amplitude and occur with greater frequency than in controls; (3) plasma FSH concentrations are higher than in controls; (4) after administration of LH-RH, the plasms FSH and LH concentrations increase more rapidly and to higher values than in similarly treated controls; (5) plasma testosterone concentrations are not significantly different from those of controls; and (6) the increase in plasma testosterone after LH administration in similar to cryptorchid and normal bulls.

Effects of Androgens

Androgens have many important co-ordinating functions, some of which are listed here:

1. They inhibit gonadotrophin secretion and also affect the type of gonadotrophin secreted. Bogdanove *et al.* (1975) found that the FSH secreted by intact male rats differed from the FSH secreted by castrated males with respect to (*a*) a change in the ratio of potency as determined by the bioassay to potency as determined by radioimmunoassay (RIA), such that FSH of castrated rats had a higher biological activity of

FSH with respect to the RIA than the FSH of intact rats; (*b*) the molecular weight of the FSH of castrated rats being higher than that of the FSH of intact rats; and (*c*) the clearance rate, and thus the survival, in the circulation of FSH from castrated rats being greater than that of FSH from intact rats.

2. In killifish, the Indian catfish, pigeons, and the mammals investigated, androgens can either maintain or restore spermatogenesis in animals.

3. Androgens play an important part in regulating the physiology of the epididymis. In the lizard *Lacerta vivipara*, the secretion of granules that are mixed with the sperm is androgen dependent. The function of these granules is not known. In bulls, Amann and Ganjam (1976) showed that androgens are transferred across the pampiniform plexus. This transfer raises the concentration of androgens in the epididymis to twice that found in the peripheral blood. In the epididymis, the spermatozoa gain in fertilizing capacity (sperm maturation). This sperm maturation is androgen-dependent and is probably exerted through secretions by the epididymal epithelium, but what these substances are and how they mediate their effect is not known.

4. In mammals, androgens are required for the maintenance of the fertilizing capacity of sperm in the ductus deferens; in chickens this does not appear to be necessary.

5. Androgens are required for maintenance of the seminal vesicles of Indian catfish and of the secondary sex glands of mammals.

6. The thermoregulatory function of the scrotum is androgen dependent.

7. The development of the anal fin, which serves as a copulatory organ in some fishes, is androgen dependent.

8. Many secondary sex characteristics, such as the nuptial colours of fish, the comb of the rooster, the development of the

thumb pads, the vocal apparatus, and the muscular development of the forelimbs of frogs, and the development of the antlers of deer are all androgen dependent. One needs little imagination to understand the adaptive value of these characteristics.

9. Androgens, at least in birds and mammals, increase nitrogen retention and thus promote muscular development. The use of metabolic androgens (which have a high metabolic potency but a low masculinizing effect) by weight lifters and other athletes and by older people who may suffer from muscular weakness is based on this nitrogen-sparing effect.

10. Androgens have profound effects in a number of classes of animals (possibly with the exception of amphibia) on aggressive and sexual behaviour.

THE OVARY

As described earlier that in vertebrates the number of ovaries is never more than one pair. However in birds only one ovary becomes functional and other rudimentary.

Like the testis, the ovary performs two main functions, gametogenesis and hormone production. There are other similarities, as well as some differences between testes and ovaries.

Oogenesis

We must first discuss the process of meiosis in the ovary. The oogonia undergo mitotic divisions and yield primary oocytes, which undergo a meiotic division that results in a secondary oocyte (2n chromosomes) and a polar body. The secondary oocyte by a meiotic division yields the ovum and a second polar body each with *n* chromosomes. This series of processes, which is concerned with the nuclear divisions, is called oogenesis, and should not be confused with growth of the follicle and with vitellogenesis (yolk formation), which may be quite spectacular, as in avian and reptilian ovaries, for example.

Over the years there has been some controversy about whether in some species (cyclostomes, elasmobranchs, a few teleosts, possibly a few reptiles, all birds, and all or most mammals), mitoses of oogonia occur after sexual maturity has been achieved (Zuckerman and Baker, 1977). For rabbits, Kennelly and Foote (1966) clearly demonstrated, in a classical paper, that neogenesis of oocytes did not occur. They labeled the oogonia by injecting [^{3}H]-thymidine systemically on the day of birth; then at different ages they collected the ovaries or, after sexual maturity, secondary oocytes after superovulation. Siunce 90 per cent of the oocytes in these rabbits between age 12 and 20 weeks and 80 per cent of the ovulated oocytes carried the [^{3H}] label, it was clear that the labeled thymidine had not been diluted. If mitosis had occurred, the labeled thymidine should have been diluted. This convincing evidence has bot been countered by other experiments.

The primary oocyte is enclosed in a follicle and, depending on the species one is considering, either follicular growth or the deposition of yolk are the major changes that occur between birth and the preovulatory changes. We shall discuss the control of follicular growth and oocyte maturation for different classes of vertebates, although our knowledge, especially for nonmammalian classes, is fragmentary, and it is speculative to state that a pattern exists. If one is willing to speculate, however, it a appears that the ovary is more autonomous in the lower vertebrates than in the higher vertebrates. Much more evidence needs to be obtained, however, especially in teleost fishes, before we can move from the realm of speculation to that of evolutionary trends.

TABLE 6.5

Comparison of the Ovary and the Testis

Characteristic	*Ovary*	*Testis*
Location	Abdominal in all species	Abdominal in inframmalian animals scrotal in many

Symmetry	In many reptis most birds, only left ovary developed in duckbil platypus andsome bats only 1 ovary developed	mammals. In most species two testes function
Origin in elasmob-ranchs and tetrapods	Cartex of indifferent gonad	Medulla of indifferent gonad
Seasonal cycles	In many species	In many species
Cyclic phenomena during breeding season	In many mammals, prono-unced cycles	No pronounced cycles
Homones secreted	Estrogens Progestins Androgens	Androgens Progestins Estrogens
Control by gonadotrophins	Follicular growth by FSH + LH Interstitial cells by LH Corpus luteum by LH (?)	Spermatogenesis by FSH or LH (?) interstitial cells by LH
Mitosis	Mitosis continues in teleosts amphibians, Most reptiles after sexual marutiry	Mitosis after sexual maturity
	No mitosis after sexual maturity in Cyclostomes, elasmobranchs, birds, and most or all mammals	

Cyclostomes

In the lamprey *(Petromyzon)*, hypophysectomy prevents the increase in dry weight of the oocytes, and the oocytes remain morphologically immature, but there is *no atresia* of the follicles. Growth of the oocytes and vitellogenesis are possible in *Lampetra fluviatilis.* In the female hagfish, *Eptatretus stouti,* has, as in the case of the male, little effect on the gonads. Ovulations and estrogen secretions are not different for fish, controls, and fish treated with hagfish pituitaries.

Elasmobranchs

Total hypophysectomy of the lesser spotted dogfish *(Scyliorhinus canicula)*does not affect the rate of vitellogenin (a yolk precursor) formation or the rate of synthesis of yolk granules; but it does delay removal of vitellogenin from the plasma. Craik (1978) obtained similar results after removal of the neurointermediate-median-rostral lobe complex as after total hypophysectomy, but ventral lobectomy did not have this effect.

Removal of the ventral lobe of the hypophysis resulted in atresia that could be detected about three weeks after the operation. Resorption of the yolk from the utretic follicles may take as long as 14 months, and Craik (1978) proposed that the ventral lobe may control the transport of yolk into the follicles. Ovulation is, according to Dodd (1975), probably under control of the hypophysis. It is not known which hormones are involved in the control of either yolk deposition or ovulation. So far only one gonadotrophic hormone has been isolated from dogfish.

Teleosts

The probably controls oogonial proliferations since reduces oogonial proliferation in the plaice *(Pleuronectes platessa)* and eliminates it in the goldfish *(Carassius auratus)*. The prolifirations of oogonia in goldfish can be restored by carp or salmon AP extracts. The growth of yolkless oocytes is apparently independent of the AP. In vitro culture of goldfish ovaries suggeests that oogonial mitoses, maturation to the meiotic prophase, and previtellogenic growth can occur, in succession, independent of pituitary gonadotrophins; however, survival of vitellogenic oocytes was improved by parabiosis of AP and ovary of goldfish, and by carp AP extract, and to a lesser extent by hCG and testosterone.

The most extensive investigations of the hypophyseal control of the ovary have been carried on with the Indian catfish *(Heteropneustes fossillis)* Hypophysectomy, during the early spawning period, when ovaries are large and the follicles are filled with yolk, results in atresia. This atresia can be prevented by the injection of as little as 5 µg of partially purified salmon GTH, or 1 µg of purified GTH from the puntions carp *(Puntius gonionotus)*.

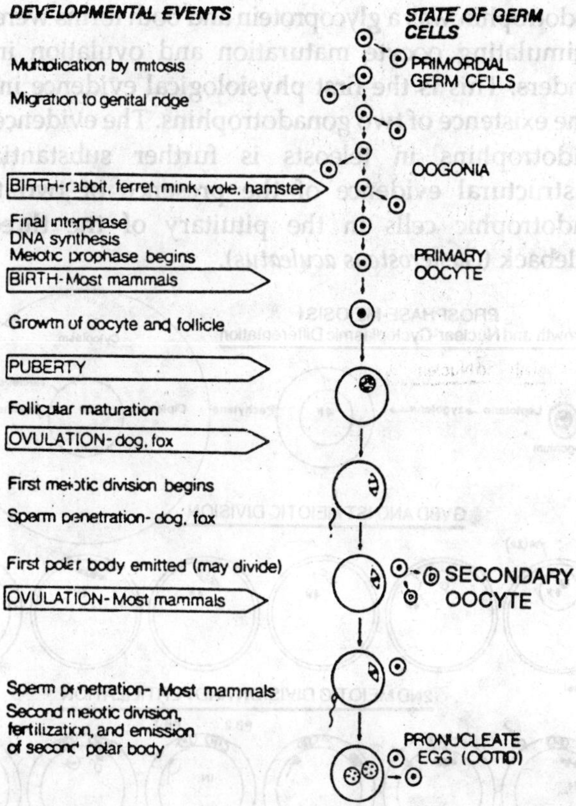

Fig. 6.14. Lite-cycle of the female germ cell *(Note :* In the insectivore genera *Setiler* and *Heincentetes* (the so-called Madagascar hedagehopst the secondary oocyte is fertilized while it is still the ovarian follicle (Mossman and Duke, 1973) in the ovarian fertilization may also occur in the shre *Blarina brevicauda.*

Ng and Idler (1978) were able to extract two gonadotrophic hormones from the pituitaries of *Hippoglossoides platessoides*, both of which existed in two forms, with molecular weight of 28,000 and 62,000. One gonadotrophin either had a low carbohydrate content or was a non-glycoprotein, and both forms stimulated vitellogenesis and ovarian growth in winter flounders (*Pseudopleuronectes americanus*). The other gonadotrophin was a glycoprotein and both forms were capable of stimulating oocyte maturation and ovulation in winter flounders. This is the first physiological evidence in teleosts for the existence of two gonadotrophins. The evidence for two gonadotrophins in teleosts is further substantiated by ultrastructural evidence of the presence of two types of gonadotrophic cells in the pituitary of the three-spined stickleback (*Gasterosteus aculeatus*).

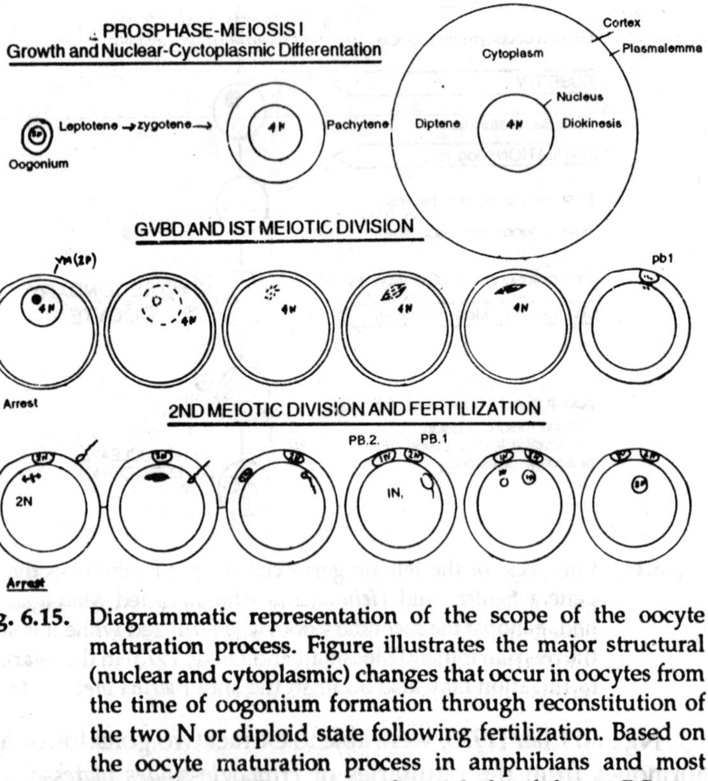

Fig. 6.15. Diagrammatic representation of the scope of the oocyte maturation process. Figure illustrates the major structural (nuclear and cytoplasmic) changes that occur in oocytes from the time of oogonium formation through reconstitution of the two N or diploid state following fertilization. Based on the oocyte maturation process in amphibians and most mammals. GVBD = germinal vesicle breakdown. P.B. 1 = first polar body, P.B. 2 = second polar body, VM = vitelline membrane, ZP = zona pellucida.

In vivo, estradiol (E_2), estrone (E_1), and estriol (E_3) can each partly maintain the follicles of gravid (with large yolky follicles) catfish, but tosterone, progesterone, and corticosteroids are marginally effective (i.e., maintenance is better than in controls, but not as good as with the estrogens), whereas mammalian LH (20 μg), hCG (25 IU) and to a somewhat lesser extent PMSG (40 IU) can maintain the follicles so that they are only slightly smaller than in intact controls.

It is extremely interesting that the gonadal hormones (pregnenolone, 17α-OH pregnenolone, 17α-OH progesterone, testosterone, E_1, E_2 and E_3) were either ineffective or only marginally effective in inducing oocyte maturation of Indian catfish in vitro, but that hydrocortisone and deoxycorticosterone were quite effective at doses of 1-10 μg/ml. Experiments by Truscott *et al.* (1978) have shown that injections of oLH or of porcine ACTH in Indian catfish with either regressed follicles or large yolky follicles, were followed by an increase in plasma cortisol concentrations. Injections of purified salmon gonadotrophin into gravid Indian catfish also increased plasma cortisol concentrations as did oLH and porcine ACTH injections in ovariectomized animals. These results suggest that the adrenal of this fish is stimulated to secrete cortisol not only by a mammalian ACTH, but also by oLH and a fish gonadotrophin. As we shall discuss later in this chapter in the section on ovulation, corticosteroids play an important role in ovulation in some teleosts.

In the Japanese rice fish (*Oryzias latipes*), cortisol is probably secreted by the ovary and is more potent than progesterone in inducing oocyte maturation. In the rainbow trout (*Salmo gairdneri*), the northern pike (*Esox lucius*), and the goldfish, 17α-hydroxy-20β-dihydroprogesterone is the strongest inducer of oocyte maturation. It is thus of considerable interest that 17α-OH progesterone (which also induces oocyte maturation in rainbow trout, northern pike, and goldfish) and 17α-hydroxy-20β-dihydroprogesterone (17α-20β Pg) have been found in the plasma of prematuring and maturing winter flounder. Either of these steroids added to plasma will induce maturation of the oocytes of this species in vitro. The changes is concentration of these steroids between premature fish and mature fish suggest that 17α-OH-progesterone is more important than 17α-OH dihydroprogesterone (Campbell *et al.*, 1976). Goetz and Bergman (1978) tested the effect of progestins, androgens, and corticosteroids on oocyte maturation in vitro on brook trout (*Salvelinus fontinalis*) and yellow perch (*Perca flavescens*). They concluded that androgens and 11-oxygenated corticosteroids were less effective than progestins and the other corticosteroids.

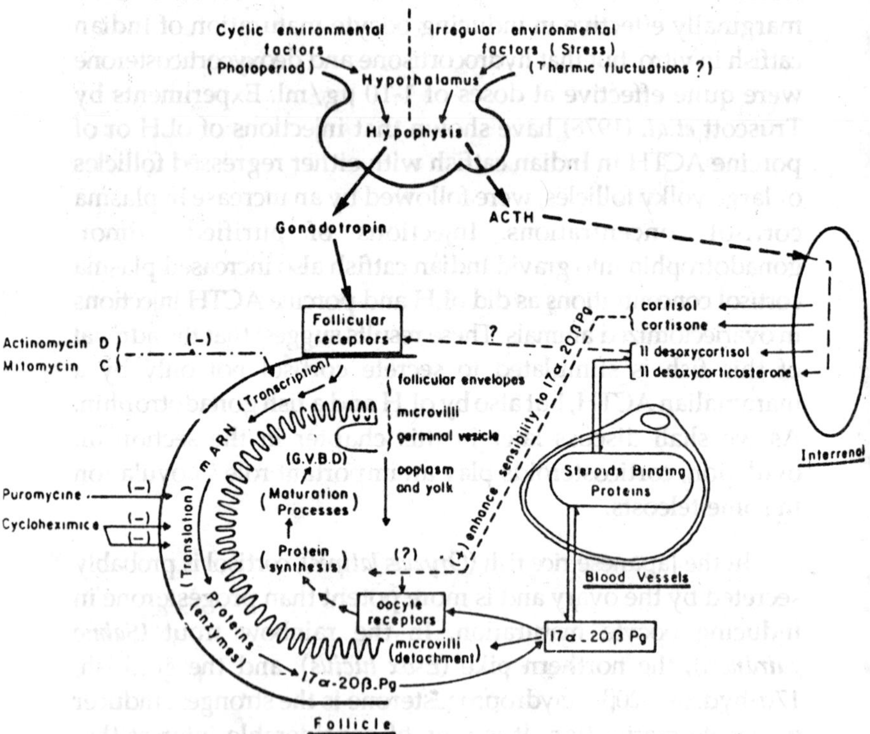

Fig. 6.16. Tentative scheme for the control of oocyte maturation in trout. GVBD = germinal vesicle breakdown.

The follicular growth in the ovary of the toad (*Bufo bufo bufo*) has been particularly well-studied, and the following account is largely taken from the publications of Barker-Jorgensen (1974, 1975). The smallest oocytes are surrounded by a single layer of flat follicle cells. This is the previtellogenic growth phase, or first growth phase (FGP), during which the cell diameter may increase ten times and the cortical cytoplasm develops microvilli toward the follicle cells. This first growth phase ends with the appearance of these microvilli.

The second growth phase (SGP) consists of yolk accumulation, during which the oocyte becomes surrounded by an epithelial, a thecal, and a granulosa layer. The oocyte may increase another 10 times in size as a result of this yolk accumulation. The hormonal control of vitellogenesis, i.e., the synthesis of yolk precursors, which is essential for yolk accumulation to occur.

The effects of hCG on the ovaries of *Bufo bufo bufo* depend on the state of the ovary. When no SGP follicles are present, hCG promotes the recruitment of FGP follicles to SGP follicles (barker-Jorgensen, 1974, 1975). If, however, SGP follicles are present, hCG maintains the SGP follicles but does not induce the recruitment of FGP follicles to SGP follicles.

In a recent publication, Browne *et al.* (1979) provided evidence that in the African clawed toad (*Xenopus laevis*), hCG regulates either the opening or the assembling of gap junctions between follicle cells and oocytes, or both. The function of these gap junctions has not been determined, but Browne *et al.* (1979) speculate that a factor that affects endocytotic uptake of vitellogenin by the oocyte may be transmitted from follicle cells to oocyte via these gap junctions.

The exact role of amphibian gonadotrophins in the regulation of follicular growth and oocyte maturation awaits experiments either with purified amphibian gonadotrophins in amphibia or experiments with specific antisera to amphibian gonadotrophins.

Maturation of amphibian oocytes will be discussed in some detail in the section on ovulation in this chapter. It is sufficient here to state that mammalian LH-induced resumption of meiosis seems to be mediated by progesterone and/or its metabolites 20β-progesterone and 17α, 20β-hydroxy-progesterone.

Reptiles

In reptiles, five patterns of ovulation can be distinguished: (1) In *monoallochronic* ovulation, one oocyte is released from an ovary and later another oocyte is related from the other ovary, but the release is so regulated that there is never more than one egg in the oviduct. This pattern is regularly found in the genera, *Anolis, Chamapleosis, Chamaelinorops, Phenacosaurus* and *Tropidodactylvs.* (2) In *polyallochronic* ovulation, more than one oocyte is released from one ovary. According to Smith *et al.* (1973), this pattern may be hypothetical because it has not been found in this class. (3) In *polyautochronic* ovulation, more than one oocyte is released simultaneously or nearly simultaneously from the two ovaries. Despite the presence of only one functional oviduct, this pattern is found in some species of *Tantilla* and all investigated species of *Ramphotyphlops, Typhlops, Leptotyphlops, Helminthophis* and *Anomalepis.* (4) In *monoautochronic* ovulation, one oocyte is released synchronously from each of the two ovaries. This pattern has been found in most geckos and in *Xantusia vigilis* and *Carlia (Leiolopisma) rhomboidalis.* (5) In *monochronic* ovulation, only one ovary is functional; I do not know whether this pattern occurs in reptiles.

These patterns of ovulation may be determined to some extent by the patters of the ovarian growth. Jones *et al.* (1978) have pointed out that each ovary contains a small region on its dorsal surface, the so-called *germinal bed*, from which growing follicles originate during embryonic development. Some species of lizards, for example, *Xantusia vigilis* and *Carlia rhomboidalis*, have two such germinal beds per ovary. In these two species, many follicles grow, but the number that might ovulate is reduced by atresia, so that the ovulation pattern is reduced to the monoautochronic pattern.

Jone *et al.* (1978) have analyzed the follicular growth of *Lepidodactylus lugubris*, a member of the Gekkonidae, which has one germinal bed per ovary. In this species, the hierarchy of follicular sizes is maintained by the recruitment from the smaller follicles, if a large follicle is ovulated or has become atretic. Atresia in this species occurs in about 17 per cent of the large follicles, whereas in *Xantusia vigilis* it occurs in 54 per cent of the large follicles.

In reptiles, the control of the ovary by the AP has probably been studied most extensively in the American chameleon (*Anolis carolinensis*). Hypophysectomy leads to atresia of the yolky follicles, which is more effectively prevented by administration of mammalian FSH than by mammalian LH. Tokarz (1978), ~~AP~~ using American chameleons during the reproductively quiescent season, found that oFSH, but not oLH, caused incorporation of [^{3}H]-thymidine into surface epithelial cells, prefollicular cells, and oogonia (in this species the oogonia undergo mitotic divisions after the animal is sexually mature. Injections of oFSH also increased the number of oogonia and the ratio of labeled to unlabeled oogonia. These effects of oFSH are apparently not mediated by estradiol, because administration of estradiol-17β did not have any of the effects of oFSH. After ~~AP~~ the small follicles of the American chameleon do not become atretic, suggesting that the maintenance of small follicles does not require gonadotrophins, but that their growth and the subsequent maintenance of the yolky condition do. In intact juvenile American chameleons, oFSH (1) stimulates differentiation and enlargement of oocytes, (2) stimulates formation of follicles and differentiation of the granulosa cells, and (3) maintains the hierarchy of the follicles. This effect of FSH on the ovary is mediated by estrogens and is the result of a differential increase in the vascularity of the largest follicles.

Birds

In the domestic chicken, the ovary undergoes a spectacular increase in weight at the time of sexual maturity; it increases from about 1 g at 120 days old to 20 g at 180 days. In the earlier-maturing modern strains of chickens, this change probably occurs earlier, and the rate at which the ovary increases in weight may be even greater. The increase in weight of the ovary is the result of yolk deposition, which is discussed in the section on vitellogenesis in this chapter. Such yolk deposition in the chicken does not occur before about 100-120 days, even when vitellogenesis has been induced by estrogens and the estrogen-treated pullet is injected with chicken pituitary extracts.

In the sexually chicken, AP results in atresia, but injections of either mammalian gonadotrophins or of chicken pituitary extract prevent the atresia, and ovulation by such follicles can be induced with either of these two types of preparations (Nalbandov, 1976). In intact birds, the effects of mammalian gonadotrophin treatment seem to depend on the species. In chickens with developed ovaries, mammalian gonadotrophin (FSH, PMSG) injections stimulate the growth of a large number of follicles, so that the hierarchy of these structures is lost and spontaneous ovulations stop. In song birds, the administration of PMSG stimulates the follicles, so that the hierarchy among the follicles normally found during the breeding season results. The reason for this difference in response between the chicken and song birds are not clear. Although purified chicken FSH and LH have been prepared, no experiments seem to have been carried out with such preparations of AP hens.

Mammals

In the mammalian ovary (at least in the cat, mink, ferret and mouse), the rete ovarii plays an important role in oogenesis and in follicle formation. The rete cells, which are probably forerunners of the follicular cells, specifically of the granulosa cells, are secretory and secrete a meiosis-inducing substance by which they presumably affect the germ cells, since the germ cells in contact with the rete cells are the first to begin meiosis. In addition to this intraovarian control of oogenesis, the control by the anterior pituitary is of particular importance. The difficulties of interpretation of experiments in which AP animals are used, and we have emphasized the advantages of using FSH and LH antisera in arriving at an understanding of the role of these hormones in regulation of ovarian function.

Schwartz (1974), in an excellent critical review, has considered (1) the control by the AP of follicular growth, maturation, and ovulation, and (2) whether FSH and LH are secreted as one gonadotrophic complex or as two separate hormones. The salient points of her findings are:

1. At the end of oogonial mitosis, the number of primordial follicles is at its maximum and the subsequent decrease in these follicles is the result of (a) atresia, which occurs after the follicle has reached the 21- to 69-cell stage or has progressed further, and (b) the maturation of the follicle to a type 8 follicle and the subsequent ovulation, after which the follicle is converted to a CL.

2. The early postnatal growth of the follicle of the mouse may be independent of the AP, but after ~~AP~~, the initial development of follicles and atresia continue, although the rate of loss of follicles decreases. Nakano et al. (1975), using [^{3}H]-thymidine labelling found that in immature ~~AP~~ rats, the growth of follicles continued and that the regressions of the number of labeled granulosa cells on the number of granulosa cells in the follicle were not different between intact and AP ~~AP~~ rats. Schwartz (1974) has suggested that GTH increases the rate of entry of follicles from the non-proliferating into the proliferating pool of follicles.

3. In the absence of gonadotrophins, no follicles grow beyond type 5, according to Schwartz (1974). Lunenfeld *et al.* (1975), using antiserum to rat gonadotrophin, found that in the mouse, few follicles grew beyond the 40-cell stage, although the total number of growing follicles was not changed; however, injections of FSH or human menopausal gonadotrophin (which has FSH and LH properties) increased the number of follicles growing beyond the 40-cell stage. These observations are in accord with the findings of Edwards *et al.* (1977) in ~~AP~~ rats treated with either FSH or hCG; FSH was effective in restoring follicular growth in the absence of stimulation of steroidogenesis.

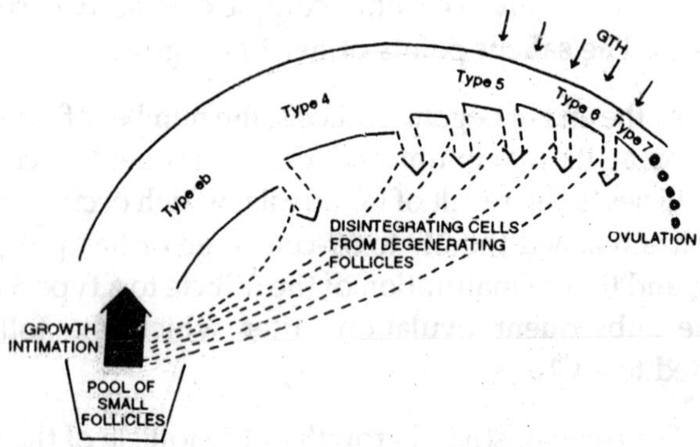

Fig. 6.17. A simplified model of follicle growth. The model postulates that initiation of follicle growth is negatively influenced by substances from degenerating larger follicles. This hypothesis provides a mechanism for an internal control of rate of follicle proliferation, in addition to the external one by means of gonadotrophic hormones (GTH).

4. The use of LH antiserum and FSH antiserum after day 5 did not permit a definite a conclusion about the role of either LH or FSH in early postnatal follicular growth.

5. In the AP juvenile prepubertal rat, oFSH can cause antrum formation, and LH can restore the interstitial cells of the ovary without inducing uterine growth or follicular growth. (FSH and LH are required for estrogen secretion.

6. Follicular growth in adult mammals requires the presence of the AP, or, in the AP animal, it requires administration of gonadotrophins. However, the exact role of FSH or LH, or the necessity of a synergism between them, has not been fully elucidated. Peters *et al.* (1975) have proposed that FSH prevents resting large follicles from becoming atretic without inducing growth in such follicles.

7. Hirschfield and Midgley (1978) have presented

persuasive evidence that the release of FSH during one estrous cycle of the rat may stimulate the growth of follicles of about 200-400 μm which are destined to be ovulated during the next cycle.

In addition to a possible direct effect of GTH on the follicles, there is probably an indirect effect, through stimulation of steroid secretion. There is evidence that gonadal hormones can affect the follicles. First, diethylstilbestrol (DES) prevents atresia and the decrease in ovarian weight after AP of rats, and it allows the formation of medium-sized follicles. In AP rats, estradiol 2 mg/day) stimulates preantral growth of follicies, and increase the estrogen receptors in the ovary and the response of granulosa cells to FSH. Injection of LH into such estrogen-treated AP rats causes follicular atresia, whereas administration of FSH + LH + estradiol to AP rats induces luteinization. Treatment of AP immature female rats with DES also stimulates the production of a protein, with the characteristics of an androgen receptor, in the cytosol of the ovaries. As mentioned below, androgens may be important in the preantral follicular development, so that DES would thus also provide for the transport mechanism of androgens. Second, this effect of DES is inhibited by injections of small doses of hCG (0.3-3 IU), LH (3 IU), but not by FSH. Third, this countereffect of LH and of hCG to DES is, in turn, prevented by the administration of antiserum to testosterone. Finally, LH or hCG injections into intact immature female rats result in a decrease in ovarian weight, presumably because the secretion of androgen in response to these hormones inhibits follicular growth.

Vitellogenesis

In species of oviparous groups (e.g., elasmobranchs, teleosts, amphibia, reptiles, birds, and monotremes), the follicles contains substantial amounts of yolk, and the spectacular increase in size of the follicles in such animals consists largely of the deposition of yolk. Yolk precursors are formed in the liver in response to estrogens, or when fat bodies are present, estrogens may mobilize fats to restructure these precursors in the liver.

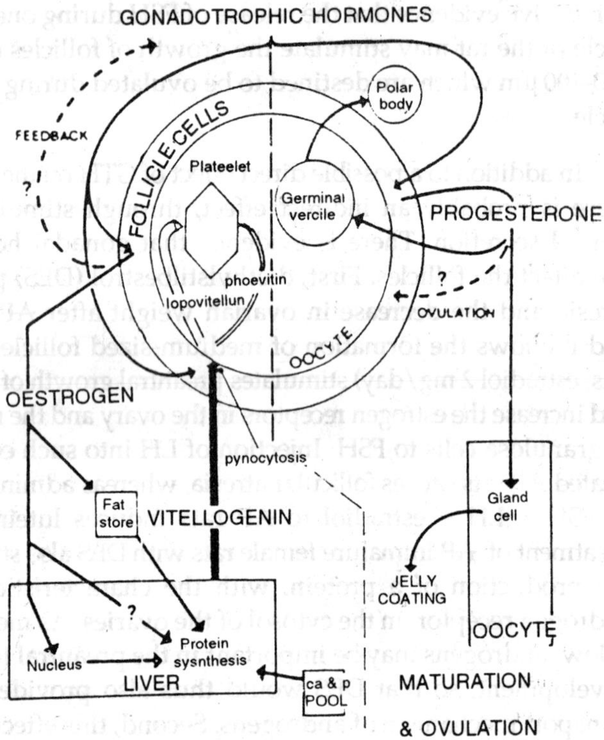

Fig. 6.18. In this generalized scheme [of female amphibian reproduction], the pituitary gonadotropins and some of the processes involved in the production of an egg suitable for fertilization are depicted. On the left side of the figur, the primary responses involved in the vitellogenesis are shown, while on the right the involvement of progestins in maturation, ovulation, and the release of oviducal jelly coatings is summarized. Although the diagram has been constructed primarily from research on *xenopus laevis* and *Buto bufo,* it is likely that in other species, part of all of the diagram is applicable. The pituitary-hypothalamic relationships are not shown.

Gonadotrophic hormones stimulate estrogen secretion by the ovary; even in immature chickens, in which follicular growth cannot be stimulated, there is evidence of ovarian estrogen secretion in response to exogenous gonadotrophins.

The estrogens stimulate vitellogenesis by their effect on the liver, and in amphibians, they promote the mobilization of lipids from the far bodies. Estrogen administration to cockerels increases the concentration of a large number of yolk precursors in the blood. Table 6.6 illustrates some of the most prominent changes in blood composition after estrogen administration. An increase in blood phospholipids has also been observed in the bass (*Paralabrax clatharus*), the bullfrog (*Rana catesbeiana*), the turtle (*Pseudemys scripta troostii*), and the mouse (*Mus musculus*). The concentration of calcium in the blood after estrogen injection increases substantially in the bass, bullfrog, turtle, but not in the mouse. This increase in the calcium concentrations of the blood in the three non-mammalian species is associated with the large amounts of yolk precursors, which are not formed in the mouse.

In ~~AP~~ winter flounders and in intact South African clawed toads (*Xenopus laevis*), yolk precursors accumulate in the blood after estrogen administration. Incorporation of yolk precursors in ~~AP~~ winter flounders is stimulated by the non-glycoprotein fraction of teleost APs; in intact *Xenopus laevis*, incorporation of yolk into the follicles is stimulated by mammalian FSH. In the ~~AP~~ lizard, *Dipsosaurus dorsalis*, estrogen administration does not induce vitellogenesis, but a combination of estrogen and ovine growth hormone does. This observation is difficult to interpret because growth hormones show a great deal of species specificity.

In captive mallard ducks (*Anas platyrhynchos*), the ovarian follicles frequently fail to develop beyond 2-3 mm because yolk is not deposited. Injections of estrogens result in the expected increase in yolk precursors, such as phospholipids, but neither estrogen alone nor estrogen pretreatment followed by injections of chicken pituitary powder bring about yolk deposition in the follicles.

It is probably safe to assume that the following sequence takes place: gonadotrophin → stimulation of steroid production (especially estrogens) → estrogens or estrogens plus AP hormones → stimulation of synthesis of yolk precursors by the liver and mobilization of yolk precursors from stores → deposition of yolk in follicles under influence of AP hormones.

Ovulation

After the follicle has reached the appropriate size, ovulation can occur. In most species, nuclear changes either precede or accompany ovulation. In most mammals, the second oocyte is released at ovulation; however, in the mare, bitch and vixen, the primary oocyte is released, and in the tenrec (*Setifer setosus*) and the shrew (*Blarina brevicauda*), the sperm penetrates the primary oocyte in the follicle before ovulation takes place. As we shall see later in this chapter, in teleosts, a number of permutations of the sequence, ovulation to fertilization to birth, occur.

TABLE 6.6

Changes in Blood Composition of Chickens after Estrogen Administration

Blood component	*Sex*	*Control*	*Estrogen-treated*
Lipids			
Total lipids (mg/100 ml plasma)	M	1,100	14,210
Phospholipids (mg/100 ml plasma)	M	162	934
Sphingomyelin (mg/100 ml plasma)	M	22	54
Cephalin (mg/100 ml plasma)	M	34	214
Cholesterol (mg/100 ml plasma)	M	235	1,136
Proteins			
Total protein (gm/100 ml serum)	M	3.90	7.40
Albumen (gm/100 ml serum)	M	1.00	0.60
Globulin (gm/100 ml serum)	M	2.90	6.80
Vitellin (dilution at which detected)	F	0	40
Hemoglobin (gm/100 ml blood)	F	9.0	5.6
Fat soluble vitamins			
Total Vitamin A (μg/100 ml)	F	5.1	46.8
Vitamin A ester (μg/100 ml)	F	0.9	42.8
Vitamin A alcohol (μg/100 ml)	F	4.2	4.0

Water soluble vitamins			
Riboflavin (p.p.m.)	F	trace	1.22
Biotin (ng/ml)	F	1.3	8.3
Minerals			
Calcium (mg/100 ml)	M	10	97
Ultrafilterable Ca (mg/100 ml)	M	6.50	8.00
Magnesium (μg/100 ml)	F	Not detected	13.8
Inorganic phosphate (mg 100 ml)	M	6.20	20.00
Total sulfate (mg/100 ml)	M	5.80	1.70
Iron (μg/100 ml)	F	100	700

Cyclostomes

As we mentioned earlier in this chapter, according to Matty *et al.* (1976), ovulations occur in the ~~AP~~ hagfish (*Eptatretus stouti*). In ~~AP~~ river lampreys (*Lampetra fluviatilis*), ovulations occur if large follicles are present at the time of ~~AP~~.

Elasmobranchs

Not enough is known about the control of ovulation in elasmobranchs, but Dodd (1975) has speculated that LH release occurs before ovulation.

Teleosts

Ovulation is teleost fishes is under the control of the AP through its regulation of steroid secretion by either the ovary or the interrenal gland. The hypophyseal-gonadotrophic control of ovulation in the goldfish is indicated by the release of GTH, as measured by a peak in plasma GTH concentration, during the late part of the light phase and the start of the dark phase in fish kept on a 16L : 18D photoperiod. The effects oi GTH administration on ovulation in the Indian catfish is probably mediated by corticosteroids secreted by the internal gland; however, the effect the GTH on the internal gland may not be a general phenomenon in teleosts, for in the Japanese rich fish (*Oryzias latipes*) the GTH appears to stimulate corticosteroid production by ovarian (follicular?) tissue. The secretion of corticosteroids by ovarian tissue is in accord with the finding of Colombo *et al.* (1973) that ovaries of *Laptocottus armatus, Gillichthys mirabilis,* and *Microgadus proximus* produced '11-

deoxycortisol and 11-deoxycorticosterone in vitro. Hirose (1976) observed that ovulation of Japanese rice fish oocytes could be induced in vitro by cortisol, but that removal of the follicle prevented the cortisol-induced germinal vesicle breakdown (GVBD), the indicator of meiosis. He proposed that cortisol has its effect on the oocyte nucleus through an agent other than cortisol, produced by the follicular wall. In the Indian carfish and the Japanese rice fish, the maturation of the oocyte (i.e., GVBD) and ovulation are intimately linked and have not been separated experimentally. In the rainbow trout, in vitro, ovulation does not occur at the end of oocyte maturation, and the two processes are controlled by different mediators. As we saw earlier in this chapter, 17α-20β Pg induces oocyte maturation in the rainbow trout, and also in the pike and goldfish, and corticosteroids enhance the sensitivity of the follicle to the effects of GTH and 17α-20β Pg.

In female *Tilapia aurea*, spawning is accompanied by large increase in the plasma concentrations of testosterone, 11-ketotestosterone, 11β-hydroxytestosterone, and deoxycorticosterone. The increase in deoxycorticosterone may be either the result of the secretion of corticosteroids required for ovulation to occur (as discussed above for the Indian catfish, rainbow trout, pike, and Japanese rice fish) or it may be the result of a release of corticosteroids associated will release of the oocytes into the water. As we shall discuss later in this chapter, egg laying in chickens is accompanied by an increase in plasma corticosterone concentration, and a similar process may prevail in *Tilapia aurea*, although this has not been investigated specifically. Hirose (1976) noted that GTH increases the water uptake by adult female ayu (*Plecoglossus altivelis*) and proposed that this water transfer is accompanied by movement of Na^+ into the animal. Corticosteroids are regulators of water and mineral metabolism, and the secretion of corticosteroids at the time of ovulation or spawning may thus be one of regulating the transfer of ions that are, in turn, regulating the process of ovulation. Different solutions have evolved for secretion of corticosteroids; in the Indian catfish, they are secreted by the internal gland in response to GTH; in the Japanese rice fish, they are produced by the ovary in response to GTH, whereas in the goldfish, pike, and rainbow trout they are produced by the internal gland but probably under the influence of ACTH.

A tentative sequence of events that follows the release of GTH and the subsequent secretion of steroid hormones, such as corticosteroids and/or progestins, for the rainbow trout. The evidence that the sympathetic nervous system plays a role in this sequence is reasonably strong, but indirect. Epinephrine, one of the neurotransmitters of the sympathetic nervous system, can induce ovulation in vitro, an effect that can be inhibited by the α site blockers dibenamine and dibenzyline, and phentolamine, whereas β site blockers have no effect. The evidence that prostaglandins play a role in ovulation in the rainbow trout is mainly derived from experiments in which indomethacin, which inhibits prostaglandin synthesis, inhibited in vitro ovulation, and from experiments in which prostaglandin $F_{2\alpha}$ ($PGF_{2\alpha}$) induced ovulation in vitro. The $PGF_{2\alpha}$ may act on musclelike structures in the follicle and thus cause expulsion of the oocyte.

Amphibians

In amphibians, the processes of oocyte maturation and resumption of meiosis are linked, so that under many experimental conditions they have not been separated. The control of ovulation has been most thoroughly investigated in *Rana pipiens* and *Xenopus laevis*. As in the teleosts, gonadotrophins induce oocyte maturation and ovulation by steroidogenesis. In this case, progesterone is secreted and it or its metabolites 17α-hydroxyprogesterone, 20α-hydroxyprogesterone or 17α, 20β-hydroxyprogesterone induce meiosis, resulting in GVBD and ovulation. Germinal vesicle breakdown required that protein synthesis occur, but this protein synthesis does not require the presence of the cell nucleus. Several factors have been isolated in the induction of meiotic maturation (Schuetz, 1977): a maturation factor; a cytostatic factor arresting mitosis and cell cleavage; a pseudomaturation factor, inhibiting protein synthesis and causing abnormal synthesis; a chromosomal condensation factor, inducing chromosomal condensation; and a DNA synthesis factor, initiating DNA synthesis.

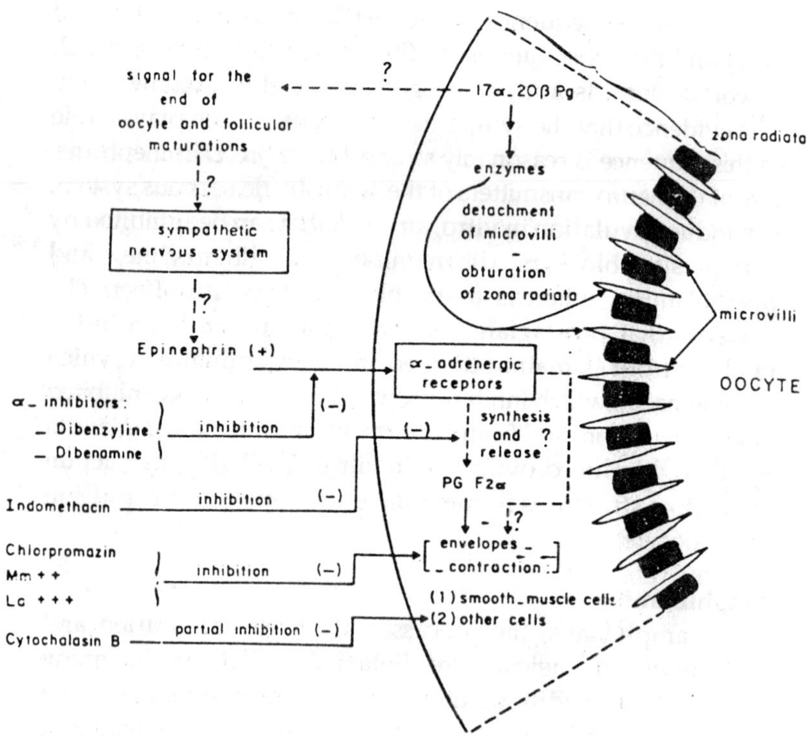

Fig. 6.19. Tentative scheme for the [endocrine] control of ovulation trout.

Biochemically, progesterone-induced maturation is accompanied by a decrease in cyclic 3'5' adenosinemonophosphate (cAMP) in the follicles, a decree which may be necessary for normal maturation to proceed. The addition of dibutyryl-cAMP (db-cAMP) to the medium inhibits progesterone-induced GVBD. Progesterone-induced meiosis is also inhibited by the catalytic subunit of cAMP-dependent protein kinase, but meiosis can be induced by injection of the regulatory subunit of this protein kinase. The inhibition of progesterone-induced GVBD by db-cAMP required the presence of Ca^{2+}.

The following tentative scheme can be proposed: gonadotrophin → progesterone synthesis → (progesterone metabolism?) → uptake of progesterone or its metabolites → increase in free Ca^{2+} → activation of phosphodiesterase → decrease in cAMP → decrease in catalytic unit of proteinkinase → decrease in phosphorylated protein that blocks meiosis.

It remains to be investigated to what extent the tentative sequence of events that has been proposed for the meiotic and ovulatory processes in the rainbow trout is also applicable to these processes in amphibians.

Reptiles

The mechanisms involved in the meiotic maturation and in the ovulation of reptilian oocytes seem not to have been investigated to any great extent. The large amount of yolk in the follicle makes it more difficult to study the biochemical events that accompany the maturation of the oocytes in reptiles than in fishes and amphibians. It is known, however, that the AP is required for ovulation and that turtle FSH is more potent than turtle LH in inducing ovulation in American chameleons.

Birds

Ovulation can be induced in chickens by injection of mammalian LH preparations. As in fishes, catecholamines are involved in the ovulatory response. Ovulation can be blocked by injection of the catecholamine synthesis inhibitor α methyl metatyrosine (MMT), and the α adrenergic blocking agents dibenzyline and phentolamine, with the β adrenergic blocking agents not being effective. The fact that the anti-adrenergic blockage could be overcome by injection of 125 μg FSH plus 125 μg LH into the follicle should be interpreted cautiously in view of the very high doses of hormones used. (For comparison, in our laboratory, 16 μg LH given intravenously is 90-100 per cent effective in causing ovulations in control hens). The phentolamine blockage could be overcome by epinephrine (E), norepinephrine (NE) and phenylephrine, an α-site stimulant. Epinephrine was the only catecholamine that overcame the effect of MMT, and none of the above adrenergic drugs mentioned overcame the effect of dibenzyline. However, cAMP injection into the follicle caused ovulation of dibenzyline-blocked follicles. In our laboratory, cAMP was effective in vitro in inducing ovulation when injured directly into the follicle.

Bahr *et al.* (1974) proposed the scheme and implied that it might be applicable to bird and mammals; however, evidence that prostaglandins are involved in ovulation in birds is scanty. Samsonovitch and Lague (1977) found that neither injection of indomethacin, which blocks PG synthesis, nor of PGE_1 or PGE_2 into the follicle affected ovulation. Day and Nalbandov (1977) found an increase in PGF from near zero to about 8 ng/ 100 mg in the largest follicle of chickens at about 10 and 6 h prior to ovulation, and an increase to about 26 ng/100 mg at 1 h after ovulation. Indomethacin, injected intramuscularly, decreased the PGF peaks but did not prevent ovulation, although it caused a delay in oviposition. Indomethacin also did not affect steroidogenesis by the follicle. In our laboratory, we have found that follicles from hens pretreated with PMSG, ruptured incompletely after the injection of PGF_{2-} into the follicle.

The pathway via prostaglandins may not be present in hens. The necessity of steroidogenesis for ovulation to occur has also not been demonstrated, nor has it been demonstrated that steroids injected into the follicle cause ovulation.

The physical processes involved in ovulation have been better studied in mammals than birds. Ischemia has been proposed as one of the causes for the weakening of the follicular wall. Support for this hypothesis, albeit indirect, is found in the experiments of Opel and Nalbandov (1962), in which they found that after ~~AP~~ follicles become more sensitive to LH, and multiple ovulations result, although in untreated hens multiple ovulations cannot be obtained by LH injections.

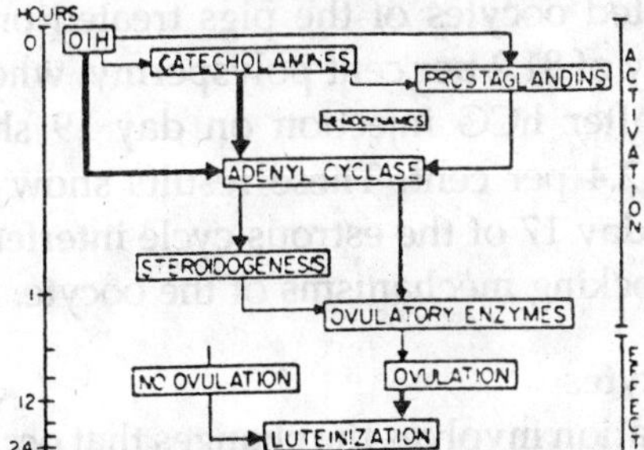

Fig. 6.20. Sequence of events postulated to occur after simulation of the ovarian follicle by LH and resulting in ovulation in birds and mammals and in luteinization in mammals. OIH = ovulation-including hormone.

Mammals

In mammals, there is appears to be a reciprocal relationship between the oocyte and the follicular granulosa cells. Oocytes removed from the follicle have resumed meiosis even when no cumulus cells were attached to the oocyte, thus suggesting that follicular cells inhibit meiosis. Furthermore, El Fouly *et al.* (1970) reported that in rabbits, removal of the oocyte (ooectomy) resulted in luteinization of the follicle.

In vitro experiments have confirmed the observation that in the presence of oocytes, granulosa cell do not differentiate into luteal cells, whereas in the presence of degenerating oocytes or in the absence of oocytes, granulosa cells do not differentiate into luteal cells, Rothchild (1979) has pointed out, however, that the evidence for this interrelationship is still controversial; further investigations are needed before definite conclusions can be drawn.

Hunter *et al.* (1976) have shown that in pigs, ovulation can be induced before meiosis in the oocytes has been resumed. Injection of hCG on day 17 of the estrous cycle resulted in a high number of ovulations (18.6/pig) with 18.2 per cent of the oocytes being primary oocytes. In contrast, injection of hCG on day 19 of the estrous cycle resulted in 12.9 ovulations/pig but only 2.1 percent of the oocytes were primary oocytes. The prematurely ovulated oocytes of the pigs treated on day 17 showed a frequency of 81.2 per cent polyspermy, whereas the oocytes ovulated after hCG injection on day 19 showed a frequency of only 2.4 per cent. These results show that the hCG treatment on day 17 of the estrous cycle interfered with the polyspermy-blocking mechanisms of the oocyte.

Maturation of Oocytes

Oocytes maturation involves the changes that occur in the nuclear, cytoplasm, and membranes of the oocyte, and in the capacity of the oocyte to be penetrated by the sperm cell and respond by appropriate changes in the nucleus, membranes, and cytoplasm after the sperm has entered the oocyte, which is then an ovum.

When the oocytes are collected from antral follicles of juvenile animals, resumption of meiosis generally fails to occur during in vitro culture; such oocytes have been called incompetent oocytes. In contract, oocytes from preovulatory follicles of sexually mature animals resume meiosis during in vitro culture; such oocytes have been called competent oocytes. Morphologically, competent oocytes differ from incompetent oocytes by their great size (60-80 μm versus 20 μm); the thicker more dense zona pellucida (ZP), the large number of microvilli per unit of surface area, the more vesicular smooth endoplasmic reticulum, round mitochondria instead of elongated and dumbbell-shaped mitochondria, a Golgi apparatus consisting of parallel lamellae instead of flattened stacks of lamellae in a parallel arrangement, and a larger, fibrillar, dense nucleolus instead of a rapidly enlarging fibrillo-granular nucleolus. In addition to these ultrastructural differences, competent oocytes show a greater protein synthesis during meiotic maturation; however, the concommitant protein synthesis at the time of resumption of meiosis is not required for GVBD.

Investigations about the endocrine control of meiosis have revealed that ~~AP~~ in rats results in resumption of meiosis in oocytes removed from the follicles, but in 98 per cent of the cases, meiosis stops at anaphase I, whereas oocytes removed from intact proestrous rats proceed to metaphase I. Erickson and Ryan (1976) have stated that FSH may be the factor required for meiosis to proceed from anaphase I to metaphase I.

Tsafriri (1979), comparing isolated rat oocytes that resume meiosis spontaneously with follicle-enclosed oocytes treated with LH to induce resumption of meiosis, found that in the follicle-enclosed oocyte, the LH-induced resumption of meiosis is probably mediated by a protein synthesized by the granulosa cells in response to the LH. As might be expected, puromycin inhibited the LH-induced resumption of meiosis, but did not inhibit the resumption of meiosis of isolated oocytes in vitro. Tsafriri (1979) contends that in the maturation of isolated oocytes "some of the regulatory events which lead to the resumption of meiosis following the LH-triggered maturation are bypassed."

Some investigators have reported that pig ovarian follicular fluid (PPF) contains a factor or factors, probably peptides, which inhibit the resumption of meiosis by isolated oocytes. Hillensj o *et al.* (1979) have provided suggestive evidence that PFF inhibition of oocyte maturation in pigs requires the presence of cumulus oophorus cells capable of progesterone secretion.

The maturity of the oocyte determines to a large extent whether sperm can penetrate the oocyte. Hamster oocytes matured in vitro are not penetrated by sperm in vitro under conditions under which oocytes recovered from the Fallopian tube are penetrated. Presumably the ZP requires maturation in the follicle, which is GTH dependent. In some species (rabbit, pig, cow), sperm do penetrate the oocyte matured in vitro, but decondensation of the male pronucleus fails to occur, presumably because an oocyte cytoplasmic factor that regulates this decondensation is either absent or present. Treatment of oocytes matured in vitro with steroids, especially E_2-17β, 17α-OH-progesterone, and testosterone, induces the conditions that allow decondensation of the male pronucleus. Rabbit and human oocytes matured in vitro without follicle cells are capable of parthenogenetic development after suitable activation.

Experimental evidence and correlations with normal events in the reproductive cycle of the mammals investigated strongly suggest that the maturation required to obtain a normal oocyte, i.e., fertilizable and capable of normal development, is dependent upon gonadal steroid actions. Schuetz (1979) has compared the maturation of oocytes in mammals with that of animals other than mammals. He concludes that in non-mammalian vertebrates there is an increasing concentration of meiosis-inducing substance(s) which eventually reaches sufficient concentration(s) to result in resumption of meiosis. In mammals, however, there is a gradual decrease of meiosis-inhibiting factor(s) required before meiosis can continue.

In both amphibia and mammals, the oocytes remain encapsulated and in contact with somatic cells. In the frog, there is a single layer of follicle cells from which macrovilli project into the vitelline membrane to the oocyte. Under the influence of the appropriate stimulus (LH, progesterone), the macrovilli retract, and loose clumps of cells form on the vitelline membrane. Such changes can also be induced by injection of a cytoplasmic maturation-promotion factor into the oocyte. Inhibition of protein synthesis prevents the maturation changes, suggesting that proteins, which may originate in the oocyte, are required for oocyte maturation. In mammals, the cumulus oophorus cells interdigitate with the ZP. During oocyte maturation, the germ cells and somatic cells disengage, an event accompanied by an increased sensitivity of the cumulus cells to the enzyme hyaluronidase and changes in the gap junctions between cumulus cells and oocyte.

Ovulation has probably been studied in more detail in mammals than in any other group. For the purpose of this discussion we will arbitrarily divide ovulation into two parts: (1) the endocrine events bringing about ovulation and (2) the rupture of the follicular wall and expulsion of the oocyte.

By use of antisera it has been shown that LH is the principal ovulation-inducing hormone in rats; in hamsters, FSH may be the ovulation-inducing hormone. However, LH, unlike FSH, induces steroidogenesis by the follicle of mammals. In most species investigated, ovulation is preceded by a surge of LH *and* FSH, although in the rabbit, and sometimes in women, there is an LH surge without an increase in FSH concentrations. In ~~AP~~ animals, either FSH or LH can induce ovulation, provided mature follicles are present. The question whether FSH and LH act in an additive or in a synergistic fashion has not been answered definitively. Lindner *et al.* (1977) state that FSH may prepare the granulosa membrane for estrogen production and that FSH plus estrogen sensitize the follicle for LH action by increasing the binding of LH.

The evidence that ovulation-inducing hormones (OIH) act via catecholamines is based on the use of blocking agents and the subsequent injection of various catecholamines to overcome the blocking effect. Most of the evidence indicates that α receptors take part in this process.

Prostaglandins (PG) are produced in the ovary, and their production is stimulated by LH; however, oocyte maturation, steroidogenesis, and luteinization proceed when PG synthesis is inhibited, although follicular rupture required PG. The PGs presumably act on the contractile elements in the cells of the theca interna cells (Lindner *et al.*, 1977); thus by causing contraction of these elements, PGF_{2} may aid in the expulsion of the egg. The contractile elements of the follicle wall also respond to sympathomimetic drugs. The effect of these drugs may explain part of the effects of catecholamines on ovulation.

Adenylcyclase may be activated directed by OIH and indirectly by PG and the catecholamines. The increase in cAMP (in amphibians, oocyte maturation and ovulation require a *decrease* in cAMP) stimulates steroidogenesis, which may or may not be required for ovulation, through protein synthesis. The evidence that steroids are required for ovulation is still controversial. Bullock and Kappauf (1973) were able to dissociate ovulation from steroidogenesis in the rat, thus providing evidence that steroids were not required for ovulation in this animal. However, Takahashi *et al.* (1977) successfully induced ovulation in AP rats by repeated injections of progesterone, thus indicating that *in the absence of gonadotrophic hormones* exogenous progesterone can induce ovulation, suggesting that this hormone is part of the first common pathway for inducing ovulation. It is possible that steroids may facilitate or cause ovulation, but that they are dispensable. We should not be surprised that in such an important process as ovulation, which is required for survival of the species, there is redundance in the processes that bring about the rupture of the follicle. Some event in the sequence can be induced in more than one manner; for example, adenylcyclase activation can occur as the result of a direct action of OIH or by an indirect action of OIH through either the catecholamines or the PGs. Similarly the synthesis of the PGs can be stimulated either directly by OIH or indirectly through the catecholamines. Moreover, the different agents can have several effects; for example, OIH stimulates adenylcyclase activation, catecholamine release, and prostaglandin synthesis.

Morphologically many changes precede and accompany and rupture of the follicular wall. Bjersing and Cajander (1974 a-f) have presented beautiful photomicrographs, scanning electron micrographs, and transmission electronmicrographs of changes in the rabbit's ovarian follicle about the time of ovulation. It is not possible to do justice to these photographs, and so the interested reader should consult these papers. In brief, they observed that after GTH release, surface epithelial cells became larger and electron-dense bodies, which might represent lysosomal enzymes, appeared. The granulosa cells became less dense, and the granulosa cell projections increased in number and size, and penetrated into a partly fragmented basement membrane. Spherical inclusions became numerous in the granulosa cells about 8 h after the GTH release and the number of gap junctions decreased from 6 hours after GTH release until the time of ovulation, 10 h after GTH release. The decrease in the number of gap junctions correlated with a decrease in the cohesive force between granulosa cells, so that expansion took place. The influx of fluid and rapid follicular growth were probably facilitated by the changes in the basement membrane and the open channels to the antrum. The theca interna contained collagen fibers and smooth muscles in the apical region of the follicle. About 8 h after GTH release, edema became recognizable below the germinal epithelium covering the follicles. The edematous area contained dead or degenerating cells, while the tunica albuginea contained degenerating fibroblasts and fragmented collagen.

The release of GTH also increased the incidence of small fenestrations of the endothelium of the blood capillaries. Close to the time of ovulation large perforations developed; consequently the fragmentation of the basement membrane allowed free passage of the fluid from the capillary to the interstitium, resulting in edema. Pressure and fluid eventually were transmitted from the capillaries to the interstitium and to the antrum. This did not occur, however, until the tensile strength of the follicular wall had decreased and the subsequent influx of fluid had caused follicular expansion and eventually ovulation.

The rupture of the follicular wall has recently been compared with an inflammatory process. On the basis of several similarities between the inflammatory processes and ovulation, a tentative model was developed. In this model, the PGs are assigned a key role: $PGF_{2\alpha}$ stimulates the activation of quiescent fibroblasts to proliferating fibroblasts, which become the source of the proenzyme procollagenase and of the serine proteases that stimulate the conversion of procollagen to collagenase. Moreover, PGE_2 acts on the preovulatory follicle, in collaboration with histamine, serotonin, and possibly bradykinin, to make the preovulatory follicle a hyperemic inflamed follicle. This inflamed follicle is also a source of serine proteases. The collagenase, formed under the action of the serine proteases, acts on the collagen components of the follicular wall, and the subsequent weakening of this wall results in its rupture. The role of collagenase in ovulation has been substantiated by experiments in which collagenase injected into the follicle facilities its rupture. In chickens, ovulation can also be induced by injection of either collagenase or non-specific proteases.

OVARIAN HORMONE SECRETIONS

Steroid Hormones

In the lamprey (*Petromyzon marinus*), the ovary contains E_1, E_2 and P_4; in the ovary of the river lamprey (*Lampetra fluviatilis*), 3β-HSD has been found in the granulosa cells. Additional, indirect evidence for estrogen secretion under the influence of the AP is found in the fact that secondary sexual characters do not manifest themselves in either ovariectomized or animals.

For the elasmobranchs, teleosts, amphibia, and amniotes, there is evidence from different sources that estrogens and other steroids are secreted by the ovary. The cellular sources for the production of different steroid hormones have been most elegantly ascertained by in vitro studies, in which the different cells have been separated and incubated either alone or in different combinations with each other. This has not been done in all cases to be reported here, but other types of evidence can be used to determine which cells produce the various steroids.

Fortune *et al.* (1975) showed that pieces of *Xenopus laevis* ovary produced P_4 in vitro, and that P_4 production was increased by the addition of either hCG or frog pituitary homogenate to the medium and also by in vivo injections of hCG. Mulner *et al.* (1978) subsequently showed that in this species vitellogenic follicles (0.8-1.0 mm in diam.) and full-grown follicles (1.2 mm or larger) transformed pregnenolone to progesterone and then further to 17α-hydroxyprogesterone, 17α, 20α-dihydropregn-4-en-3-one, androstenedione, and testosterone. Aromatization of androstenedione and testosterone was 10-50 times higher in vitellogenic than in mature follicles. The presence of 3β-HSD and aromatase activity in the follicular walls only provide evidence that it was indeed the follicular wall that had the steroid-producing capacity. Colombo *et al.* (1977) have reported that upon incubation of ovarian tissue of the newt, *Triturus alpestris alpestris*, no steroid was synthesized from acetate; however, when progesterone was used as a precursor, 11-deoxycorticosterone, 11-deoxycortisol, 17α-hydroxyprogesterone, androstenedione, and testosterone were formed, with the yields of 11-deoxycorticosterone and 11-deoxycortisol being higher than those of the androgens. Upon superfusion of ovarian tissue, 11-deoxycorticosterone was produced for endogeneous precursors, with the production increasing after the addition of hCG to the medium.

Dispersed follicular cells from the ovary of the turtle, *Chrysemys picta*, produced progesterone and testosterone in vitro, a production that was increased by the addition of either oFSH and oLH. Estradiol was also produced, but an increase in production was only obtained after addition of FSH, not of LH. Corpora lutea cells secreted P_4 in vitro, a steroid production that was increased by the addition of db-cAMP.

The walls of the preovulatory follicle of the chicken contain progesterone, testosterone, and estrogen, whereas the preovulatory follicle contains progesterone, androstenedione, testosterone, estradiol and estrone.

Flushg and coworkers (Fauang and Naatraandov) have studied in the contributions that chicken thecal and granulosa cells make to the production of different steroid hormones. Thecal cells from either not or preovulatory follicles, women incubates alone, failed to produce any of the steroid hormones (P_4, T, E) assayed. The addition of either of FSH or stimulated a small amount of T production, similar to those produced by the granulosa cells of the same (second and third largest) follicles. Ovine FSH and cLH also increased the estrogen production by the thecal cells of these follicles, but progesterone production, which was less than 1 per cent of that production by granulosa cells, was not stimulated by either of these two gonadotrophic hormones.

The granulosa cells of the three largest follicles all were capable of T and P_2 synthesis, with the granulosa cells of the second largest follicle showing the highest production of T and P_4 at 15-16 h ovulation of the largest follicle. Granulosa cells of the largest follicle showed their highest P_4 production capacity at the time of the preovulatory LH urge. The T-producing capacity of the granulosa cells of the largest follicle were not significantly affected by endogenous LH. Granulosa cells of postovulatory follicles were capable of T and P_4 synthesis in vitro, suggesting that in the chicken, production of these hormones does not require luteinization. The progesterone-producing capacity decreased significantly at 6-7 h after ovulation, but the T-producing capacity showed a bimodal distribution, with a decline at 4-5 h and a subsequent increase at 11-12 h, followed by a sharp decline at 19-20 h after ovulation.

When testosterone was provided as a substrate, thecal cells of the second and third largest follicles (F_2 and F_3) were capable of secreting estrogens, but thecal cells of the follicle destined to ovulate (F_1) and thecal cells of the postovulatory follicle lacked the aromatizing capacity. The thecal cells of F_2 and F_3 follicles produced testosterone when incubated with progesterone as a substrate.

Thecal and granulosa cells lincubated together produced less P_4 then granulosa cells alone. The addition of oLH to the medium increased the Py production by thecal and granulosa cells when compared with thecal plus granulosa cells without oLH added; however, this P_4 production remained less than that of granulosa cells alon. Thecal and granulose cells incubated together produced more T and E than either granulosa or thecal cells alone, and the T production was stimulated by oLH. On the basis of these findings.

TABLE 6.7

Diploid Concentration or Content in the Granulosa and Thecal Layers of Present Post-ovulatory Follicles of the Chicken.

	Steroid				
Follicle tissue	*P* (*ng/g*)	E_1 (*ng/g*)	E_2 (*ng/g*)	*T* (*ng/g*)	*A* (*ng/g*)
Preovulatory theca	3.13 ± 6.7	0.71 ± 0.08	4.76 ± 0.08	7.7 ± 1.3	<1
Preovluatory theca	64.9 ± 11.2	4.3 ± 0.85	11.2 ± 3.8	1.6 ± .03	5.0 ± 2.9
	ng/foll.	ng/foll.	ng/foll.	ng/foll.	ng/foll.
Preovulatory granulosa	2670 ± 1364	0.42 ± 0.05	4.5 ± 1.9	<1	3.0 ± 1.3
Preovulatory granulosa	59.7 ± 6.7	0.76 ± 0.29	1.5 ± 0	1.48 ± 0.33	5.6 ± 2.2

P = progesterone, E_1 = estrone, E_2 = estradiol, T = testosterone, A = androstenedione.

The interstital cells of the ovary produce androgens. Woods and Domm (1966), using immunofluorescent techniques, showed that androgens were present in the theca interna and granulosa layer of the developing and postovulatory follicle and in interstitial cells of the ovary. They also demonstrated and presence of androgens in the cells of the medullary cords and tubules, the rete testes, and the interstitial cells of the hypertrophied right gonad of sinistrally ovariectomized hens.

In mammals there is evidence in several species (pig, horse, cow, sheep, rabbit and rat) that granulosa cells can produce estrogens, provided that androgen is supplied as a substrate. Luteinizing hormone stimulates androgen secretion by thecal cells in the rat, and FSH stimulates the granulosa cells to convert androgen to estrogen. The thecal cells in monkeys and humans seem to be capable of estrogen secretion in the absence of granulosa cells, although the presence of granulosa cells increases the production of estrogens (Channing, 1979a).

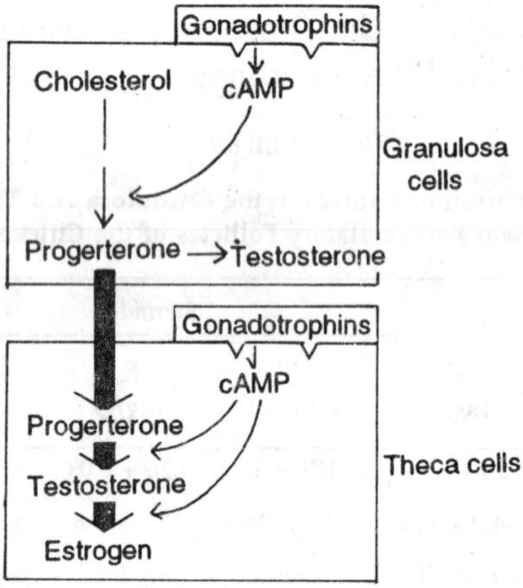

Fig. 6.21. Model of the two-cell-type hypothesis for [chicken] follicular testosterone and estrogen synthesis. Thick arrow represents major route and thin arrow represents minor route.

Thecal cells of cattle ovaries produced progesterone. Estradiol-17β (E_2) added to the medium in concentrations of 0.001 μg/ml more than doubled the P_4 production at 24-48 h of culture, but this effect was not as pronounced at 48-72 h of culture. At concentrations of 0.1, 1.0 and 10 μg/ml, E_2 inhibited P_4 secretion.

In vitro, granulosa cells from pig follicles and rats can also secrete progesterone. This secretion of P_4 is enhanced by androgens, FSH, LH, FSH plus LH, and by follicular fluid from preovulatory porcine follicles. Progesterone secretion by follicular granulosa cells of the pig is inhibited by a fraction of the fluid from 1-2 mm pig follicles. Rothchild (1979) has pointed out that the findings from in vitro experiments that follicular granulosa cells can synthesize progesterone and estrogens are not in accord with ultrastructural and cyto-chemical evidence that these cells lack the systems presumably required for steroid biosynthesis.

The interrelationships between the thecal and granulosa cells in steroid production, and the inhibitory and stimulatory effect of these steroids and of follicular fluid (to be discussed below) suggest that in addition to regulation by gonadotrophic hormones, intrafollicular and intraovarian regulation is important in modifying follicular function. The quantitative aspects of these regulatory interrelationships in vivo need to be established before it is feasible to formulate a quantitative model of the physiology of the follicle at different stages of development.

The mammalian CL secretes principally progesterone, but also other hormones. For instance, bovine CL secretes progesterone plus 20β-hydroxy-4-pregn-3-one and pregnenolone, but no estrogen, whereas the human CL secretes, in addition to progesterone, 17-hydroxyprogesterone, 4-androstenedione, pregnenolone, 20α-hydroxypregn-4-en-3-one, estrone, and estradiol-17β. The CL of the rabbit, pig, sheep, horse secrete progesterone, 17α-hydroxyprogesterone, and 4-androstenedione, but not the estrogens, estrone and estradiol. The CL of the estrous cycle of the hamster during day 2 of the cycle elaborates progesterone and androstenedione in about equal amounts.

The CL of pseudopregnant and pregnant rats secrete progesterone, 20α-hydroxypregn-4-en-3-one, and estrogens. The evidence for estrogen secretion by the CL is based on: (1) in vivo experiments, in which indirect evidence of estrogen secretion was obtained, e.g., cornified smears and increased uterine weight in rats with an autotransplant after LH administration and (2) blastocyst implantation in AP rats provided with an exogenous or endogenous source of progesterone. However, in experiments, in which the follicles were destroyed by X-irradiation, morphologically normal CL did not secrete estrogen, as determined by vaginal smears. The presence of estradiol in the CL of pregnant rats at the end of pregnancy is further evidence of estrogen secretion by CL, as is the in vitro production of estrogen (at 1/20-1/200 of the rate of non-luteal ovarian tissue of the same rat ovaries that yielded the CL) by the CL of pregnant rats. The fact that the CL can aromatize testosterone may indicate that testosterone serves as the substrate for estrogen formation by the CL at the end of pregnancy.

Steroid hormones in the ovary are produced not only by the follicles and CL but also by the cells of the interstitial gland. In the rabbit, the interstitial gland, which is well-developed, produces 20α-hydroxypregn-4-en-3-one. In the human and rat ovary, the interstitial cells secrete androgens or contain them.

Non-steroid Hormones

Besides the steroid hormones, non-steroid hormones, such as relaxin, follicular hormones, and prostaglandins, may affect reproduction in mammals.

Relaxin. This is a protein hormone with a molecular weight of about 6300 consisting of two subunits linked by disulfide bonds. The concentrations of this hormone, which can be determined by RIA varies during the estrous cycle of sheep. It is high during day 15 and 16 and at 24 h after the onset of the LH peak and at the time of ovulation. In some species, the relaxin concentration in the blood increases during gestation, e.g., sheep, pigs, humans, hamsters and rats, mice, and guinea pigs; and relaxin is present in high concentrations in pregnant dogs and monkeys. In sheep, there is a sharp peak in the relaxin concentration at the time of parturition, when the lamb's head is out of the vagina of the ewe.

The hormone is secreted by the CL, specifically in the lutein cells. However, the ovary is not the only source of relaxin, because substantial concentration of relaxin are present in the blood of ovariectomized pregnant sheep. The rat's uterine material gland is, however, not a source of relaxin.

The significance of the fluctuations in relaxin concentrations during the estrous cycle is not at all clear. The high concentrations of relaxin during pregnancy are probably related to the effects of relaxin of reproductive tract, e.g., softening and relaxation of the cervix in hamsters and relaxation of the pelvic ligament in guinea pigs, which permits the large young to pass through the pelvis.

Relaxin also inhibits spontaneous utreine motility, increases the uterine water and glycogen content, and, together with estrogen and progesterone, stimulates uterine growth.

Follicular Fluid Factors. The follicular fluid of pigs (PFF) contains several factors that affect the physiology of the follicle and oocyte.

1. *Oocyte maturation inhibitor*. This inhibitor has a MW of about 2,000 daltons. The concentration of this factor is higher in the PFF of small (1—2 mm) folicles than in the PFF of large (6—12 mm) follicles. The factor inhibits the spontaneous maturation of isolated pig oocytes surrounded by their cumulus, as well as progesterone secretion by the cumulus cells, and both effects are dose dependent. The effect of the inhibitor on oocyte maturation and progesterone secretion by the cumulus cells is reversible. This inhibitor also causes the outgrowth of the cumulus cells in vitro.

2. *Luteinization inhibitor*. This factor, present mainly in PFF of small follicles has a molecular weight greater than 100,000. It inhibits the spontaneous luteinization of granulosa cells cultured in vitro and the progesterone secretion by such cells. It also inhibits the stimulation of cAMP production by LH of these cells, possibly by decreasing the receptors for LH and hCG.

3. *Folliculastatin or ovarian inhibin.* This factor, which is present in higher concentrations in small follicles than in large follicles, has a MW greater than 10,000 daltons. In rats and pigs, it prevents the secondary rise in FSH concentration that follows ovulation; in rats, it prevents the rise in FSH that follows castration (a rise that cannot be inhibited by estrogen) and the LH-RH-induced increases in FSH concentration, although it does not inhibit the LH-RH-induced rise in LH concentration. Administration of folliculastatin results in a decrease in FSH concentration in the intact rat and in the intact and the ovariectomized rhesus monkey.

Prostaglandins. Earlier in this chapter we mentioned that gonadotrophins induce ovulation, probably by stimulating PG synthesis. Recently, Dodson and Watson (1979) have reported that in superfused preovulatory pig follicles, FSH addition to the medium resulted in a threefold increase in production of PGE and $PGE_{2\alpha}$ whereas LH addition resulted in about a twofold increase in the production of thesePGs. Estradiol-17β addition increased PGE_2 production fourfold and $PGF_{2\alpha}$ production nearly tenfold. Since FSH and LH act synergistically in stimulating estrogen production, the results reported above suggest that FSH and LH each stimulate estrogen production to a suboptimal level and that gonadotrophin-induced PG production requires the intermediate estradiol production.

Functions of Ovarian Steroids. We wish to mention here some of the important functions of estrogens, progesterone, and androgens in female reproduction. The physiological functions of these hormones in the chicken may serve as an example. Just prior to sexual maturity, defined as ovulation or laying of the first egg, the plasma estradiol concentration rises from about 70-80 pg/ml 10 days before the first egg is laid to about 350 pg/ml at two days before the first egg is laid and then drops to about 125 pg/ml. Estrogens have several effects important for the initiation and maintenance of sustained egg production.

1. Estrogens increase food intake from 180 calories/day to about 300 calories/day, which is clearly adaptive, because it ensures a sufficient caloric intake to sustain egg production. (A hen laying 270 eggs/year produces 540 g of calcium, 1900 g of protein, 5400 g of solids — 2 times its body weight — 1600 g of lipids, and 21,600 calories.) Of the 300 calories consumed. 80 are deposited in the egg. The mechanism by which estrogens stimulate appètite is not established. It can either be a direct effect on the hypothalamic centers that regulate food intake, or it can induce changes in the blood constituents, which might then change the hypothalamic regulation, and, of course, a combination of these mechanisms might operate.

2. Estrogen in concert with progesterone is necessary for the priming of the hypothalamo hypophyseal system, so that progesterone can induce LH release in hens.

3. Estrogens are probably involved in the regulation of female sexual behaviour, e.g., crouching.

4. As we discussed under vitellogenesis, estrogens change the blood composition by acting of the liver, thus making precursors for yolk formation available.

5. Estrogens stimulate the deposition of calcium as medullary bone in the long bones. This calcium serves as a reserve during the periods of high egg production, when calcium cannot be take in with food in enough quantity for egg production.

6. Estrogens stimulate growth and cytodifferentiation of the oviduct, priming it for the action of progesterone. Administration of estrogen to immature chicks results in the secretion of ovalbumin, conalbumin and lysozyme. The excellent reviews by O'Malley *et al.* (1975) and Schimke *et al.* (1975) should be consulted for extensive discussions of the molecular aspects of estrogen (and progesterone) action on the cellular metabolism of the oviduct.

7. Estrogens affect the plumage, so that both colour and feather structure can be used as indications of estrogen secretion; ovariectomized hens and castrated and intact males all have the same type of feathers, i.e., the neutral type.

Progesterone administered alone causes differentiation of the surface epithelium of the oviduct, with formation of goblet cells and ciliated epithelial cells, but protein is not secreted. When estrogen administration stops, ovalbumin synthesis stops, but it can be reinitiated by the injection of either estrogens or progesterone. Progesterone is essential for the induction of avidin synthesis by the estrogen-primed oviductal goblet cells. Administration of progesterone to laying hens between 22 h and 12 h before ovulation induces LH release and premature ovulation. Progesterone administered in high doses can precipitate a molt in laying hens, probably by stimulation of the feather papillae, and cause atresia of the large follicles.

The chicken ovary secretes androgens, as is externally evident from the growth and reddening of the comb and wattles at the time of sexual maturity. Plasma androgen concentrations show a peak in concentration two hours prior to ovulation and so might to thought to be involved in the induction of ovulation. However, injections of testosterone at doses comparable to effective doses of progesterone (0.1-1.0 mg) are not as effective, i.e., about 50 per cent of the hens ovulate after injection of 1 mg of TP and 16 mg need to be injected to obtain nearly 100 per cent ovulations. Androgens induce protein synthesis is estrogen-primed oviducts, but the mechanism is not as well-studied as for progesterone.

Several reviews are helpful in understanding the integrative action of ovarian hormones in bird species other than chickens or in other classes. The concept of the integrative actions of hormones, together with a knowledge of the natural history of the species of interest, usually makes it possible to deduce the integrative action of these hormones for that particular species.

7

Ovum Transport

The germinal epithelium of the ovary undergoes oogenesis. The mature Ova are given out into the body cavity by a process of *Ovulation*. The released ova are then taken up by the oviducal funnel of the oviduct to transport them to the uterus.

The ovum is fertilized in the oviduct, and during the first days of pregnancy the oviduct is the physiological home of the early embryo. Also, remarkably enough, the oviduct is capable of transporting the male and female germ cells in opposite directions at practically the same time. It is only in the last few years that we have come to recognise clearly the complex function of this organ in instituting and regulating the earliest phases of pregnancy and, as a result of this knowledge, we are now in a better position to explain the causes of certain cases of sterility and infertility in women, and also to understand some peripheral mechanisms of action of hormonal contraceptives.

The oviduct or uterine tube consists of two parts, the lateral, wider ampulla and the medial, narrower isthmus. At the ovarian end, the ampulla ends in the fimbriated infundibulum, and medially the isthmus opens into the uterus via the pars isthmica interstitialis. In the human uterine tube, which is about 12 cm long, the transition between ampulla and isthmus is between the inner and middle third; in the uterine tube of the rabbit, which is about the same length, and in most laboratory animals it is in the middle of the organ. The wall of the uterine tube consists of three layers which merge into one another. The subperitoneal longitudinal muscle layer runs in the direction of the tube. In the central layer, which contains blood vessels, muscle fibres and blood vessels run around the tube. The inner, autochthonous muscle layer is spiral in form. The three layers of muscle form a coherent system which is closely linked both from the anatomical and functional point of view. In the ampulla, the muscular wall of the tube is delicate and thin, but in the isthmus it is very much more powerful. In some species of animal the tube is rather tortuous.

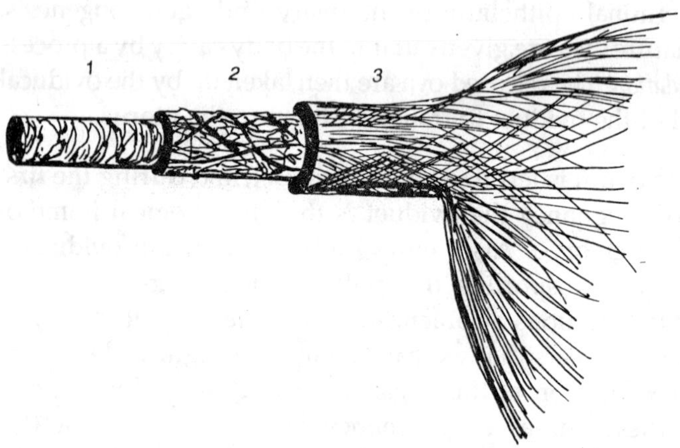

Fig. 7.1. Muscular layers of the oviduct. *1.* circular layer, *2.* vascular layer, *3.* longitudinal layer (H. Erb, 1969).

The mucosa of the tube is thrown up into many folds and villi which almost fill the lumen of the tube. However, the size of these folds may vary at different points in the same tube and also between the right and left tube.

The mucosa is lined with a simple epithelium which undergoes cyclic changes. This epithelium consists of secretory cells, which produce the tubal secretion, and ciliated cells, the bunches of cilia of which all beat in the same direction. The cilia conform to the standard structural plan of kinocilia, whereby nine pairs of fibrils are always arranged concentrically around a central pair of fibrils.

The function of the oviduct in ovum transport begins when it picks up the ovum which has been released from the ovary at ovulation. In some species of animal, e.g., mouse, rat and hamster, the fimbriated end of the tube is connected to the ovary by a specially shaped peritoneal fold, the bursa ovarica, and this close contact between the two organs facilitates transfer of the ovum. In man, rabbit, guinea pig, sheep and other animals there is no permanent close contact between the two organs. Thus, in these species the oviduct is not directly, connected to the ovary and for a long time it was not understood how the ovum reached the tube. After earlier, strange and rather exotic theories, made observations by laparotomy on rabbits and by laparoscopy on rhesus monkeys and later also in man, and was able to demonstrate the complicated interaction between oviduct and ovary. These observations revealed that the fimbriated end of the tube which, at the time of ovulation, is wide open and engorged with blood, is brought into close contact with the surface of the ovary by muscular activity of the mesotubarium, particularly of the M, atrahens tubae which runs in the fimbria ovarica. At the same time, the ovary is

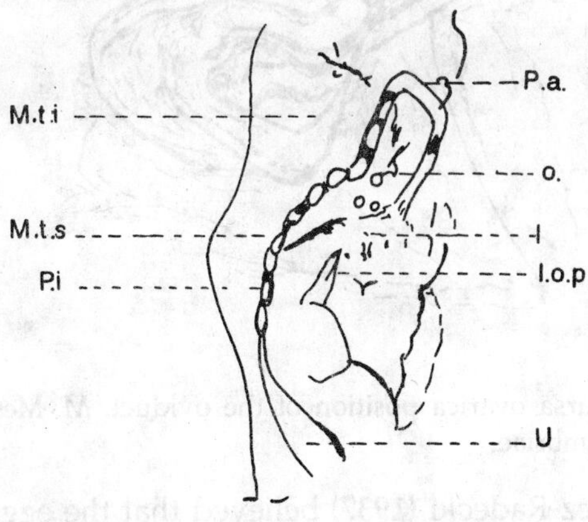

Fig. 7.2. Rabbit oviduct during estrus (A. Westman, in: F. v. Mikulicz-Radecki, 1931). *O.* Ovarium, *I.* Infundibulum, *U. Uterus, P.a.* Pars ampularis, *P.i.* Pars isthmica, *M.t.i.* Mosotubar um inferius, *M.t.s.* Mesotubarium superius, *L.o.p.* Ligamentum ovarii propium.

moved slowly to and fro and about its longitudinal axis by contractions of the ligamentum ovarii proprium so that the fimbriated infundibulum is constantly feeling all over the surface of the ovary.

Fig. 7.3. Ovum pickup in man.

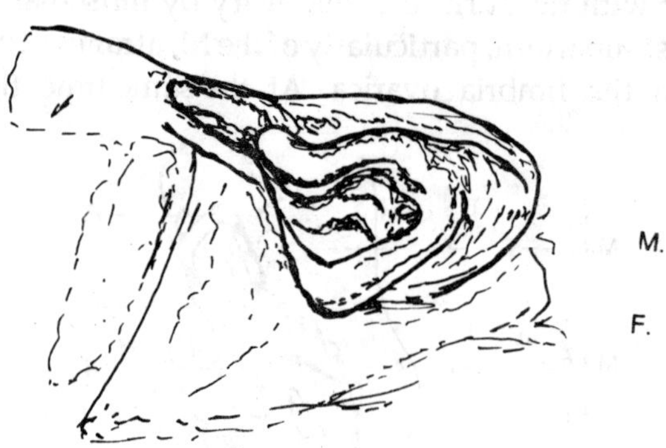

Fig. 7.4. Bursa ovarica position of the oviduct. *M.* Mesosalpinx, *F.* Fimbriae.

Mikulicz-Radecki (1937) believed that the egg was taken up in this way. His results from animals experiments and X-ray studies show that, at the time of ovulation, the end of the tube takes up a special position. It surrounds the ovary in such a way that, along with the mesotubarium inferius and superius, it forms a pocket into which the ovum slides and from where it can be taken up by the fimbriated infundibulum. Since this arrangement corresponds to the anatomically permanent bursa ovarica in other species of animal, Mikulicz Radecki (1929) called it the *"bursa ovarica position."*

Elert (1947) examined women by laparoscopy and he found that the two mechanisms occur side by side. He observed that the exploration of the ovarian surface by the fimbriated infundibulum lasts for only a few minutes at a time and is probably repeated frequently. During the intermenstruum, the tubes sink temporarily into the pouch of Douglas in the "bursa ovarica position" and remain there for a time. This evidently represents an additional safety measure.

Dubreuil (1944) and Kneer (1949) suggested that the ovum was drawn into the fimbriated infundibulum by a stream of liquid passing into the tube. However, in 1958 Clewe and Mastroianni found that in rabbits the ovum consistently passes into the fimbriated infundibulum even if the tube is tied off directly behind the infundibulum. Harper (1961) used rats to film the mechanism by which the ovum is taken up. He showed that once the ovum has been slowly forced out of a small aperture in the follicle, a process which takes five to seven seconds, the cilia on the fimbriae are sufficiently powerful to waft the ovum and cumulus oophorus, which is still attached, rapidly and steadily into the fimbriated infundibulum. The same happens with starch grains and, in the mouse, with minute carbon particles. The ovum was still taken up even when, as an experiment in rabbits, the ovary was placed two to three centimetres away from the fimbriated end. Ova were not taken up if they had been injected intraperitoneally and hence were too far from the fimbriated infundibulum.

On its migration through the twelve centimetres or so of tube, the ovum is very much under the control of its environment, since, unlike sperms, it is non-motile. However, surprisingly enough, there is still not complete agreement as to whether the ovum is moved by contractions of the tubal muscles or by the action of the epithelial cilia. Many investigators believe it is most likely to be the tubal muscles, although this does not explain one or two of the features of ovum transport. In the meantime, the nature of the muscular contractions has been examined in man, in many species of animal, including rabbits and pigs, cows, sheep, horses, mice, rats and monkeys. There were no great differences of tubal motility, and the type of movement appears to be the same in all species. At various points in the tube contraction-rings of longitudinal muscle from rapidly, and these occasionally run for a short distance in the direction of the uterus and then disappear, either rapidly or slowly. Antiperistaltic movements are observed only occasionally. Contractions in the ampulla are relatively weak, but they are more powerful and more regular in the isthmus. Contractions in the isthmus occur at intervals of seven to fifteen seconds and last for two to four seconds. Since the inner circular fibres contract first, the tube stretches at the beginning of the contraction; it then shortens, owing to contraction of the outer fibres, before finally it relaxes again. When the contraction has subsided, the previously contracted segment becomes very engorged with blood, and this may cause that section of tube to become double the normal thickness. Thus the whole tube appears to be very active, although, because of the many types of contraction which are possible, the individual movements often appear to be rather irregular and they are very difficult to analyse.

At the time of ovulation, under the influence of estrogens there is a sharp increase in the intensity of contractions, but by the second, third and especially the fourth, post-ovulatory day they again become-weaker and more irregular as the progesterone level increases. They increase again at the end of pregnancy. Thus, the tube goes through its first period of major activity during migration of the ovum.

This temporal association is certain the reason for the functional connection assumed by many authors. After Parker

(1931) had produced definite evidence that the cilia are capable of transporting the ovum, Harper (1961) produced what is, in my opinion, conclusive evidence of the effectiveness of ciliary action. He was able to show that the stream produced in the tube by the cilia is capable of moving the ovum even after the motility of the muscles has been eliminated experimentally. Wintenberger (1955) excised segments of ampulla and isthmus from sheep and sewed them back in the same or in the reverse direction. After healing, 82% of ova were retained by the reversed segments in which, of course, the cilia were also working in the reverse direction, whereas segments which had been replaced in the normal direction retained only 32% of all ova.

Thus, not only do the contractions of tubal muscles not appear to have the same effect as intestinal peristalsis in pushing forward the contents, but they actually seem to cause a slight delay of transport. From the nature of the muscular contractions described, it would appear that they produce thorough mixing of the tubal contents, promote fertilization and help to denude the ova. It is, however, the cilia which appear to be responsible both for uptake of the ovum by the fimbriated infundibulum and also for transport of the ovum.

Using a method developed by Dalham for measuring ciliary activity, Borell, Nilson and Westman (1957) were able to produce precise data on ciliary activity during passage of the ovum. They found that in the rabbit, the cilia beat towards the uterus at a rate of 1500 beats per minute during oestrus. After ovulation, as the level of progesterone rose rapidly, the frequency increased steadily and became 20% faster. However, this rise did not occur until 48 hours after ovulation, i.e., long after the ova had reached the isthmus, and when the intensity of the muscular contractions had already begun to decline. This increased ciliary activity was maintained until after implantation of the embryo.

The effect of strong tubal activity on the contents of the tube is a constant source of interest to research workers, mainly because, in spite of powerful ciliary movements and tubal contractions, the ovum needs a surprisingly long time to complete its journey. In most mammals, the ovum spends most of its pre-implantation phase of about three to four days in the tube, in which time the fertilized ova may develop to the morula stage or, in some species, even the blastula stage.

TABLE 7.1.

Length of Time Taken for Transport of Ovum Through the Oviduct in Some Mammals

Animal	*Time (hours)*
Dog	168
Cat	148
Horse	98
Rat	96
Rhesus monkey	96
Cow	90
Guinea pig	82
Sheep	72
Mouse	72
Rabbit	60
Pig	50

After ovulation, the ciliary activity rapidly and steadily wafts the ova through the fimbriated infundibulum and past the first loop of the tube, which is usually twisted. The forward movement of the ovum caused by the cilia then stops, and the ovum is constantly pushed backwards and forwards between the rings of contraction. Occasionally, the ovum will get past one of these contraction-rings and then it will again begin the rather violent "pendulum" movement in the next segment of tube.

The ovum is propelled at very different speeds in different sections of tube. It migrates through the ampullary part of the tube in a few hours or, according to Harper (1961), in a few months. It is arrested for a time at the ampullary-isthmical junction. Delaying of the ova at this point has also been observed in the pig, in the cat and in the guinea pig. Wintenberger (1953) was able to demonstrate that, in sheep, ova reached the junction of ampulla and isthmus after two-and-a-half hours but entered the isthmus only after 49 hours, and then passed through this part of the tube very quickly. Black and Asdell (1958) found that although droplets of oil and Indian ink were distributed rapidly in the ampulla by pendular contractions of the tube, they entered the isthmus only slowly. According to Greenwald (1961), rabbit ova take only two hours to pass through the first half of the tube, but they can be detected in the isthmus only after 70 hours, and they pass through the last part of the tube very rapidly. Harper, Bennett, Boursnell and Rowson (1960) and Harper (1964) observed small radioactively-labelled glass beads the size of ova at the ampullary-isthmical junction eight hours after insertion, but they passed slowly into the isthmus only after 40 hours. The ova enter the isthmus only after a fairly long delay at this point in the tube, and they pass through the last part of the tube so rapidly that they can hardly ever be detected there (Greenwald, 1961).

It was suggested long ago that the utero-tubal junction, i.e., the place where the tube opens into the uterus, has a regulatory function. Black and Asdell (1958) were able to demonstrate just how far this is true with a very simple and impressive experiment. Using the rabbit as the experimental animal, they ligated the tube at the fimbriated end immediately after ovulation. After only 24 hours they found that the less muscular part of the oviduct was extremely distended, indicating that no fluid was able to pass out of the oviduct through the utero-tubal junction. Although the tube was grossly dilated, the fluid pressure developed within it was evidently insufficient to overcome the resistance. Andersen (1927) stated this pressure to be 18 mm Hg. Only after 60 to 72 hours, i.e., the usual time required for the ova to enter the uterus, did the resistance to fluid pressure fall sharply, evidently as a result of opening of

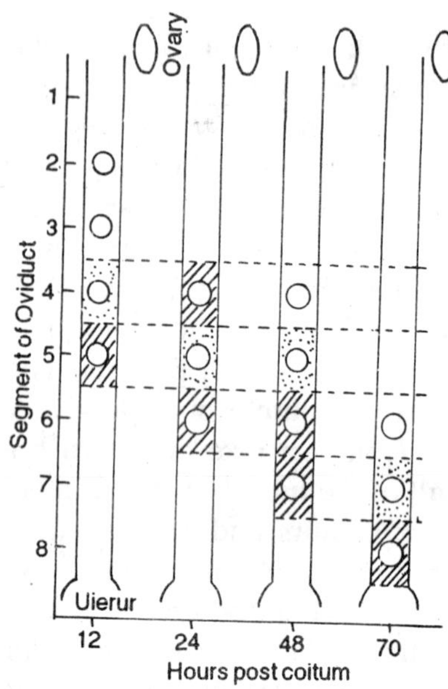

Fig. 7.5. Distribution of rabbit ova during the normal 3-day passage through the oviduct. Numbers indicate the number of ova within each segment of the oviduct

the utero-tubal junction. If a short plastic tube was placed in the terminal segment, the secretion did not accumulate. This experiment illustrated the control function of the utero-tubal junction and it also demonstrated a fact was not previously known, that during migration of the ovum tubal secretion does not flow towards the uterus but in the opposite direction, back through the infundibulum into the peritoneal cavity. Moreover, this shows that ova move against the stream whilst sperms move with the stream. Since then the same mechanism has been demonstrated in sheep and in the cow. Rauscher (1968) has recently reported similar findings, obtained during surgery, in human beings. In rabbits, the utero-tubal junction opens only on the third and fourth day and on 30th and 31st day of pregnancy, i.e., when the embryos pass out of the tube into the uterus and at the time of parturition. Thus, it would seem that during these days the stream of fluid is at least partially reversed and the fertilized ova travel the remaining short distance into the uterus with the stream, whereas on their way through the tube they move against it.

Andersen (1928) examined the morphology of the utero-tubal junction in 25 different mammals from ten different orders, and she found that the restrictive effect varied. Working with rabbits, Black and Asdell (1959) concluded that estrogen-induced oedema was responsible for the occlusion, but they did not actually succeed in demonstrating it. Oestrogen-induced oedema has been shown in sheep in the subserous and muscular coasts of the utero-tubal junction, and the lumen at this point was kinked so that the folds of the mucous membrane formed a valve-like occlusion.

However, studies in the cow indicate that the muscle tissue in this region also has an important part to play. In fact, if segments are removed from this point, the tube is no longer occluded, and this is also the case if the musculature of the tube is paralysed with nicotine.

Using serial cinephotography of large histological sections of specimens removed during operations of human, Rocker (1964) was able to shed further light on the morphology of the utero-tubal junction and describe a probable functional mechanism. He showed that the musculature of the uterus continues into the utero-tubal junction, where the interlaced muscle fibres of the inner layers of the uterus become parallel and are arranged like a sphincter muscle. The lumen of the tube is completely straight as far as the vascular layer. Here the tube is sharply kinked and the lumen is only very small (100 γ). Thus, at the utero-tubal junction the tube has an outer layer of longitudinal muscle, an inner layer of sphincter, and inside this another layer of longitudinal muscle which is part of the uterine muscle. Thus, each contraction of uterine muscle can close the tube.

Thus, together with the muscular contractions, ciliary movement and tubal secretion, the utero-tubal junction is one of the factors which determine the movement and the timing of the movement of the ovum. Hafez (1963) and also Harper, Bennett, Boursnell and Rowson (1960) and Harper (1964) take the view that the utero-tubal junction is controlled by changes in the oestrogen-progesterone ratio, and thus can easily be regulated. A constantly-changing relationship between these two hormones would appear to be necessary for this mechanism to function smoothly, but the exact quantitative scale of this relationship is still not known. Thus, after ovulation the ova are retained in the oviduct, as a result of closure of the utero-tubal junction by the effect of oestrogen produced in the ovary, until the uterus and endometrium have been properly prepared for implantation of the ovum by the increasing levels of progesterone. When, at the right moment, oestrogen-progesterone ratio reaches a certain level, the path through the tube is opened up.

However, we still cannot explain the arrest of the ova in the middle of the tube, at the junction between the isthmus and the ampulla. Research workers were so surprised at the retention of the ova there, that Greenwald (1963) suggested that there might be, at that point, an even more effective closure of the tube than the block shown by Black and Asdell (1968) to occur at the utero-tubal junction. As a result of her experiments on sheep, Winten Berger-Torrés (1961) was also forced to assume that the occluding mechanism at the mid-point of the oviduct was more important.

In recent years several studies have set out to explain this observation. Edgar and Asdell (1961) suggested that there was occlusion of the tubal lumen owing to oedema of the subserosa and muscles both at the mid-point and also at the utero-tubal junction, but they were unable to demonstrate this type of oedema with any degree of certainty. Harper (1964) postulated that this type of oedema occurs throughout the isthmus, and that peristalsis of the ampulla is weaker. Black and Asdell (1958) considered that it was due to a decline in or failure of the activity of the sphincter fibres at this point, whereas Winten

Berger-Torrés (1961) considered that powerful antiperistaltic waves in the isthmus are responsible for the 48-to-68-hour delay of the ova. Greenwald (1961) made a vain attempt to demonstrate a muscular sphincter at this point by cutting serial sections; Humphrey (1969) also assumed that there was a muscular sphincter. Older explanations of this phenomenon were also referred to: that there is a special arrangement of the muscle fibres at this point or that the amplitude and frequency of the muscular contractions in the ampulla and isthmus vary.

Greenwald (1963) studied the internal pressure in rabbit tubes and failed to find any change in the amplitude or frequency of the contractions in the ampulla during the three-day passage of the ovum. In the more muscular isthmus, under the effect of increasing levels of progesterone the power of contractions declined. In oophorectomized animals, estrogen administration restored the power of the contractions to the level seen during oestrus and the first day of migration of the ovum. This was achieved with a dose of 250 γ and also with only 25 γ. Since earlier experiments had shown that whereas with the higher dose ova can be seen at the ampullary-isthmical junction for many days (the phenomenon described as "tube locking") but with the lower dose passage of the ovum is accelerated, yet both doses increase motiiity, Greenwald (1963) believed that tube motility could be excluded as the cause of the block in the middle of the tube. As yet there has been no explanation of this "sphincter" mechanism at the ampullary-isthmical junction.

Therefore, we set out to discover whether, in addition to tubal motility and ciliary activity, tubal secretion is also an important factor in ovum transport. This appeared to us important, in that Black and Asdell (1958) had discovered that the ovum migrates against the direction of fluid flow.

It was not until 1956 that Bishop demonstrated that the tubal fluid is produced by a genuine secretory process of the tubal epithelium. Bishop (1956) and later also Mastroianni *et al.* (1961) were able to show that the rate of secretion undergoes cyclic fluctuations. In sheep, monkeys and other species of animal, it was possible to demonstrate that at the time of ovulation there was a sudden and sharp rise in the rate of secretion, and that as the corpus luteum developed the rate fell again. There was no increase of secretion of anovulatory cycles, and in costrated animals the amount of secretion fell by two-thirds. In rabbits in estrus, secretion is 0.92 to 2.31 ml/day (average 1.57 ml/day), and this falls by 50% by the third day after ovulation. Thus, secretion is the oviduct is controlled hormonally, i.e., increased by estrogen and inhibited by progesterone.

It has been seen that the secretion of tubal fluid alters in each particular section of the oviduct as the position of the ovum alters, the secretion being most pronounced around the site of the ovum at any given time. In other words, the most powerful secretory activity travels like a wave with the ovum from the ovarian end to the uterine end of the oviduct. To investigate this theory we carried out experiments on rabbits.

In order to demonstrate a change in the type of secretion, the oviduct was examined 14 hours and 62 hours after ovulation, i.e., on the first and third days of the three-day passage of the ova through the oviduct. To determine the rate of secretion of various sections of the oviduct, we employed parenteral administration of anorganic radioactivity-labelled sulphur, which is practically selectively deposited in the mucopolysaccharides which represent a substantial part of the tubal secretion. A few hours after administering sulphur, an autoradiograph indicated activity in the secretory cells of the tubal epithelium, but not in the remaining tissues. Thus, the level of this radioactivity could be taken as a parameter of secretory activity. Seven hours after administering $Na_2{}^{35}SO_4$, the oviduct was divided into five equal segments and, after ignition, the radioactivity was assayed by a standardized automatic process.

In order to compare the activities in the different segments of tube, the radioactivity for such segment was related to the mean activity of the complete tube. The resulting figures for "relative segmental activity" were examined statistically. In this way we were able to compare a sufficiently large number of tubal segments with each other. The results showed us that tubal secretion did not move in a wavelike pattern from the ovarian to the uterine end to the tube. Secretory activity was always highest in the isthmus and lowest in the ampulla. This was because the epithelium of the isthmus comprises mainly secretory cells, with isolated ciliated cells, whereas the epithelium of the ampulla is composed mainly of ciliated cells with few secretory cells.

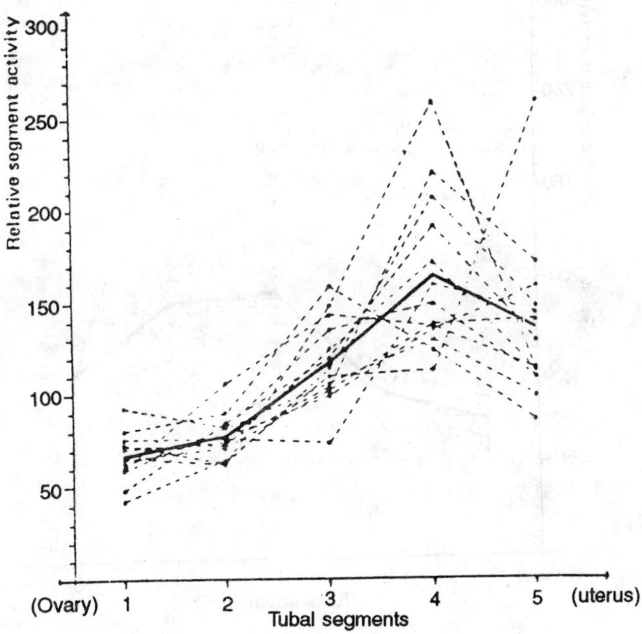

Fig. 7.6. Relative segmental activity at first day of normal pregnancy.

An audiograph of the isthmus taken on the first day after ovulation shows intense, darkening of the film over the whole epithelium. In an autoradiograph of the ampulla there are only view few activity secretory cells and graphs of the results reveal the same picture. The relative segmental activity of segments 4 and 5 of the isthmus is significantly higher than that of segments 1 and 2 of the ampulla.

On the third day after ovulation, i.e., on the last day on which the ova are passing through the tube, autoradiographs demonstrated that secretory activity is definitely drying up.

There is hardly any secretory activity in the isthmus. Secretion has disappeared in the ampulla. In the graphs the relative activities of ampullary and isthmic segments approach each other.

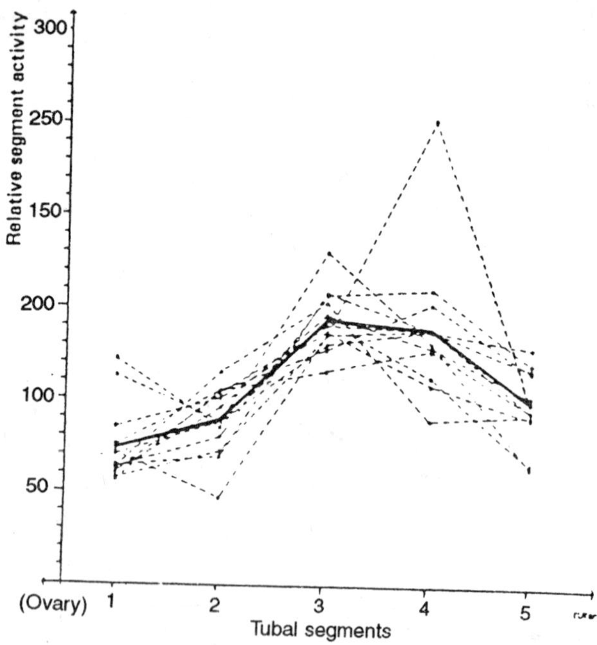

Fig. 7.7. Relative segmental activity at third day of normal pregnancy.

Comparison of the summated curves reveals that there is a definite difference between the first and the third day. There is a difference between the first and third day in that, at the end of ovum transport, the relation secretory activities of the two parts of the tube approach each other. The high density of secretory cells in the isthmus makes the fall of secretory output in the third day after ovulation, indicated by a fairly sharp fall in radioactivity over this section (as described by Bishop and other authors), even more apparent. Thus, because there is a decline of total secretion, the relative activities of each segment approach each other.

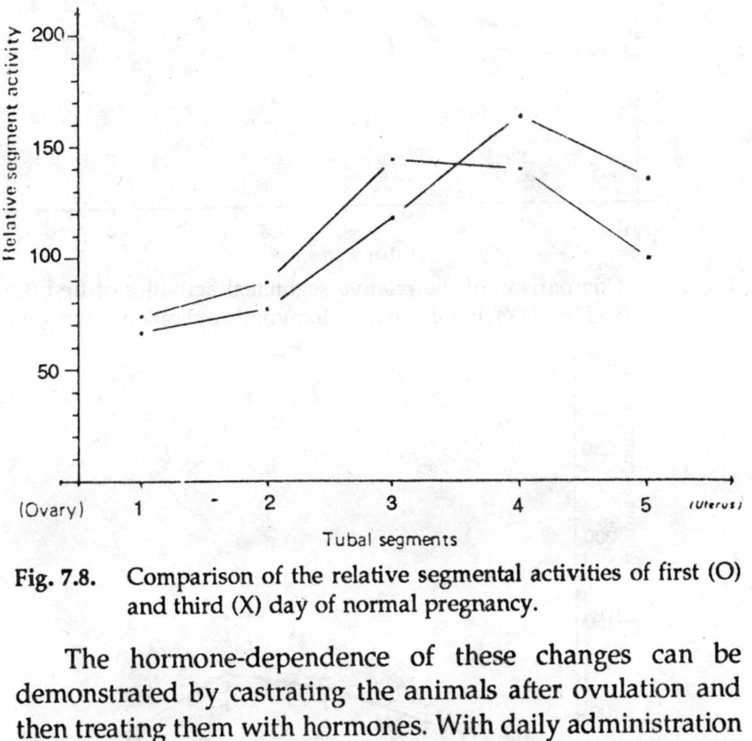

Fig. 7.8. Comparison of the relative segmental activities of first (O) and third (X) day of normal pregnancy.

The hormone-dependence of these changes can be demonstrated by castrating the animals after ovulation and then treating them with hormones. With daily administration of 100 γ 17-β-estradiol secretion is just as powerful even on the third day. Comparison of the summated curves also shows that the type of secretion on the first day is maintained by giving an estrogen.

If 10 mg progesterone is given daily after oophorectomy, an autoradiogram reveals hardly any secretion in the isthmus on the third day. This is manifest in the graphs, in which the curves for ampulla and isthmus approach each other.

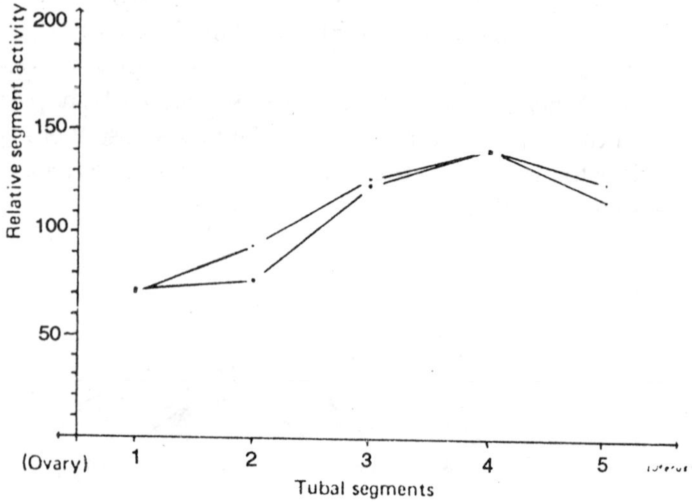

Fig. 7.9. Comparison of the relative segmental activities of first (O) and third (X) day after ovariectomy and application of estrogen.

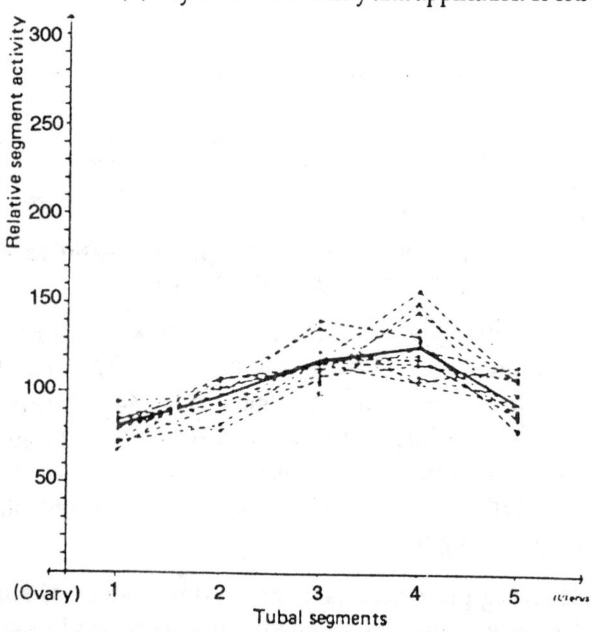

Fig. 7.10. Relative segmental activity at third day after ovariectomy and application of progesterone.

Hormone combinations, such as those used to inhibit ovulation, also elicit this effect. In rabbits which, after oophorectomy, were treated with a combination of 5 γ ethinylestradiol and 50 γ norgestrel (Eugynon, Schering), there was a typical progesterone curve on the third day in spite of the estrogen component of the hormone mixture. Here again secretion dried up and the relative activities of the isthmus and ampulla changed and approached each other.

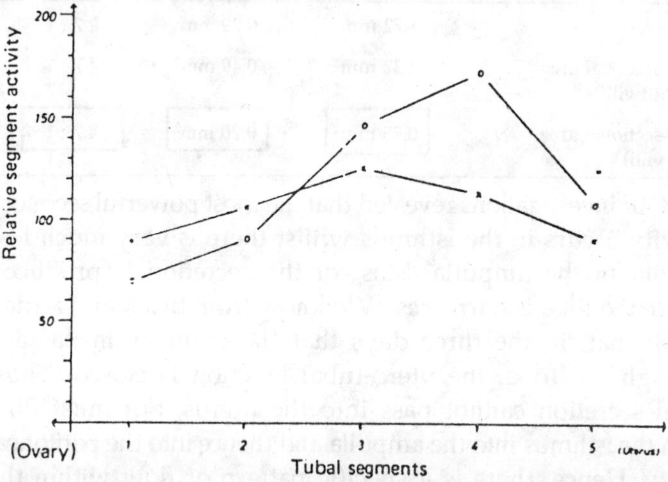

Fig. 7.11. Comparison of the relative segmental activities at third day after ovariectomy and application of ethinyl-estradiol (O) and a compound of ethinyl-estradiol with norgestrel (X).

Now, what is the significance of these observations in respect of the ova which are in the oviduct for these three days? The secretory picture described here may help to explain the as-yet-unexplained problem to the halting of the ovum at the ampullary-isthmical junction. It is possible to explain this if the secretory activity of each segment of the tube is related to its lumen.

The tubal lumen is irregular and narrowed by mucosal villi. Planimetric measurement of the actual tubal lumina of histological sections of the segments bordering on the ampullary-isthmical junction gave a figure of 0.95 mm^2 for the ampulla and 0.20 mm^2 for the isthmus. Thus, the cross-section of the ampulla is 4.7 times greater than that of the isthmus.

TABLE 7.2

Measurements of Tubal Lumen (Rabbit)

	Ampulla	*Isthmus*	*Ratio*
Diameter	1.72 mm	0.79 mm	2.2 : 1
Cross-sectional area (without villi)	2.32 mm^2	0.49 mm^2	4.7 : 1
Cross-sectional area (with villi)	0.95 mm^2	0.20 mm^2	4.7 : 1

Our investigations revealed that the most powerful secretory activity occurs in the isthmus whilst there is very much less activity in the ampulla. Most of the secretion is produced where the tube is narrowest. We know from Black and Asdell (1958) that, in the three days that the ovum is in passage through the tube, the utero-tubal junction is closed. Thus, tubal secretion cannot pass into the uterus, but must flow from the isthmus into the ampulla and thence into the peritoneal cavity. Hence, there is a specific pattern of flow within the tube.

Because of the different diameters of the tubal lumen in the isthmus and ampulla there are different flow-rates for tubal secretion, and this is determined by the continuity-equation

$$\text{const.} = q_1 \cdot v_1 = q_2 \cdot v_2.$$

That is to say, the rate of flow (v) is determined by the cross-section (q) of a tube, and so the same volume of fluid flows faster in a narrow tube than in a wide one.

If the tube was smooth, the rate of flow of secretion produced in the isthmus (isthmus segment 4) would, on the basis of the calculated cross-sections, be reduced to one-fifth after it entered the ampulla.

In collaboration with the Department of Experimental Physics of the Giessen University, we are now attempting to determine the viscosity of tubal secretion. The final values for the various days after ovulation are not available at the moment, but we can say that the results have indicated that the viscosity of tubal fluid is practically the same as that of water.

Although the absolute volume of tubal secretion produced on each of the days is known, the rate cannot definitely calculated, since the tubal lumen is not smooth but is narrowed by mucosal villi and is constantly being changed by contractions. Even though the continuity-equation can only be regarded as an "approximation", it seems that we are justified in using it to explain the flow picture:

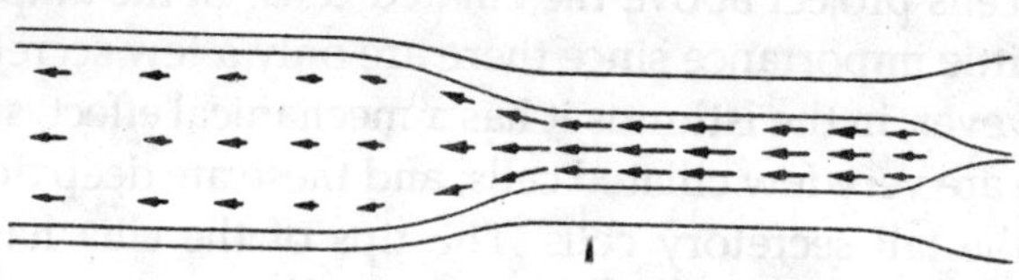

Fig. 7.12. Schema of tubal flow.

Accordingly, tubal secretion flows faster in the narrower isthmus and slower in the wider ampulla, at least in those segments which border on the junction between isthmus and ampulla. If this fact is seen in the light of ciliary movements, which are also an important factor in ovum transport, we have a possible explanation for the irregular rate of ovum migration in each of the segments, and, in particular for the stopping of the ova at the ampullary-isthmical junction.

The epithelium of the ampulla is made up mainly of ciliated cells, with scattered, isolated secretory cells. Here, the beats of the cilia rapidly move the egg cell forwards against the flow of tubal secretion. However, the tubal secretion is moving so slowly in the ampulla that it is overcome by the power of the cilia. At the ampullary-isthmical junction the tubal lumen narrows and there is a sharp increase in the rate of flow. At the same time, there is a sharp fall in the number of ciliated cells as compared with secretory cells. Thus, the power of the cilia driving the ovum forward is less effective against the counter-current of secretion at this point. All this occurs under the predominant influence of estrogens produced during estrus and on the first day after ovulation.

As progesterone activity begins to increase towards the end of the third day of the ovum's passage through the tube, the rate of beating of the ciliated cells increases by 20% and there is a sharp declined in secretion in the isthmus. Both these factors encourage the ovum to enter the isthmus.

This mechanism is backed up by a morphological change of the tubal epithelium. Under the influence of estrogens, the secretory cells project above the ciliated cells. In the ampulla this is of little importance since there are only a few secretory cells. However, in the isthmus it has a mechanical effect, since here there are very few ciliated cells, and these are deep down between the tall secretory cells. The tips of the cilia hardly reach the tubal lumen and thus exert no effect.

In the second phase of ovum transport, the secretory cells drop back under the influence of progesterone, leaving the ciliated cells free. The cilia then reach far into the tubal lumen of the isthmus and they are in a position to drive the ovum forwards.

It can, thus, be confirmed that ovum transport is controlled both by the secretory process in the tube and by the effects of hormone on tubal secretion, and also by a blockade mechanism at the junction between isthmus and ampulla, but not by muscular tubal activity. This does not occur as a result of one of the many suggested closure mechanisms at this point, but is evidently governed by the balance between ciliary beats and the counter-flow of secretion, the rate of which is determined by the different widths of the tubal lumen and by production of secretion in the isthmus, which is controlled hormonally.

This mechanism which regulates migration of the ovum is dependent on closure of the ultro-tubal junction which is determined by a specific estrogen-progesterone ratio. Obviously, it can work when the utero-tubal junction is closed. The rapidly changing ratio of the two hormones after ovulation appears to be the fundamental factor in controlling migration of the ovum, since ovum transport through the tube occurs normally only during the first three to four days after ovulation, during which progesterone production rises sharply.

Thus, migration of the ovum is hormonally controlled, and hence can be disrupted hormonally. The results of many animal experiments have shown that ovum transport can be influenced by hormones. Both oestrogens and also progestogens are capable of delaying or accelerating ovum transport, depending on the species, dose, time and duration of administration. Many of the observations appear to be contradictory, particularly if muscular activity of the tube is assumed to be the main mechanism responsible for transport. However, if we consider the secretory picture within the tube, it provides a possible explanation for some of the observations reported in the literature. For example, Greenwald (1963) was able to produce retention of the ova for many days at the ampullary-isthmical junction by giving rabbits a single high dose of a depot-estrogen (250 γ estradiol-cyclopentyl-propionate). The same thing happened with daily injections of low doses of estrogen (10 γ estradiol benzoate). However, after a single dose of only 25 γ of a depot-estrogen, the opposite result was obtained. In rats and rabbits, ovum transport was accelerated and many ova appeared in the uterus prematurely (i.e., within 48 hours) and were discharged into the vagina. Similar observations were made in guinea pig by Deanesly (1963), Chang and Yanagamachi (1965) and in rabbits and hamsters by Noyes, Adams and Walton (1959). In rabbits, with high doses of estrogen they were able to cause retention of ova in the tube, while low doses accelerated the passage of the ova and hence their entry into the uterus. Using similar doses of hormone, Harper (1961) was able to achieve the same effect with radioactively-labelled glass beads of the size of ova. This qualitatively-different, dose-related effect had already been described by many authors.

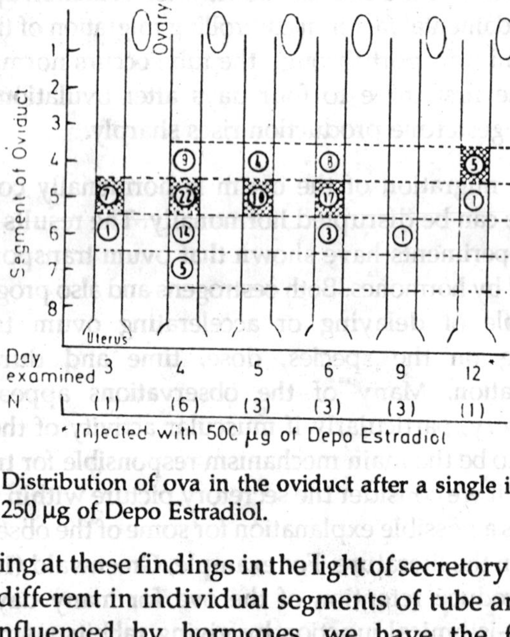

Fig. 7.13. Distribution of ova in the oviduct after a single injection of 250 µg of Depo Estradiol.

Looking at these findings in the light of secretory function, which is different in individual segments of tube and which can be influenced by hormones, we have the following explanation: with high doses of estrogen, there is a high output of secretion in the isthmus, as compared with the ampulla, and the utero-tubal junction remains closed, so that ova cannot move into the isthmus. Whilst the hormone is exerting its effect, ova are retained in the middle of the tube by the consistently high level of secretion and the accompanying morphological changes of the epithelial cells.

With a single low dose of estrogen, the estrogen-progesterone ration which is necessary for closure of the utero-tubal junction is evidently disrupted, and the ova can enter the uterus prematurely through the open utero-tubal junction. Premature opening of the tube to the uterus also alters the direction of flow of the secretion, so that ova are not retained at the junction of the ampulla and isthmus.

Reports in the literature of changes of ovum transport after treatment with progesterone also accord with a functional mechanism mediated by tubal secretion. By administering progesterone or other progestational compounds, it is possible to prevent ova from being arrested at the ampullary-isthmical junction, so that the ova enter the isthmus prematurely, although often their entry into the uterus is only a little premature. This has also been observed in the cow and in sheep (Robinson, 1951). In oophorectomized estrogen-treated rabbits, it is again possible to prevent arrest of the ova in the middle of the tube by concurrent administration of progesterone.

These observations can be explained by the reduction of secretory activity, particularly of the isthmus, under the influence of progesterone, since, when this happens, the cilia begin to exert a more powerful effect on the counter-current of secretion.

Oophorectomy will also prevent arrest of ova at the ampullary-isthmical junction, because of the resultant sharp reduction of secretion, and the migration of the ova will not be delayed very much even though tubal motility is disturbed; however, the ova will not develop properly in the uterus because of the absence of progesterone. This degeneration can be prevented by giving corpus luteum extract or progesterone. Passage of the ovum can be disrupted by giving natural sex hormones and also by synthetic steroid hormones used for inhibition of ovulation.

Non-steroid estrogens, prostaglandins and also compounds such as, for example, reserpine, chlorpromazine and tetrabenazine, can have an effect on ovum transport. They either cause acceleration of the ova in the tube or retention ("tube locking"), usually at the junction of ampulla and isthmus.

These findings are of practical significance. Ova in the

tube can develop further only if they pass out of the tube into the uterus at a specific point in time (specific for each species of animal). If the ova pass into the uterus a few hours too early, they may perish there because the endometrium will not be properly prepared for them. On the other hand, if the ova are delayed in the tube for more than one-and-a-half days, the already developing ova will degenerate and will not implant in the uterus.

In view of the complicated nature of ovum transport and the hormonal regulation of this process, it is astonishing that the mechanism usually works without any trouble. We certainly have here one of the intangible causes of female infertility. But above all, the mode of action of hormonal contraceptives which do not inhibit ovulation, for which there is as yet no satisfactory explanation, may well be sought in the effect they have on the interaction between muscular activity, sphincter system, ciliary current and secretion of the oviduct, and in disrupting thereby the complicated mechanism and synchronization of ovum transport.

8

Expulsion of the Oocyte, Embryo, or Fetus

The oocyte, the egg with the developing embryo in it, or the fetus must be expelled for the maternal reproductive tract. (In animals with external fertilization it is the oocyte, and in oviparous animals with internal fertilization it is the egg with the developing embryo or the fetus.) In pipefishes and seahorses, the embryo develops in the brood pouch of the male and thus the embryos have to be expelled from the male's pouch. In this chapter, we will discuss mainly the endocrine control of oviposition or parturition, that is, giving birth. We will not discuss the physiology of the muscles that expel the oocyte, egg, or fetus.

Cyclostomes

In this group male and female spawn with their genital openings in proximity. This means that behavioural and spawning mechanisms are closely synchronized, but this does not mean that the same hormones are necessarily involved in both processes. The endocrine mechanisms involved in spawning are not understood.

Elasmobranchs

La Pointe (1977) has recently reviewed the literature on the mechanisms that control contractions of the oviduct of elasmobranches and has also reported the results of some new studies of his own. In general, the isolated oviducts of the skate (*Raja erinacea*), the spiny dogfish (*Squalus acanthias*), the smooth dogfish (*Mustelus canis*), the blue shark (*Prionace glauca*), the dogfish (*Scyliorhinus canicula*), and the smooth hound (*Mustelus californicus*) respond to low doses of acetylcholine with clearly discernible contractions, but they do not respond in the same way to various posterior pituitary hormones such as oxytocin, arginine vasotocin, aspartocin, valitocin, glumitocin, and elasmobranch neural lobe extracts. The neural lobes of sharks have been found to have aspartocin and valitocin and those of rays have glumitocin. The postpartum oviduct of the leopard shark (*Triakis semifasciata*) responded to aspartocin, valitocin, oxytocin, vasotocin, and glumitocin, but the immature and the gravid oviduct did not. Epinephrine in low doses caused contractions of isolated oviducts of the skate, *Raja ernacea*, the smooth dogfish, the spiny dogfish and the blue shark. Dodd (1975) has speculated that there may be "a neural or hormonal pathway between oviducts and either the hypothalamus or the ovaries or both which controls ovulation and oviposition." This hypothesis is based on observations rather than experimental evidence and thus needs further investigation.

Teleosts

In oviparous teleosts, ovulation and subsequent oviposition are operationally similar to parturition in aplacental viviparous teleosts. Aṣ stated already ovulation in some oviparous teleosts such as the rainbow trout (*Salmo gairdneri*) and the Indian catfish (*Heteropneustes fossilis*), is regulated by catecholamines, progestins, corticosteroids, and prostaglandins.

In the aplacental viviparous guppy (*Poecilia reticulata*), Injection of 20 ng/g deoxycortisol induced premature parturition in 1w-2 days in 2 to 7 fish. Deoxycortisol also tended to reduce the latency for parturition induced by the neurohypophyseal hormones, oxytocin and vasotocin, and by an extract of carp neurohypophyses. The deoxycortisol may have promoted the breakdown of the follicular membrane, as it does in vitro, and so would make the young more susceptible to expulsion by contractions of smooth muscle induced by the neurohypophyseal hormones. These hormones are implicated in the expulsion of the egg in birds and the fetus in mammals; in teleosts, they cause the spawning reflex in the killifish *Fundulus heteroclitus*, oviposition in the mekada (*Oryzias latipes*), movements related to expulsion of the young from the brood pouch of the (male) seahorse (*Hippocampus* sp.), and premature parturition in *Gambusia* sp; and the guppy. If corticosteroids secreted by the embryo were to affect the breakdown of the follicular wall and were to induce increased sensitivity of this structure of maternal posterior pituitary, then a mechanism would exist for the young to control its own destiny.

Amphibians

Investigations of the effect of neurohypophyseal hormones on isolated oviducts in amphibians have revealed that the oviducts of some species respond by contraction to arginine vasotocin (an amphibian pituitary hormone), oxytocin, lysine vasopressin, isotocin, mesotocin (amphibian posterior pituitary hormone) and arginine vasopressin. Examples of such species are bullfrog (*Rana catesbeiana*), the toads *Bufo bufo* and *B. marinus*, the South African clawed toad (*Xenopus laevis*), the mud puppy (*Necturus maculosus*), and the salamanders *Triturus cristatus*, *Salamandra salamandra*, and *S. atra*. Diakow (1977, 1978) has made the interesting observation that the gravid, but not yet sexually receptive female frog (*Rana pipiens*) give a release call in response to attempts by the male to embrace her before she is ready to spawn. This release calling in inhibited by distension of the female with water and by injection of arginine vasotocin, which acts as an antidiuretic hormone and causes retention of water, which is necessary for reproduction in frogs with external fertilization.

Reptiles

In the lizards *Zootoca vivipara* and *Sceloporus cyanogenys*, removal of the ovary or of the CL is followed by abnormal parturition, whereas in the snakes *Natrix* sp. and *Thamnophis* sp. ovariectomy late in gestation does not affect parturition. It thus appears that progesterone or other ovarian hormones affect parturition. Unfortunately, no data seem to be available on the changes in hormone concentrations in reptiles before, during, and shortly after parturition.

Injections of neurohypophyseal extracts induce premature expulsion of the egg or the embryo in the lizards *Lacerta vivipara, Sceloporus undulatus, Uta stansburiana,* and *Zootoca vivipara* and in the snakes *Natrix sp., Thamnophis* sp. and *Storeria* sp.

Hypophysectomy, causes abnormal parturition in the water snake *Natrix sipedon* and the garter snake *Thamnophis sirtalis.* Although results in a decrease in the cellular size of the CL it seems that this effect of is independent of the CL since CL removal does not affect parturition.

Birds

Much of the information about oviposition in birds has been obtained with the domestic fowl. The chicken is well suited for such investigations because the time expected oviposition can be accurately predicted and because hypophysectomy and neurohypophysectomy can be performed by a skilled experimenter without affecting the survival of the animal. Furthermore, serial blood samples can be obtained conveniently and in fairly large numbers from the hen.

It has already been pointed out that except for the last oviposition of a sequence, there was a close temporal relationship between oviposition of one egg and the ovulation of the next sequence. This phenomenon led to the discovery by Fraps (1942) that LH-induced premature ovulation was accompanied by premature expulsion of an oviductal egg. Other hormones have been implicated in the control of ovulation, i.e., posterior pituitary hormones, hormones from the ruptured follicle, and steroid hormones. The plasma vasotocin concentration increases more than 40-fold immediately prior to and during oviposition. Injection of avian posterior pituitary hormones, i.e., arginine vasotocin, and oxytocin can induce premature expulsion of the egg from the shell gland. Electric stimulation of the hen's preoptic area of the brain results in premature oviposition. Presumably, this stimulation results in the secretion of the hypothalamic hormones that are stored in posterior pituitary. It should be noted, however, that neurohypophysectomy affects neither the pattern nor the timing of oviposition.

Rothchild and Fraps (1944) observed that removal of the recently ruptured follicle resulted in a delay of oviposition, sometimes for several days. In 1978 Gilbert et al., showed that removal of the gramulosa cells from the ruptured follicle was as effective as removal of the follicle itself. There are two schools of thought about the nature of the hormone secreted by the recently ruptured follicle, which is a rapidly regressing structure. On the one hand, Tanaka and Goto (1976) found that the active fraction of extracts from recently ruptured follicles was a small peptide with disulfide bridge, which thus resembled the posterior pituitary hormones. On the other hand, evidence by other investigators strongly suggests that prostaglandins are the hormones in the ruptured follicle that cause expulsion of the egg. First, plasma concentrations of prostaglandin E(PGE), are higher when there is an egg in the shell gland than shortly after oviposition. Second, shortly before oviposition the PGF concentration in the ruptured follicle rises sharply. Third, administration of indomethacin, which blocks the synthesis of PG, delays oviposition in the chicken, and this effect can be overcome by injection of PGE into the shell gland. Indomethacin also inhibits the normal preoviposition increase in electrical activity of the uterine muscle. Finally, PGE induces premature oviposition in chickens and Japanese quail. Prostaglandin $F_{2\_}$ ($PGF_{2\_}$) injections into laying hens resulted in increased uterine contractions, as measured by the electrical activity of the uterus. Injection of $PGF_{2\_}$, 2 h prior to the expected time of oviposition was followed by premature oviposition (Shimada and Asai, 1979). The evidence is thus quite persuasive that prostaglandins are involved in oviposition, but this does not mean that the recently ruptured follicle is the exclusive source of prostaglandins. Recently, Ogasawara and Koga (1978) have presented evidence that prostaglandins are also produced in the shell gland and that these prostaglandins affect oviposition.

Among the steroid hormones, exogenous progesterone (P_4) delayed oviposition between 0 h and 6 h after ovulation, sq that oviposition followed p_4 injection by about 31h, an interval similar to the interval between the preovulatory P_4 peak and oviposition. Incidentally, this exogenous P_4 also delayed ovulation, so that the original C_3 follicle became the C_1 follicle of a new sequence. When P_4 was administered 6-15 h after ovulation, it also delayed the C_3 ovulation, so that it became the C_1 ovulation of a new sequence, and the P_4 also delayed oviposition by 10 h, so that ovulation occurred at about the same time as the delayed oviposition. Injection of P_2 just prior to the occurrence of the preovulatory P_4 peak (9-12 h prior to ovulation) resulted in premature ovulation and oviposition of the oviductal egg, so that ovulation and oviposition occurred rarely simultaneously. The results obtained with injection of P_4 between 6 h and 15 h after ovulation, and between 9 h and 12 h prior to ovulation provide an explanation of the finding of Fraps (1942) that LH injection induced oviposition. If we assume that the LH-induced P_4 secretion and that the bursting of the follicle resulted in PG secretion, then P_4 and PG would be present in concentration similar to those found at normal oviposition. No data appear to available about the concentration of posterior pituitary hormones around the time of LH-induced ovipositions.

The sharp peak in plasma corticoserone concentration at the time of oviposition is accompanied by a number of physiological and behavioural changes. Hens that are held in floor pens and that are provided with nests will make egg-laying calls, examine nests, and assume the egg-laying posture. Hens that are kept in cages will show restlessness and so-called panic behaviour and may give egg-laying calls. These behavioural changes are accompanied by an increase in body temperature and by uterine contractions prior to oviposition. Although normally associated with oviposition, the changes in behaviour, body temperature, and corticosterone concentration can be disassociated from oviposition experimentally. After removal of the recently ruptured follicle, oviposition is delayed, but laying behaviour, rise in body temperature, and increase in corticosterone occur at the time of the normally expected oviposition, yet fail to occur when the egg is eventually laid. When oviposition is advanced by the injection of vasopressin, mere is an increase in corticosterone that is probably the result of the vasopressin itself, and the corticosterone peak may or may not be the result of the expulsion of the egg. The important point however, is that at the time that the egg would normally have been laid, the hen shows normal egg-laying behaviour and an increase in corticosterone concentration. It appears that the laying behaviour corticosterone peak, and temperature increase are dependent on the presence of the recently ruptured follicle and that the actual oviposition contributes only slightly to the increase in corticosterone and body temperature. Similarly, contractions of the uterus occur at the time of expected oviposition 26-28 h after ovulation, even when there is no egg in the oviduct, as for example when the yolk falls in the body cavity and is resorbed, instead of being engulfed by infundibulum of the oviduct.

Although not concerned directly with oviposition, uterine contractions, as determined by measurements of electrical activity, occur not only at the time of oviposition, but also when the egg arrives in the uterus and at about 1 h before ovulation of the C_1 follicle, at a time when there is no egg in the oviduct. The C_1 follicle apparently produces a hormone, which acts either directly on the uterine masculature or acts indirectly though the neurohypophyseal hormones. Shimada *et al.* (1981), in a series of elegant experiments, in which they ligated the stalk of the C_1 follicle at various times before ovulation, demonstrated that this hormone was released at 1.5 h to 2 h before ovulation. The chemical nature of the hormone remains to be determined.

In the hen, oviposition appears to the independent of the adrenergic and cholinergic nervous system; neither denervation nor specific blocking agents affect the timing or the process of oviposition in a significant manner.

In birds, all embryos are in about the same stage of development at the time of oviposition, because all eggs are in the oviduct for about the same length of time after fertilisation in the infundibulum of the oviduct. The initiation of oviposition is probably mainly controlled by the hormones produced by the ruptured follicle. Is mammals, the initiation of parturition is, at least in some species controlled by the secretion of hormones by the fetus, thus ensuring that the young are born at about the same stage of development for the particular species.

Mammals

Various external factors and the genotype are known to affect the duration of gestation in mammals, and thus to affect the approximate timing of the onset of parturition. On of the external factors that has an effect on parturition is the time in the breeding season when insemination occurs. Sheep bred early in the breeding season, for example, have a longer period of gestation than sheep bred late in the season. Horses bred between September and December have an average gestation period of 328.0 d, but horses bred between March and June have a gestation period of 352.2 d, the difference being significant. In cattle, which unlike sheep and horses, do not have a breeding season, the effect of month of insemination on duration of gestation is not significant. The hour of the day is important in horses and alpacas; 86 per cent of foals are born between 1900 and 0700, with the highest frequency of births occurring between 2200 and 2300; in alpacas, 79 per cent of the deliveries occur between 0700 and 1100 h. George (1969) found that during the winter, 63 per cent of Dorset horn ewes lambed between 0800 and 1800, but that only 37 per cent of Merion ewes lambed during that period; during the summer, 23 per cent of Dorset horn and 39 per cent of Merion ewes lambed during those hours. Thus both season are breed affected the distribution of the time of parturition during the day. A less pronounced, but definite effect of time of day on the distribution of births is reported for pigs, mice, Chines hamsters, and humans, but not for cattle.

The effect of the genotype of the fetus on gestation length can be illustrated most dramatically by the genotype of horse-donkey crosses. The foal (stallion X mare) is carried 340 days, the hinney foal (stallion X jennet) 345 days, the mule foal (jack X mare) 355 days, and the donkey foal (jack X jennet) 365 days. Less dramatic, but nevertheless significant, is the effect of certain bulls on the length of time their offspring are carried. In dairy cattle, the heritability of the effect of the bull on gestation length as a characteristic of the fetus is estimated to be 0.42; for horses the corresponding figure is 0.36 for swine 0.30, and for sheep, in which the fetal and maternal effects were not separated, 0.40 to 0.50.

The sex of the fetus can affect the length of gestation to some extent. For example, the duration of gestation is 1.97 days longer for bull calves than for heifer calves, and male foals are carried longer than mare foals. In sheep, however, the fetal sex has no apparent effect on the length of gestation.

In mice, the size of the litter is negatively correlated with the duration of gestation. In experiments, the weight of individual offspring was increased by crossing inbred lines of mice, suggesting that the total fetal mass (total weight of the litter) was negatively correlated with the length of the gestation period; however, the weight of the placenta played no important part in determining the length of gestation, although the placental weight, of course, contributed to the complete contents of the uterus. Similar correlations between fetal mass and the length of the gestation period have been found in rabbits, guinea pigs, cattle, goats, sheep, mink, and humans, but apparently not in pigs.

Before parturition, the fetus undergoes changes that aid its survival after birth. Surfactants appear in the alveoli of the lungs, which aid in the maintenance of the air-distended state of the lungs after birth. Glycogen is deposited in the liver, where it acts as readily available reserve to carbohydrate, and in the heart muscle, where it serves as a source of energy during hypoxia, which may occur during parturition. Finally, while and brown fat are deposited, the brown fat being particularly important for thermogenesis.

The mother also undergoes several changes prior to parturition, including dilation of the cervix and changes in the pelvis. Fitzpatrick (1977) concluded that in sheep, prostaglandins probably are required for the activation of enzymes that degrade the collagen in the cervix. According to Naaktgeboren and Slijper (1970), it is relaxin that causes the softening of the cervix in pigs, cattle, rats, monkey and humans. However, Fitzpatrick (1977) has stated that studies done with respect to relaxin and cervical dilation are "inadequate and should be repeated in relation to known hormonal changes." Recently, Hollingworth *et al.* (1979) reported that in pregnant rats the extensibility of the cervix increased sharply at day 22. Pregnant rats, ovariectomized on day 16 and treated with estradiol benzoate (EB) and P_4 to maintain pregnancy, showed little increase in cervical extensibility; however, a combined treatment of EB, P, and relaxin resulted in an increase in extensibility to about 50 per cent of the value found in day 22 in normal pregnant rats.

It is also possible that prostaglandins (PG) cause release in relaxin. In pigs, there is a peak in the plasma relaxin concentration at delivery. This peak occurs prematurely if labour is induced prematurely by PG infusion, and it is delayed if delivery is delayed by indomethacin treatment.

The symphysis pubis in several species (e.g., the mole, some bats, and the Pinnepedia), consists of loose connective tissue. In some other species, it consists of cartilage, which near parturition, under the influence of relaxin, is changed into a soft connective tissue that permits the passage of large young, as, for instance, in guinea pigs. Even in some animals with a calcified symphysis pubis, this structure becomes soft, as in the pocket gophers (*Geomys bursarius* and *Thomomys bottae*), the mole rat (*Spalax leucodon*), and mongooses. The onset of parturition is under different control mechanisms depending upon whether the CL is required throughout gestation or whether the placenta is producing its own gestational hormones. Much of the initial break-through in the understanding of the regulation of the initiation of parturition has come from pathological conditions. In the CL-dependent cattle, there are some breeds, e.g., Friesians, in which an autosomal recessive gene when present in the homozygous condition in the calf, prevents the normal delivery of that calf, and at the expected time of parturition, the dam does not show the relaxation of the pelvic ligaments, the softening of the cervix, and the decrease in plasma progesterone concentration that normally occur at the time of parturition. Such calves have an adrenal insufficient. In Guernseys, a homozygous recessive gene, when present in the calf, results in a failure to grow after the seventh month of gestation and a failure in the onset of parturition at the normally expected time; such calves lack an anterior pituitary and, as a result, also have adrenal insufficient. Among Angora goats, some does habitually abort. This abnormality was traced to hyperplasia of the adrenal, suggesting a role of the fetal adrenal in initiation of parturition. In sheep, a species in which ovariectomy does not lead to abortion, the eating of skunk cabbage (*Veratrum californicum*) between day 1 and day 15 of gestation resulted in abnormal fetuses, characterized by cyclopia, fetal adrenal and thyroid hypoplasia, and fetal adrenals that in vitro produced more testosterone, but less cortisol, cortisone, and corticosterone than normal fetal adrenals. In South Africa, the eating by pregnant ewes of the shrub, *Salsola tuberculata* Schinz var. *tomentosa*, during the last 50 days of the normal gestation period resulted in gestations that were as long as 213 d instead of the normal 140 d, lambs that were up to three times the normal size and that showed pituitary and adrenal atrophy. This information on cattle, goats, and sheep suggested that the fetal pituitary and adrenal might have a key function in the initiation of parturition. Such a key role for the adrenal is attractive also because it is known that corticosteroids induce the formation of lung surfactants, which are necessary for adequate functioning of the lungs immediately after the umbilical cord is severed. As the result of the considerable skill of animal surgeons, it became possible to test this hypothesis by ~~AP~~ performing and adrenalectomy on fetuses in utero.

For the sake of convenience, we will discuss in the account that follows the regulation of parturition in species in which the CL is required of the entire or nearly the entire period of gestation, i.e., CL-dependent species, and species in which the CL can be removed relatively early during gestation without affecting the duration of gestation significantly, i.e., CL-independent species.

Initiation of Parturition in CL-Dependent Species

We will discuss as examples goats, cattle, swine, rabbits and rats.

Goats. About 48 h before the onset of parturition, the maternal plasma progesterone (P_4) concentration starts to decrease sharply and maternal plasma estradiol-17β starts to increase. The normal gestation period is about 148 d, and ovariectomy leads to immediate abortion. These data suggested that the decrease in P_4 concentration might cause the onset of uterine contractions. Such an effect might act through P_4 inhibiting oxytocin release or by inhibition of myometrical contractions by high P_4 concentrations. Roberts (1971) obtained evidence that P_4 inhibited oxytocin release by acting on the nervous system; however, Blank and DeBias (1977) have stated that progesterone infusion during pregnancy does not always inhibit oxytocin release.

The signal for regression of the CL and thus for diminished P_4 secretion is probably of fetal origin. The evidence to support this hypothesis comes from several sources. First, fetal ~~AP~~ prolongs gestation, and the kids, when delivered, have small adrenals. Second, fetal plasma corticosteroid concentrations increase from near zero at 15 d before parturition to about 40 ng/ml at 5 d before parturition and then further to about 300 ng/ml at parturition. Moreover, infusion of ACTH in physiological amounts (Nathanielsz, 1978) results in a sharp drop in maternal plasma P_4 concentration, an increase in the prostaglandin F (PGF) concentration, and premature delivery of the kids. Finally, administration of corticosteroids to the fetus induced premature parturition. There is no evidence, however, that the ACTH concentration in fetal plasma increases before parturition or before fetal plasma cortisol concentration increases. There and other consideration led Nathanielsz (1978) to conclude that there is no proof that the signal for parturition is of fetal pituitary origin. However, much of the evidence justifies considering the fetal AP-adrenal axis as the initiator of parturition as a working hypothesis. High fetal plasma corticosteroid concentrations stimulate placental enzymes, which are needed for the synthesis of estrogens from pregnenolone. The estrogens probably serve two functions: (1) they promote the dissolution of the feto-maternal connections, and (2) they result in luteolysis and thus lower P_4 secretion. This may be a direct effect of the estrogen on the CL, or the estrogens may act by increasing the synthesis and secretion of PGF. Evidence that fetal corticosteroids are implicated in estrogen production is provided by the findings (1) that delivery of ~~AP~~ kids was not preceded by an increase in maternal plasma estrogen concentrations, and (2) that after administration of dexamethasone (a synthetic corticosteroid), the maternal plasma concentration decreased and that of estrogen increased, as they did during spontaneous and during premature delivery of kids. The role of estrogens in parturition is indicated by the results of estrogen injection into pregnant goats. Injection of estrogen induced premature delivery of kids, which did not survive, presumably because corticosteroid secretion was not sufficient for secretion of surfactants in the fetal lungs.

Infusion of PGF_{2} into the tributaries of the uterine vein ipsilateral to the CL causes luteolysis, the PGF concentrations rise about 24 h prior to parturition at the time that luteolysis starts. The PGF_{2} may thus also serve two purposes: (1) it may induce luteolysis and thus withdrawal of progesterone from the myometrium, and (2) it may induce contractions of the myometrium. The importance of the role of P_4 is shown by the fact that progesterone administration can prevent ACTH-induced parturition.

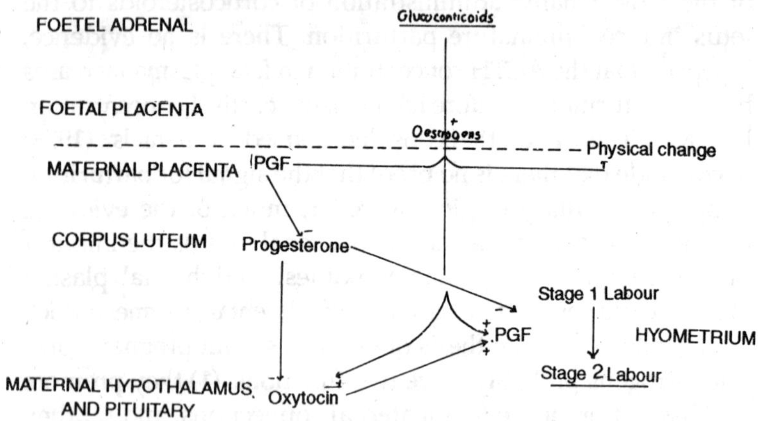

Fig. 8.1. A model for initiation of parturition in the goat.

Cattle. The plasma P_4 concentration in dairy cattle decreases about 30-40 h before the onset of parturition at the end of the gestation period of about 282 d. Before day 215 ovariectomy generally results in abortion; after day 215 it generally does not result in abortion, perhaps because there may be enough P_4 secreted by the maternal adrenals or release from fat stores to maintain pregnancy.

Maternal plasma estradiol and estrone concentrations are 3- to 10-fold higher at the time of parturition than at 26 d before parturition. Immediately before delivery, the concentration of plasma estrogens in the umbilical vein rises sharply; this has been interpreted as rapid degradation of estrogens by the fetus. After day 270 of gestation, there is a definite increase in fetal plasma corticosteroid concentration and an increase in the ratio of fetal cortisol to fetal corticosterone.

Parturition can be induced prematurely by four methods: First, it can be induced by fetal infusion of either ACTH, to stimulate fetal corticosteroid secretion, or of glucocorticosteroids, such as the naturally occurring cortisol or the synthetic dexamethasone. Such experiments suggest that, as in goats, the fetal corticosteroid induce parturition. Second, it can be induced by administering synthetic glucocorticosteroids, dexamethasone, flumethasone, and triamcinolone acetonide to the dam. Glucocorticosteroids administered to cows do not induce expulsion of the fetus if it has died; in vivo they do not inhibit P_4 synthesis by CL. These observations suggest that glucocorticosteroids do not induce parturition mainly as the result of their luteolytic effect. Third, parturition can be induced by the administration of $PGF_{2\alpha}$, and finally, by administration of estrogen, especially of diethylstilbestrol. This evidence, although not as convincing as that obtained with goats, suggests that the fetal pituitary, by its stimulation of fetal corticosteroid production, induces parturition in a manner similar to that suggested for the goat. Experiments by Hoffmann *et al.* (1979), however, have shown that the fetal cotyledons are the main source of estrogens and that estrogen production is not affected by the presence of a fetus, the precursors for estrogen production coming from the mother and not the fetus. These findings suggest that neither fetal glucocorticosteroids nor fetal precursors for estrogen production are the indirect luteolytes agents.

The mechanisms for luteolysis at the end of gestation in cattle are not well enough understood to support the idea that parturition in goats and cattle is mediated by the same sequence of events.

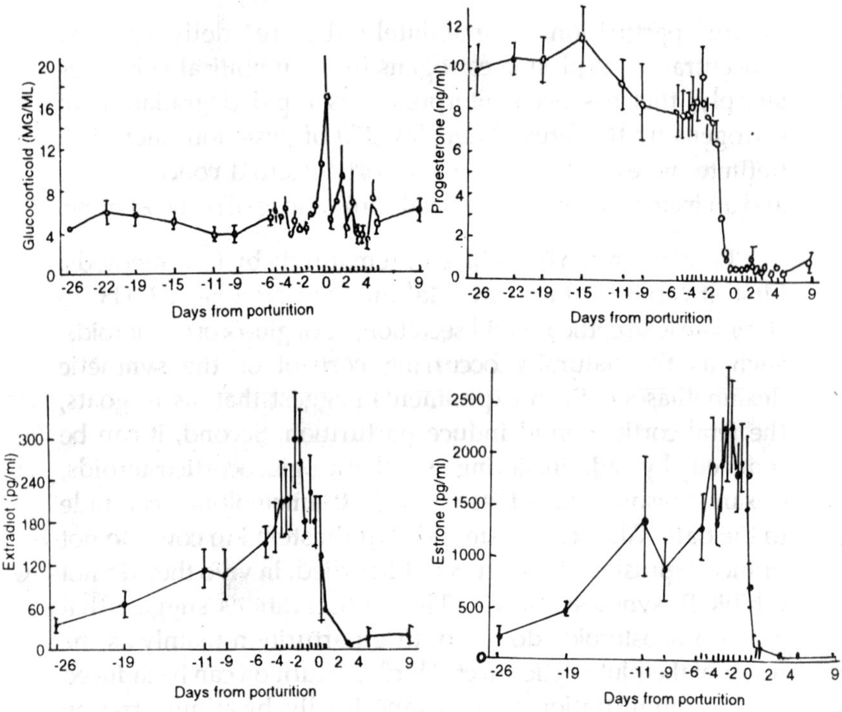

Fig. 8.2. Serum glucocorticoid serum progestin (expressed as progesterone), serum estradiol, and serum estrone in 10 cows from 26 days before to 9 days after parturition.

Swine. The gestation period for swine is about 114 d; parturition is preceded by a sharp increase of fetal plasma corticosteroid concentration about 4 d before parturition, but maternal plasma corticosteroids do not increase significantly until the day of parturition. Maternal plasma estrogen concentrations, specifically those of estradiol and estrone, start to increase at about the same time that the fetal glucocorticosteroids start to increase. These estrogens are produced by the placenta and not by the material ovaries and adrenals or fetal ovaries and adrenals. At about 24 h before the onset of parturition, the maternal plasma P_4 concentration starts to decrease sharply, possibly as the result of an initial rise in PGF_{2a}. At the time of delivery, there is a sharp rise in PGF concentrations and in the PGF metabolite, 13,14-dihydro-15-oxo prostaglandin in the amniotic fluid, and in maternal plasma PGF_{2a} metabolite concentrations. The maternal plasma PGF_{2a} metabolite concentrations. The maternal blasma relaxin concentration rises just before delivery and remains high during spontaneous delivery and during PGF_{2a} -induced delivery. Kertiles and Anderson (1979) found that injections of porcine relaxin from day 105 to day 110 of gestation caused premature cervical dilation in gilts. Just before delivery, maternal plasma corticosteroids concentrations increase sharply. Hypophysectomy and ovariectomy at any time during gestation are followed by abortion. The evidence thus suggests that the decrease in maternal P_4 secretion may control the onset of parturition, and that, as in goats and cattle, a signal from the fetus may control the maternal P_4 secretion. This idea of fetal control of the onset of parturition in swine is supported by the several sources of evidence. Hypophysectomy of 50 per cent or more of the fetuses in the litter prevents the decline of the CL and of maternal plasma P_4 concentration and delays the onset of parturition; decapitation of fetuses results in prolonged gestation. Administration of ACTH to fetal pigs induces premature parturition as does the administration of dexamethasone, which is followed by a rapid decline in maternal plasma P_4 concentration. Administration of dexamethasone to mothers does not result in premature parturition when newly induced CL are present, suggesting that dexamethasone is itself not luteolytic. When PG synthesis is inhibited by administration of indomethacin, however, dexamethasone does not induce premature parturition and, of course, plasma PG concentrations do not increase. In the absence of indomethacin treatment, dexamethasone administration results in an increase in maternal plasma PGF_{2a} metabolites and in induction of parturition. These results suggest that dexamethasone induces parturition by stimulating the production of PGF_{2a}. It appears that in pigs, the fetal pituitary regulates the onset of parturition through adrenal glucocorticosteroids production, which, in

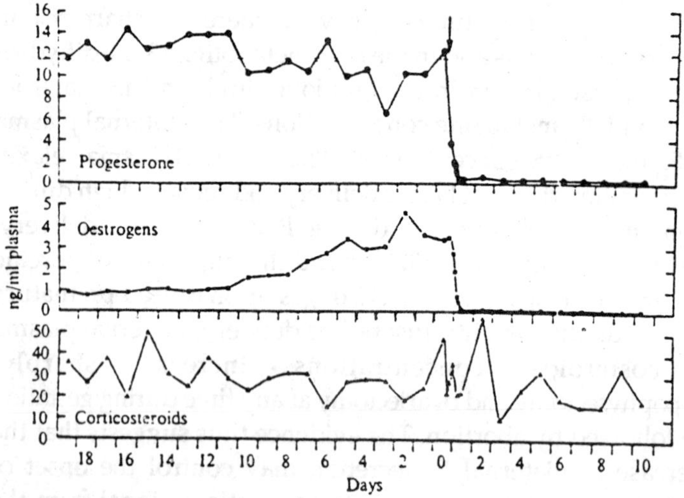

Fig. 8.3. Plasma concentrations of progesterone, total unconjugated estrogens, and corticosteroids in sow (5th pregnancy) during late pregnancy, parturition (14 piglets), and early lactation. Day 0 and the arrow denote the day of parturition.

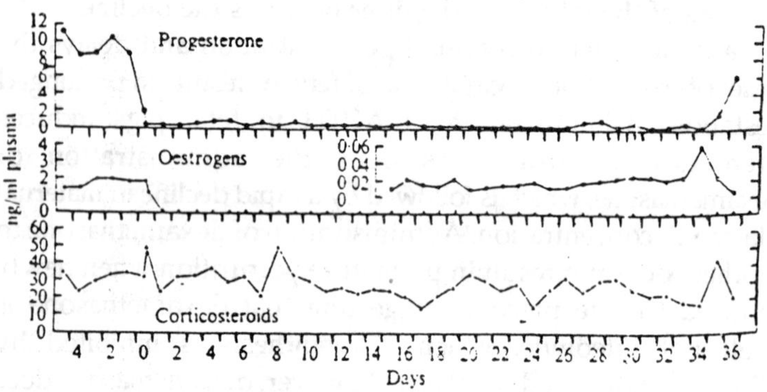

Fig. 8.4. Plasma concentrations of progesterone, total unconjugated estrogens, and corticosteroids in sow (5th pregnancy) during late pregnancy, parturition (11 piglets), and lactation and after weaning. Parturition occurred on day 0; weaning was completed by day 31, and behavioural estrus was observed on day 34.

turn, increases PG production in the uterus, with the PG causing luteolysis and a decline in P_4 production; the PGs possibly also cause uterine contractions of the now estrogen-dominated uterus. The available evidence suggests that in swine, a change in the fetal pituitary-adrenal axis may initiate the sequence of events that results in parturition.

Rabbits. Rabbits have a gestation period of 28 d. Both AP and ovary are required for maintenance of pregnancy. The signal for parturition in the rabbit is probably not of fetal origin, for (1) fetal decapitation does not prolong gestation and may, in fact, shorten it, and (2) removal of the entire fetus, leaving the placenta in place, does not affect the length of gestation. These observations suggest that the control of delivery is maternal. The site of this control may be the hypothalamus, as is suggested by experiments of Lincoln (1971) with electrical stimulation of the infundibulum and of the median eminence. Such stimulation induced labour in pregnant rabbits and milk ejection in lactating does, indicating that in the species, oxytocin release is involved in induction of labour. Contractions of the uterus induced by oxytocin are inhibited by progesterone. Prior to parturition, the plasma progesterone concentration decreases, and administration of progesterone or hCG, which induces new functional CL, delays the onset of labour. This suggests that oxytocin release, to be effective in causing expulsion of the fetuses, needs to be preceded by a decrease in progesterone secretion. The question then remains as to what causes the decrease in progesterone secretion. In view of the dependence of rabbit CL on the pituitary, it seems reasonable to assume that the maternal hypothalamus controls the fate of the CL and thus of parturition. The nature of the signal for the regression of the CL is, however, not known. The expansion of the uterus is relation to the myometrical progesterone concentration, may determine the onset of parturition.

Although the fetus is apparently not the source of the signal for the initiation of parturition, there is evidence that cortisol administered to either the doe or the fetus results in a decrease in plasma progesterone and in early delivery, which can be prevented by administration of progesterone to the doe. Chiboka (1977) found that dexamethasone administered on day 21 did not cause expulsion of the placenta in fetectomized does, but administered on day 25, it caused prematured parturition, whether a fetus was present or absent. Furthermore, he found no evidence for a local utero-ovarian pathway for the maintenance or termination of gestation in hemiovariectomized does.

Laudánski *et al.* (1979) reported that infusion of $PGF_{2\alpha}$ in pregnant does at day 20-21 and in pseudopregnant does on day 11-12 induced luteolysis and increased myometrial activity in pseudopregnant, but not in pregnant females. These results suggest that a factory in the conceptus other than progesterone may inhibit myometrial activity and thus control the onset of labour. These results and those of Chiboka (1977) suggest that change of $PGF_{2\alpha}$ synthesis is not the main factor in inducing parturition in the rabbit, so that the exact mechanism for the timing of the onset of parturition in this species remains an enigma.

Rats. Gestation in the rat normally lasts 22 d. Until day 12, pregnancy is dependent upon the presence of the maternal AP to maintain the CL; after 12, the CL are maintained by a fetal luteotrophin. During gestation, the maternal plasma progesterone concentration starts to decrease at day 17, and the estradiol-17β concentration starts to increase sharply at day 16. In fetal plasma, the ACTH concentration reaches a peak between day 16 and 19 and then decreases, and the corticosterone concentration reaches a peak on day 18 of gestation and then decreases. These findings suggest that the fetal AP-adrenal axis may not be the determining factory in the initiation of parturition. This suggestion is further supported by the observations that neither fetal decapitation nor fetectomy affects the duration of gestation, which in the case of fetectomized rats is judged by the expulsion of the placentas on day 22.

The timing of parturition is affected by photoperiod. In rats kept under 12L : 12D, with lights on from 0600 to 1800,

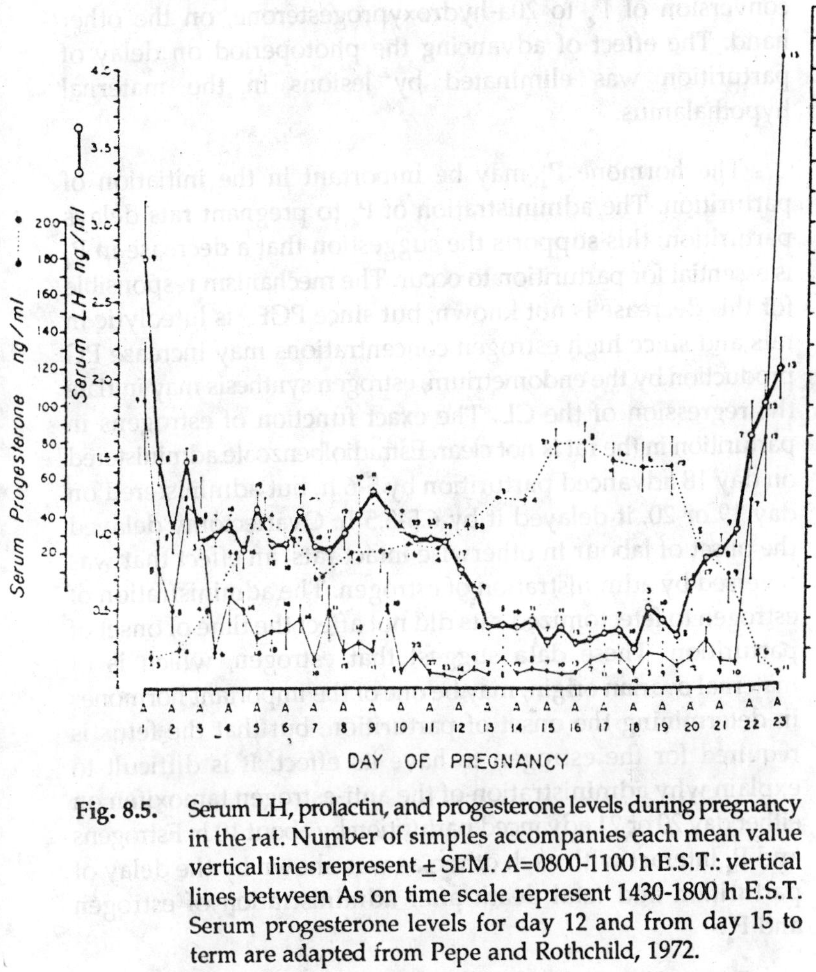

Fig. 8.5. Serum LH, prolactin, and progesterone levels during pregnancy in the rat. Number of simples accompanies each mean value vertical lines represent ± SEM. A=0800-1100 h E.S.T.: vertical lines between As on time scale represent 1430-1800 h E.S.T. Serum progesterone levels for day 12 and from day 15 to term are adapted from Pepe and Rothchild, 1972.

most births occurred between 0800 and 1400; when the light period was advanced by 8 h before midpregnancy, parturition was delayed by 12 h. In groups of rats kept under the control and the advanced photoperiod conditions, there was a correlation between the time of parturition on the one hand and the decline in maternal plasma P_4 concentration, the rise in maternal plasma prolactin, and the appearance of the enzyme 20a-hydroxysteroid dehydrogenase, which catalyzes the conversion of P_4 to 20a-hydroxyprogesterone, on the other hand. The effect of advancing the photoperiod on delay of parturition was eliminated by lesions in the maternal hypothalamus.

The hormone P_4 may be important in the initiation of parturition. The administration of P_4 to pregnant rats delays parturition; this supports the suggestion that a decrease in P_4 is essential for parturition to occur. The mechanism responsible for this decrease is not known, but since $PGF_{2\alpha}$ is luteolytic in rats and since high estrogen concentrations may increase PG production by the endometrium, estrogen synthesis may initiate the regression of the CL. The exact function of estrogens in parturition in the rat is not clear. Estradiolbenzoate administered on day 18 advanced parturition by 5-6 h, but administered on day 19 or 20, it delayed it by 6.5-8.5 h. Ovariectomy delayed the onset of labour in otherwise intact rats, an effect that was reversed by administration of estrogen. The administration of estrogen to fetectomized rats did not affect the time of onset of parturition. These data suggest that estrogen, which is of maternal ovarian origin, may be one of the important hormones in determining the onset of parturition, but that the fetus is required for the estrogen to have its effect. It is difficult to explain why administration of the anti-estrogen tamoxifen on either day 20 or 21 advanced parturition by about 10 h. Estrogens and P_4 interact with each other, as was shown by the delay of parturition and fetal death after administration of estrogen and P_4.

Flower (1977) has discussed ways in which PGs may affect parturition. They may affect the uterine tone, depending on the hormonal influences on the uterus. Their effect on the response of the uterus will be discussed in more detail later in this chapter under human parturition. Prostaglandins may have an effect by vasoconstriction, thus causing local ischemia, or they may act as luteolytic agents and cause a drop in progesterone concentrations, thus making the uterus more sensitive to the contraction-inducing effect of PG. The following evidence supports this interpretation: (1) Injection of $PGF_{2\alpha}$ at days 19 and 20 of pregnancy was followed by a drop in progesterone concentration in peripheral blood and by premature delivery of the pups. (2) Injection of $PGF_{2\alpha}$ antibodies on day 17 of pregnancy delayed parturition in some rats until day 24. (3) Administration of prostaglandin synthase inhibitors, such as aspirin, indomethacin, and fenclozic acid, delayed parturition. (4) Abortions were caused by compounds that increased concentration of PGs either by blocking their metabolism or stimulating their synthesis. (5) Administration of indomethacin and $PGF_{2\alpha}$ together did not affect the length of the gestation period. (6) Administration of estrogen (which can induce parturition) increased the activity of PG synthase and the concentration of PG in intact and in ovariectomized rats. Dukes *et al.* (1974) and Fuchs *et al.* (1976) stated that $PGF_{2\alpha}$ acts principally as a luteolysin and does not in itself cause contractions. In a clever series of experiments using unilaterally ovariectomized rats with unilaterally pregnant horns, Cavaillé and Maltier (1978) found evidence that a substance produced by one uterine horn was transmitted to the ipsilateral ovary and induced luteolysis, and subsequently parturition. The luteolytic substance may well be a PG. Further research is required both in vitro and in vivo with rat uteri under different hormonal conditions to determine whether PGs can induce contractions that are comparable to those observed at parturition.

Parturition can be induced by the release of maternal oxytocin. The periparturient uterus is sensitive to oxytocin, and although the receptors for oxytocin and PG receptors in the myometrium are separate entities these hormones interact with each other. Oxytocin increases PG synthesis in the pregnant uterus more than tenfold and such PG could then have a dual function: (1) causing luteolysis and so resulting in a decrease in progesterone and (2) causing contractions of the uterus. If oxytocin and PG each caused contractions, they would enhance delivery. Inhibitors of PG synthesis do not affect the response of rat uteri to oxytocin in vitro, and oxytocin inhibitors do not affect the uterine response to PG. In vivo, indomethacin does not affect the induction of parturition by oxytocin (Fuchs *et al.*, 1976). It thus appears that the interaction between oxytocin and PG described by Chan (1977) is a redundant mechanism, which ensures that parturition will take place efficiently. It is not known, however, what signal would trigger the release of oxytocin from the posterior pituitary. As will be discussed later in this chapter, in several species, oxytocin release occurs after the fetus has distended either the cervix of vagina, that is, after parturition has been initiated and has progressed. If oxytocin release indeed initiates parturition, then the signal for its release needs to be determined.

Although it thus appears that no signal from the fetal pituitary or adrenal is necessary for induction of labour, plasma corticosterone concentrations in the fetal rat increase sharply just before delivery, an increase that is prevented by fetal decapitation. This increase in fetal plasma corticosterone coincides with the increase in the ACTH content of the fetal pituitary. These changes in corticosterone may be important for the fetus, increasing the surfactant secretion in the lungs for example, buy they may not constitute a signal for the onset of labour.

Initiation of Parturition in CL-Independent Species

We will use sheep, rhesus monkeys, and humans as examples.

Sheep. The duration of gestation in sheep is about 140 d. and after day 50 the AP and ovary are dispensable, because the placenta produces sufficient amounts of P_4 for maintenance of pregnancy. Maternal corticosteroid concentrations increase 4 d prior to parturition, at about the same time that P_4 starts to decrease. The decrease in maternal P_4 is not a requirement for the initiation of parturition, as it is in goats. Estrone, estradiol, and prolactin concentrations in the maternal plasma increase about 2 d before parturition, whereas LH concentrations do not change significantly.

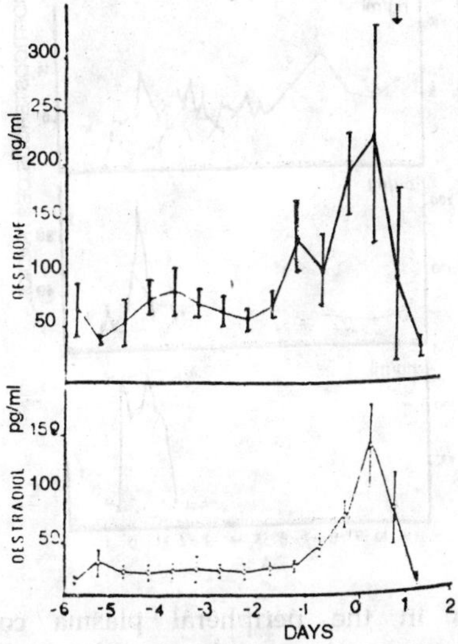

Fig. 8.6. Changes in concentration of estrogen (■) and estradiol-17β (□) in peripheral plasma [of ewes] (mean±SEM) until just before parturition, N=12; in subsequent collections N=6-8. The mean hour of lambing is indicated by the arrow.

The evidence that the fetal AP and adrenal initiate parturition in sheep in persuasive. Hypophysectomy or adrenalectomy of the fetus, but not of the mother, prolongs gestation. Infusion of synacthen ($ACTH_{1-24}$) into fetuses resulted in parturition, despite fetal corticosteroid concentrations being only about 50 ng/ml and in one case 20 ng/ml, in contrast to the average of 100-120 ng/ml in intact control fetuses. Synacthen infusion into intact fetuses at 125 d resulted also in premature parturition within 4 d and was accompanied by suppression of endogenous ACTH secretion during the first 24 h, followed by large fluctuations, indicating stimulation of ACTH secretion. Fetal corticosteroid concentrations fluctuated widely after the first 24 h of infusion. Jones *et al.* (1978a) speculated that maturational changes in the central nervous system stimulate secretion of ACTH and make it less sensitive to inhibition by high corticosteroid concentrations. Robinson (1979) has proposed that in addition to ACTH, another factor of AP origin is required

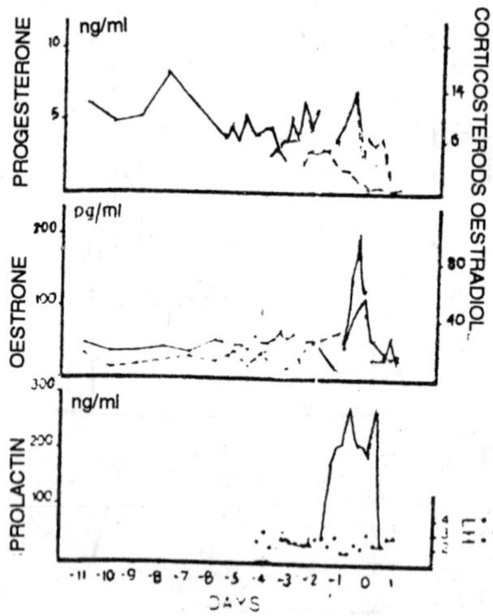

Fig. 8.7. Changes in the peripheral plasma concentration of progesterone, corticosteroids, estrone, estradiol-17β, LH, and prolactin in one ewe which delivered a single lamb.

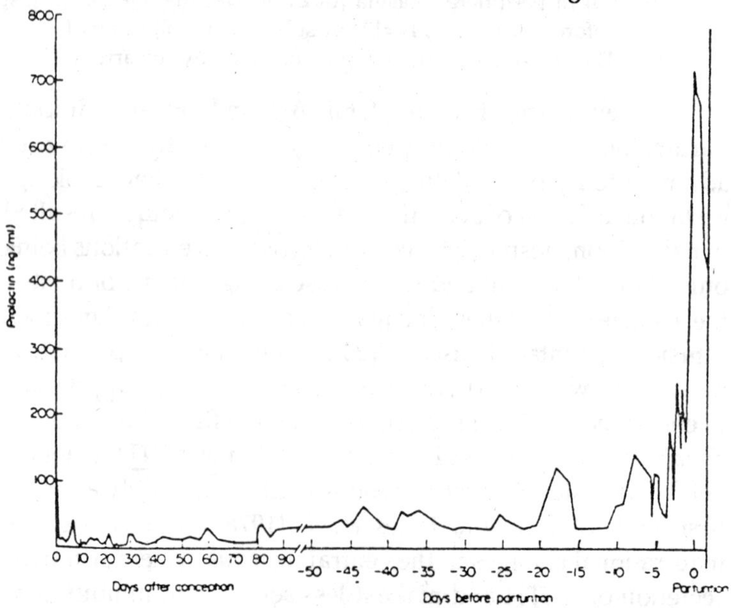

Fig. 8.8. Average prolactin levels in plasma of seven ewes during gestation.

to mimic the increase in fetal corticosteroid concentration and in maternal plasma estrogen concentrations before parturition in intact ewes.

These observations, together with the findings that there is an increase in corticosteroids several days before parturition, but an increase in ACTH that *follows* the corticosteroid increase, show that although the fetal pituitary is required for initiating parturition, it is not the cause for the increase in corticosteroid secretion. The increase in corticosteroids may, however, be caused by prostaglandins. Louis *et al.* (1976) found that fetal infusion of $PGF_{2\alpha}$ into the carotid artery resulted in about a threefold increase in corticosteroid concentrations. It remains to be determined whether the fetal pituitary is required for this response to $PFG_{2\alpha}$. The high corticosteroid concentrations, that are associated with the onset of pituitary are apparently not a requirement for initiation of this process, as the experiments by Jones *et al.*, (1978b) with fetuses, cited above, indicate.

The high cortisol concentrations about the time of parturition are correlated with a decrease in progesterone concentrations in the uterine venous blood and an increase in the concentration of estrone and estradiol-17α in fetal and maternal blood. Liggins *et al.* (1977a) proposed that cortisol might stimulate the activity of 17α-hydroxylase, 17-20 lyase, and possibly aromatase in the placenta. Anderson *et al.* (1978b) found that cortisol infusion indeed induced 17α-hydroxylase, but it did not stimulate either 17-20 lyase or aromatase activity.

The decrease in progesterone concentration and the increase in estrogen concentration may, together, stimulate $PGF_{2\alpha}$ release, or the decreasing progesterone may stimulate, and the increasing estrogens facilitate, PGF_{2} release from the maternal placenta. A single intramuscular injection of 20 mg estradiol benzoate in ewes between day 142 and day 148 resulted in a high incidence of parturitions within 96 h.

The prostaglandins may be important in inducting and

sustaining parturition. They may initiate contractions of the myometrium, or they may lower the threshold of the myometrium to oxytocin, which also induces uterine contractions. Oxytocin and PGF complement each other in sustaining uterine contractions, with oxytocin stimulating PGF synthesis by the maternal placenta and PGF reducing the threshold for oxytocin-induced contractions. An increase in $PGF_{2\alpha}$ release preceded by an oxytocin release occurs upon distension of the vagina when the fetus is expelled.

Railways and Ward (1978b) found that in the ewe, at the onset of spontaneous increased uterine activity, progesterone concentrations were still high and that fetal and maternal estrogens rose before parturition. The maternal estrogen concentration was correlated with fetal estradiol concentration (r=0.50) and fetal corticosteroid was negatively correlated with maternal progesterone concentration (r= -0.50). Maternal $PGF_{2\alpha}$ concentrations rose before parturition and remained high, showing a high correlation with uterine activity (r=0.79) and with plasma estradiol (r=0.79), and a negative correlation with plasma progesterone (r= —0.54). These correlations agree with the account presented above.

Infusion of either synthetic ACTH or of dexamethasone into either intact or fetuses resulted in an increase in maternal plasma progesterone and 13, 14-dihydro-15-oxo $PGF_{2\alpha}$ concentrations and in parturition. Maternal estradiol concentration did not increase in the absence of the fetal AP. This suggests that the fetal AP is necessary for the increase in estradiol before parturition, but that it is not required either for the increase in maternal plasma P_4 or PG or for parturition. The initiation of parturition in the sheep and goat is similar in that in both species, the fetal pituitary initiates the secretion of fetal corticosteroids, which stimulate placental enzymes that convert progesterone to estrogens. The higher estrogen concentrations increase $PGF_{2\alpha}$, which is the goat is luteolytic and thus causes a decrease in P_4 secretion. In sheep the CL is not required for gestation, and the decrease is maternal plasma P_4 is not required for parturition to occur, whereas such a decrease in P_4 concentration is required to the goat.

Rhesus Monkeys. In this species, which has a gestation period of about 164 d, evidence for the fetus having control over its delivery is not clear cut. In one report, 2 of 8 fetuses, in which the AP was destroyed by radioactive yttrium implantation, were carried beyond normal term, but Chez (1973) found that 4 of 5 surviving fetuses were carried beyond term. Decapitation at 73-78 d of gestation of 18 fetuses resulted in 4 abortions within 10 d, delivery of 2 partially absorbed fetuses 30-65 d after surgery, and 2 fetuses dead and eventually delivered by Caesarean section. Of the 10 remaining fetuses, 2 were delivered by Caesarean section at 177 d and 188 d, and the other 8 were delivered between 137 d and 186 d, with an average length of gestation of 159.6 d, and controls having an average length of gestation of 167 d. These data do not support the concept that the fetal AP is required for normal delivery. Adrenalectomy of the fetus results in too high an incidence of death to draw any valid conclusions about its effect on length of gestation.

Lanman (1977) found that fetectomy frequently resulted in retention of the placenta beyond normal term. After fetectomy, plasma estradiol decreased to about one-third of that of controls, and there was no increase in estradiol concentration prior to expulsion of the placenta, whereas in normal parturition, there was an increase in maternal estradiol and estrone.

In the rhesus monkey, when 90 per cent of the gestation time has passed, the maternal progesterone concentration is low (2-4 ng/ml) in contrast with the chimpanzee (about 100 ng/ml), marmoset (about 150 ng/ml) and humans (about 180 ng/ml). Maternal progesterone concentrations do not decrease, but P_4 concentrations in the uterine vein and umbilical vein decline gradually towards the time of parturition.

Fetectomy does not affect maternal progesterone concentrations, probably because the maternal ovary compensates for the loss of placental progesterone production. The mother's endocrine system does not compensate, however, for the loss of estrogen production as a result of fetectomy (which deprives the placenta of precursors for estrogen synthesis). It would be attractive to proposed that the decrease in progesterone in the uterine vein and the increase in estrogen concentration are responsible for the uterine contractions that lead to delivery of the fetus, but that administration of estrogen does not result in parturition in the rhesus monkey, because progesterone secretion does not decrease.

In primates, dexamethasone and cortisol permeate the placenta and thus inhibit the fetal adrenal glands. This inhibition results in a loss of precursors for estrogen synthesis by the maternal placenta and, therefore, a decrease in maternal plasma estrogen concentration. Corticosteroid administration to either mother or fetus does not affect the length of gestation.

Prostaglandins are implicated in the delivery of the fetus. Exogenous $PGF_{2\alpha}$ can induce premature parturition, and oral administration of indomethacin results in delivery at 180 d instead of at 161 d. At parturition, but not prior to its onset, there is a sharp increase in 13, 14-dehydro-15-keto $PGF_{2\alpha}$ in the maternal plasma and in the amniotic fluid. During administration to ACTH to the fetus, a procedure that induces premature delivery, there is a significant increase in the concentrations of estrone, PGF, and PGE in the amniotic fluid, and in the concentrations of estrogen and estradiol in the maternal and fetal plasma. It remains to be determined, however, what the mechanism is that signals the initiation of parturition.

Humans. Much of the evidence pertaining to initiation of parturition depends on clinical data and is fragmentary, and so open to various interpretations. It is, therefore, not surprising that there are strong differences of opinion about the controlling mechanism in parturition in humans.

Swaab *et al.* (1973) have reported that anencephalic and normal fetuses are born at the same fetal age and that fetal injections of ACTH plus oxytocin do not cause earlier delivery. Anencephaly results in labour being prolonged. These and other data suggest that the fetal pituitary and the fetal adrenal do not play crucial roles in the initiation of parturition. The fact that the synthetic corticosteroid betamethasone, administered to pregnant women, caused a sharp decrease to cortisol concentrations in fetal plasma and in amniotic fluid, but no change in the time of spontaneous delivery suggests that the fetal adrenal does not provide the signal for the onset of parturition. However, Turnbull *et al.* (1977) interpret the sudden rise in cortisol secretion by the adrenal in normal pregnancies as the possible final trigger for parturition in women.

There seems to be reasonably good agreement that toward the end of the nine-month gestation period, maternal peripheral plasma progesterone concentrations decline and estrogen concentrations increase. Liggins *et al.* (1977b) state, however, that there is no "readily measurable changes in placental hormone metabolism before labour starts." They do, however, cite evidence that there is an increase in plasma estradiol-17β prior to onset of labour in women at risk of premature labour.

Csapo (1977) has proposed that placental progesterone suppresses myometrial activity and that labour is the result of a sharp drop in the P_4 concentration in the uterine vein. As a result of the decrease in P_4 concentration, prostaglandins may induce contractions of the myometrium. Csapo has named this the "see-saw theory." It is his contention that the fetus determines the maintenance of pregnancy (the death of a fetus generally induces labour). He implies that the role of the fetus in initiating labour may be the stretching of the uterus. This may result in interference with the blood supply of the placenta, leading to a decrease in uterine progesterone concentrations. This increases the amounts of prostaglandins (which are synthesized all the time in the rabbit, considered by Csapo as a model for human parturition) toward the end of gestation and so induces myometrial contractions. Oxytocin presumably maintains labour once it has been initiated.

General Considerations on Initiation of Parturition

It is apparent from the account just given that there is good evidence that in the goat, sheep, and to a lesser extent the cow and pig, parturition is initiated by a signal from the fetal pituitary. This signal, by way of the adrenals and their corticosteroids production, causes a decrease in progesterone concentrations and an increase in PGs. The combination of low progesterone and high PG starts myometrial contractions, which force the fetus through the cervix, which has become dilated and softened under the influence of PG and/or relaxin. As the fetus enters the vagina and stretches it, it initiates an increase in the concentration of PGF in the utero-ovarian vein, and, in sheep, an increase in uterine contractions. There is also a release of oxytocin in sheep, as well as in goats, pigs, horses, and rabbits. Section of the pelvic nerves in lactating goats prevents the oxytocin release induced by stretching of the vagina, the so-called Ferguson reflex. In rabbits, spinal cord transaction between thoracic vertebrae 2 and 3 did not affect the onset of parturition, but labour was prolonged, and a high proportion of the fetuses were dead at birth, probably because the spinal cord transaction interfered with oxytocin release. Oxytocin helps to stimulate uterine contractions which help in the expulsion of the fetus, and which may also stop uterine bleeding when the young nurses immediately after it is born, since this nursing results in oxytocin release.

The hormones involved in expulsion of either the egg or the embryo or fetus in teleosts, birds, and mammals are oxytocin, PGs, progesterone and corticosteroids. The relative importance of each differs, but similarities, though interesting, cannot be used to draw general conclusions, since our knowledge is too fragmentary.

Parturition

The activity of the myometrium that expels the fetus or fetuses has received much less attention than the endocrinology of the initiation of parturition. Uterine contractions are either in a tubocervical or in a cervicotulvi direction, and during parturition, propagation of both types of contractions occurs in the rabbit, the rat, and the pig.

During parturition in the sow, myometrial activity appears to be initiated most frequently either tubal or at the cervical end of the uterus and less frequently in the remainder of the uterine born. Before and during expulsion of a piglet, cervicotubal propagation of contractions dominate tubocervical propagation. Once a uterine born in emptied, the contractions decrease. During the peak of uterine contractions in sheep, abdominal contractions begin and increase once the cervix is dilated. These abdominal contractions override the intra-uterine pressure waves, and the lambs are delivered.

Two effects of hormones of the myometrium are discussed: (1) the effect of steroid hormones and oxytocin on contractions of the smooth muscle and (2) the effect of steroid hormones on the ultrastructure of the myometrium. The study of the contraction of strips of rabbit myometrium in vitro has shown that: (1) there was little spontaneous motility or development of tension either after electric stimulation or after oxytocin administration when the strips were obtained from either ovariectomized or immature rabbits; (2) strips from estrogen-treated rabbits showed spontaneous motility and a response to either electric or oxytocin stimulation; and (3) progesterone administration to estrogen-primed rabbits eliminated the spontaneous motility and also the responsiveness to electric and to oxytocin stimulation (van Tienhoven, 1968). This effect of progesterone may be mediated by the binding of activator calcium (A-Ca) necessary for the actinomycin-adenosine triphosphate (ATP) interaction which causes muscle cell contraction. Calcium ion transport is smooth muscle is cyclic-

AMP-dependent. Currie (1980) has found the strips of myometrium from pregnant rabbits can be stimulated by addition of 10^{-3} to 3^{-2} molar imidazole to the bathing medium. The imidazole probably acts by stimulating phosphodiesterase activity, thus lowering cyclic-AMP and allowing an influx of Ca^{2+} into the contractile cells.

In vivo studies with pregnant and parturient rabbits have shown that P_4 suppressed mechanical and electrical activity of the myometrium and that in the absence of P_4, action potentials were generated, with the electric activity being synchronous, so that pressure cycles of great amplitude were generated.

In rats, P_4 does not inhibit uterine contractions, but it restricts the contractions to local ones, which are not as rapid as the contraction waves found in estrogen-dominated uteri.

Immediately prior to, during, and immediately after parturition, gap junctions (nexuses) can be found between myometrial smooth cells in rats, guinea pigs, sheep, humans and golden hamsters. These gap junctions, by facilitating electric coupling, improve the synchronization of muscle contractions required for parturition. In rats, ovariectomy at day 16-17 of pregnancy results in abortion and the formation of gap junctions; administration of progesterone prevents normal parturition and the formation of gap junctions. In sheep the increase in estrogen concentration and decrease in progesterone concentration just prior to parturition is correlated with the appearance of gap junctions. Old golden hamsters deliver their young at day 16-18 of pregnancy while young hamsters deliver theirs at day 15. This delayed parturition is accompanied by a delayed appearance of gap junctions. In general there has been so far an excellent correlation between onset of parturition and the presence of gap junctions and their absence in non-parturient uteri.

This review of the initiation of parturition and of the control of uterine contractions shows that both the control of the initiation of parturition and the effect of a particular hormone on the contractions of the uterus vary considerably among species. This adaptive significance of each of these mechanisms is not clear.

9

Care of the Embryo

After fertilization, the zygote may be either retained by the mother (or in the case of seahorses and pipefishes in the pouch or the father) or released into the environment to take care of itself without receiving additional food, so that it lives on the yolk of the egg. However, parents may protect the embryo against predation and furnish it with hear or oxygen, or remove waste products from its environment. The main categories we will use for patterns of care of the embryo (fetus) are: (1) *oviparity*, in which the young emerge from the egg outside the dam's reproductive system and (2) *viviparity*, which is aplacental viviparity the young are retained in their mother's reproductive system, but the embryos depend (*a*) solely on yolk reserves, (*b*) on oophagy — a form of intrauterine cannibalism, or (*c*) on placental analogues, such as uterine milk and trophonemata (to be discussed later in this chapter under Teleosts). (The term *aplacental viviparity* replaces an older term, *ovoviviparity*.) In *placental viviparity*, various forms of placentas provide for the embryo. Wourms (1977) discusses in some detail the evolutionary and ecological factors that may account for the apparent independent evolution of viviparity. We will discuss the different forms of care for the different classes.

Cyclostomes

Prior to mating and spawning, lampreys build nests over sandy bottoms. They construct small walls from rocks downstream from the nest so that water runs relatively calmly over the nest. During spawning, they stir up sand to which the adhesive eggs may become attached, thus increasing their chance for survival by decreasing predation. Beyond this anticipatory care of the young, there is no further care of the embryos since the parents die after spawning.

Elasmobranchs

While Table 9.2 shows the gradation between oviparity and viviparity. The details of embryonic development will not be discussed here; instead we refer the reader to the excellent recent account by Wourms (1977). Table 9.3 gives a summary of the time it takes for embryonic development to be completed, i.e., until the young are hatched or born.

Aplacental species of elasmobranches have different ways of caring for the embryo. When the yolk with which the embryo is provided serves as the entire source of nourishment, then the care is essentially not much different from that found in oviparous species except, of course, for the protection provided by the mother. A second method for aplacental species to obtain food during gestation is the eating of yolks or of embryos that follow the oldest or strongest embryo in the uterus, e.g., in porbeagle shark (*Lamna* sp.), the sand shark (*Odontapsis taurus*), and possibly some mackerel sharks (*Lamnidae*) and thresher sharks (*Alopiidae*).

The embryo of *Lamna* sp. absorbs its own yolk sac early and then proceeds to eat yolks of oocytes that are ovulated later. The cardiac stomach becomes distended by such yolks, and this distension gives the stomach the appearance of an enormous yolk sac. The sand shark embryos swim open-mouthed in the oviduct and appear to eat everything that is not oviduct wall. This pattern of nutrition has a double advantage for the young shark; it feeds itself, and it also receives training in finding food, so that it is born an accomplished predator.

The aplacental viviparous sharks with placental analogues have two different strategies for providing the young with nourishment; (1) the secretion of uterine "milk," which may

TABLE 9.1

Changes in Weight and Composition of Selachian Embryos During Development in Oviparous, Aplacental Viviparous, and Placental Viviparous Species.

Characteristics in relation to embryo	*canicula (dogfish)*	*Scyliorhinus ocellatasmooth (electric ray)*	*Mustelus vulgans Torpedo(common violacea (nursehound)*	*Dasyatis (Trygon) laevis (sting ray)*	*Mustelus (nursehound)*
Egg status	Oviparous	Aplacental viviparous	Aplacental viviparous	Aplacental viviparous	Viviparous viviparous
Gestation (Months)	6	4	10	2	5
Weight of ovulated egg (g)	1.31	6.78	3.93	9	5.54
Weight of "completed" embryo (g)	2.69	13.37	60.16	118.0	189.6
Percent weight change in embryo	+105%	+97%	+1432%	+6105%	+3326%
Percent change in water in embryo	+213%	+215%	2490%	+10412%	+5608%
Percent change ash in embryo	+292%	+157%	+2800%	+10250%	7609%
Percent change organic sub. stances in embryo	-21%	-23%	+356%	+1628%	+1064%
Placental type or type of nutrition	None	Milk from villi on uterine wall, into gut of embryo	Uterine secretion copious	Villi thick, milk-secreting and long into gut of embryo	Intimate yolk-sac

have an organic matter content of 13 per cent and a total fat content of 8 per cent and (2) tufts of uterine epithelium or trophonemata (long glandular appendages), which enter the embryo through the spiracles and pass into the esophagus. Nutrients secreted by the trophonemata enter the gut. This pattern of feeding, which is found in *Dasyatis violacea*, *Myliobatus bovina* and *Gymnura micrura*, allows for a 17-50-fold gain in organic matter by the embryo.

Aplacental nutrient transfer is not always as efficient as just illustrated. In the electric rays, *Torpedo ocellata* and *T marmorata*, the uterine milk is watery (1.2 per cent organic matter, 0.1 per cent fat), and the embryos show a net loss of 22 percent organic matter and 34 per cent fat during development.

Placental development is found within two families, the requiem sharks and the hammerhead sharks, but not all genera, or species within genera, have a placental way of nutrient transfer; e.g., *Mustelus canis* has a placenta, but *Mustelus antarcticus* does not. in *Mustelus canis*, the embryo is dependent on its yolk during the first three months of its development and then establishes a placental connection. The placenta that develops incorporates the egg capsule and consists of yolk sac and uterine epithelium, which eventually interdigitate with each other. Such interdigitation occurs also in most other species, such as, *Prionace glauca*, *Sphyrna tiburo*; however, in some, e.g., *Carcharhinus dussumerieri* and *C. falciformis*, there is no interdigitation, and the placenta rests on a vascularized region of the uterine wall. In *C.falciformis*, there are five layers involved in the exchange between mother and fetus: (1) maternal endothelium, (2) maternal epithelium, (3) egg capsule, (4) fetal epithelium, and (5) fetal endothelium, although some of these layers may be reduced or lost. After hypophysectomy of *Mustelus canis* at the beginning of gestation, development of the young is not disturbed. Placental transfer can be quite efficient, as attested by the 8 to 10-fold increase in organic matter in the embryo during its development.

TABLE 9.2

Duration of Development in Elasmobranchs from Fertilization to Independence of Embryo from Mother

Pattern and species	*Duration of development (months)*
Oviparous	
Skates	
Raja clavata	4.5-5.5
R. naevus	8
R. marginata	15
R. eglanteria	$2^{1}/_{4}$-3
Sharks	
Heterodontis sp.	9-12
Scyliorhinus canicula	6-8
Viviparous	
Sharks	
Mustelus canis	8-12
Squalus acanthias	22-24
Rays	
Myliobatus bovina	3
Urolophus hallen	3
Dasyatis (Trygon) violacea	2

Teleosts

After external fertilization, the zygotes may simply be abandoned (e.g., herrings), or they may be protected, or protected and nourished.

Protective Strategies

Protective strategies include nest building, mouth breeding, skin incubation, and incubation in other animals.

Next building. Nests may be made in a slit in a rock or an empty shell, as found in many littoral fishes, e.g., the Blennidae and Cottidae. The fish *Pholis gunnellus* may lay her eggs in such a hiding place, but if she cannot find one, she will coil around the big ball of fertilized eggs, which, of course, contain developing embryos, and guard them in this manner. Nests may also be hollowed out in gravel or stone, as is done by trout and salmon, or they may be made of plant material that is glued together by special secretions of the kidney, as is done by the three-spined stickleback (*Gasterosteus aculeatus*). Floating nests may be made from air bubbles, as in *Macropodus* spp. Bertin (1958) gives a rather complete account of the various nests used by fishes. In some species, the parents not only build a nest, but they may also fan the developing embryos to facilitate gaseous exchange and removal of wastes, e.g., *Crenilabrus ocellatus*.

Mouth breeding. In a number of Cichlidae, the females incubate the eggs in their mouth (in *Tilapia macrocephala*, the male incubates the eggs in his mouth). Females also incubate eggs in some species of Cyprinodonidae, Osteoglossidae, Osphrominidae and Siluridae. In representatives of the Ariidae, Cyclopteridae, Apogonidae, Cichlidae, Opisthognatidae, and Belontiidae, the male incubates the eggs. Mouth breeder parasitism, similar to the nest parasitism of cuckoos, is found among cichlid fishes. The female *Haplochromis polystigma* cares for her own offspring and those of a sympatric species. *H. chrynosotus*. Two other species, *H. macrostoma* and *Serranochromis robustus*, also guard broods consisting of a mixture of their own and another species.

Skin incubation. In the banjo catfishes (Bunocephalidae), the fertilized eggs are attached to the skin in various ways, although it is not clear whether this is only for protection or also for nutrition in some cases.

Incubation in another animal. This is found in the bitterling (*Rhodeus amarus*). In this species, fertilization occurs within a mussel, with the young growing inside the mussel's gill chamber. It is also found in *Aulichthys japonicus*, in which the eggs are incubated in the peribranchial cavity of ascidia and in *Careproctus* sp., which uses the gill chamber of the crab. *Paralithodes camtschatica*, as a place of incubation.

Protective and Nutritional Strategies

Strategies that include protection and nourishment are feeding on skin secretions, pouch breeding, and nutrition from pseudoplacentas.

Feeding on skin secretions. Under the influence of prolactin, the male and female secrete a mucus that the young use as nourishment, e.g., in *Symphysodon discus* and *Aequidens latifrons*. In both of these species, prolactin administration results in the parents fanning the eggs in the nest.

Pouch breeding. In pipe fishes and seahorses, the female deposits her eggs in the pouch of the male; indeed, several females may lay eggs in the brood pouch of one male. The brood pouch maintains an environment of different composition than sea water, which is harmful for the embryos of *Hippocampus erectus*, according to Linton and Soloff (1964). Hypophysectomy of the sea-horse (*hippocampus hippocampus*) leads to an increased frequency of malformation and premature expulsions from the pouch; however, some embryos are unaffected, and these are "born" at the normal time. The proliferation of the epithelial lining of the brood pouch of seahorses, which occurs when he is incubating the young is under the apparent control of prolactin; mammalian prolactin administration results in stimulation of this epithelium in intact and in AP seahorses.

After internal fertilization, the zygotes may be discharged and abandoned, or they may be taken care of in the body of the female, depending upon whether they are oviparous or viviparous. In oviparous species, the young emerge from the egg outside the reproductive system and live on the yolk provided in yolk sac and on nutrients from the environment, as in *Trachycorystes* sp., in which the zygotes are discharged shortly after fertilization, and in *Sebastodes* sp., in which some development takes place inside the ovarian cavity. Since emergence occurs outside the body of the mother, we still classify this pattern as oviparous. Since *Sebastodes* sp., carries about 2 million embryos, there is a problem of providing sufficient oxygen and nutrients. These are provided, however, by a specialized double vascularization, in which anterior and posterior ovarian arteries are continuous with each other and thus form an arterial loop that provides an adequate oxygen supply.

In viviparity, which in teleosts is aplacental, the embryo may be dependent mainly or solely on yolk reserves, e.g., in Poeciliidae, such as *Xiphophorus helleri*, or the embryo may use dead embryos, dead sperm, and/or desquamated cells which are absorbed through a hypertrophied yolk sac, as in the sea perch (*Cymatogaster aggregata*). Gaseous exchange occurs through spoon-shaped enlargements of the dorsal fin.

Nutrition from pseudoplacentas. The embryo may be served by several types of pseudoplacentas, four of which are described briefly. (1) Portal circulation of the yolk sac may serve principally for gas exchange since the yolk seems to provide sufficient nourishment, e.g., in the guppy (*Poecilia reticulata*) and in *Heterandria formosa*; (2) The surface of the follicular wall contains villi, which are in contact with the embryo's ectoderm. The estoderm is separated from the wall of the gut by a yolk sac space, and the gut is enormously expanded to form the belly sac, which, presumably, serves for absorption of nutrients. (3) Trophotaeniae, which are found in, e.g., the Goodeidae, are fetal processes that increase the absorbing surface and so enhance the absorption of nutrients from the follicular fluid. (4) Trophonemata similar to those described for elasmobranchs are found in the Jenynsiidae. In teleosts these are extensions from the ovary, but in elasmobranchs, they are extensions from the uterus, which enter the gills and pharynx of the young. Not much is known about the endocrine influences on gestation in fishes. It is impossible to study the effect of ovariectomy, because gestation is in the ovary. Hypophysectomy does not affect embryonic mortality in *Poecilia latipinum* and increases it in *Gambusia* sp.

Amphibians

Among the Anura, there are two species with internal fertilization. In *Nectophrynoides vivipara*, more than 100 embryos are present in the viviparous female. The embryos have long slim tails, which "function as so many pipe lines which bring oxygen to the larvae kept away from the uterine wall by the bodies of their brothers and sisters."

In one of the Urodeles (a salamander, *Salamandra atra*) with internal fertilization, the embryos use their own yolk, but also practice oophagy and hemophagy (ingestion of blood present in the uterus as a result of hemorrhages in the uterus). Only one embryo out of a total of 40-60 ovulated fertilized oocytes survives, and it eats its other sibs.

In Anura with external fertilization, the young are protected and nourished in some interesting ways. In the Surinam frog (*Pipa pipa*) embryos are incubated in chambers or a pouch on the back of the mother and the marsupial toad (*Gastrotheca marsupiata*), for example, the larvae of *Pipe pipa* have very vascular, large tails, which act as a placenta by making contact with the vascular lining of the chambers. In the marsupial load, the larval gills have vascular bell-shaped extensions, which spread out under the blood vessels of the pouch. The male Chilean toad (*Rhinoderma darwinii*) incubates the eggs in two vocal pouches lateral to the esophagus, and the young frogs leap out of the father's mouth when they have sufficiently developed. The female Australian frog, *Rheobatrachus silus*, swallows the recently fertilized eggs and incubates them in her stomach where they are protected, although they live on the yolk nutrients, until they are mature enough, when they leave the mother through the mouth. It is not understood how the female's digestive juice is prevented from digesting the eggs and embryos. The male midwife toad (*Altes obstetricans*) catches the eggs which are laid in strings and then twists them around his legs. He carries these eggs for several weeks until the larvae are ready to hatch and he then delivers them to a pond, where there are no other tadpoles present.

Peptiles

Although most reptiles are oviparous, nearly half of the extant families contain some live-bearing species, and five of these families are completely viviparous. Viviparity is not necessarily confined to living forms, however, some fossil specimens of *Ichthyosaurus*, an extinct marine reptile of the Mesozoic, contain skeletons of unborn young in the body cavity. Reptilian viviparity is often the aplacental type, although representatives of at least nine families of lizards and snakes are reported to form yolk sac or chorio-allantoic placentas. According to Black-burn (1981), the minimum number of times viviparity could have evolved in reptiles is 75 (compared to 10 times in the Chondrichthyes, 12 times in the Osteichthyes. 4 times in Amphibia, and once in Mammalia). In lizards, viviparity has evolved at least 30 separate times in snakes, at least 17 times, and in the amphisbaenids, probably only once. All crocodiles, turtles and the tuatara are oviparous.

Care of the embryos or young is not common in reptiles, with some exceptions, however. The skinks of the genus *Eumeces* lay their eggs and protect them, and also warm them by basking in the sun and then coiling around the eggs to transfer heat to them. Some pythons, e.g., *Python reticulatus*, coil themselves around the eggs and brood them for about 100 days, leaving them only to eat, drink and molt. Cobras and the American mudsnakes (*Farancia abacura*) also guard their eggs. The American alligator guards the nest and, when the young hatch and make a gruning noise, the mother may help them out of the nest.

Among reptiles with internal fertilization, there are nine families that have placentas. Amoroso (1952) has classified these placentas as yolk-sac placentas and chorio-allantoic placentas.

Yolk-sac placenta. This type of placenta is found in some Australian lizards, in the Italian lizard (*Chalcides tridactylus*), and in the garter snake (*Thamnophis sirtalis*). Fetal tissue does not invade the maternal tissue, although the omphalochorionic membrane and endometrium may show interlocking folds. The yolk-sac placenta becomes isolated as the extra-embryonic mesoderm develops in a manner peculiar to placental reptiles. The chorio-allantoic region of the placenta is simple in structure, and the fetal and maternal capillaries remain separated by layers of uterine and chorionic epithelium and a thin shell membrane. In lizards, the omphalo placenta, once formed, regresses gradually and disappears as the allantois expands around the yolk. In snakes, the yolk-sac placenta does not increase in size during development. The allantois spreads beneath the isolated omphalopleure and adheres to it.

Chorio-Allantoic Placentas. These are of three types. In one type, maternal circulation and fetal circulation may be close to each other as the result of reductions in uterine and chorio-allantoic tissue. This type is found in the skink, *Chalcides ocellatus*, and the Australian snakes, *Denisonia superba* and *D. sata*. In *Denisonia*, the allantoic membrane fuses with the isolated yolk endoderm of the yolk-sac placenta to form a chorio-omphalo allantoic membrane as in *Enhydrina* and *Hydrophis*. In garter snakes, the allantois fuses with the yolk-sac placenta at its margins only. This vascularization of the yolk-sac placenta is (indirectly) allantoic. Na^+ and glycine pass to the fetus from the mother. The yolk-sac placenta apparently allows glycine to pass through it, whereas its permeability to P and Fe is low. The shell membrane seems to function as a dialyzing membrane.

In species of skinks of the genus *Lygosoma*, maternal capillaries bulge at the surface of folds of the uterus. Between these folds there are grooves lined with a glandular epithelium. Enlarged epithelial cells of the chorion interlock with the uterine folds, and the cytoplasmic processes of the chorionic epithelium penetrate between uterine epithelial cells, thus providing a relatively intimate contact between fetal and maternal tissues.

In the skink, *Chalcides tridactylus*, the folded maternal tissues overlie the main longitudinal uterine blood vessels. These

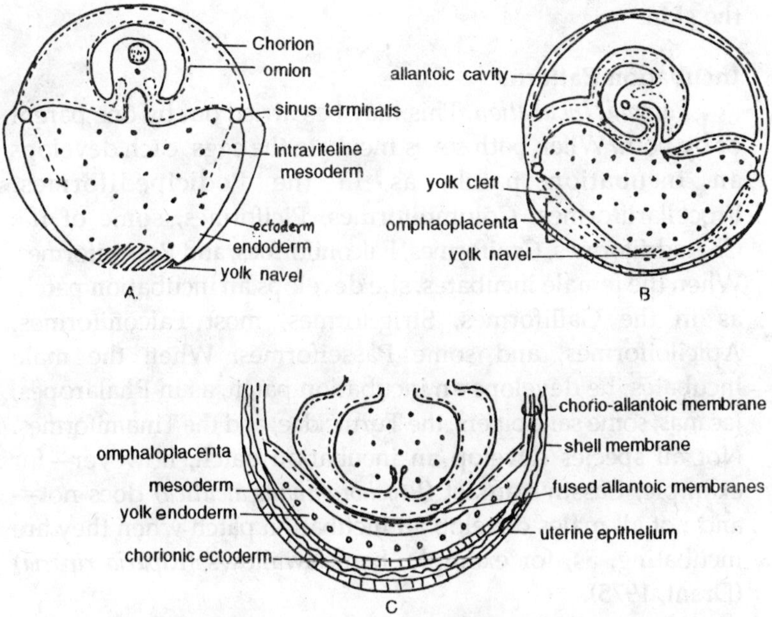

Fig. 9.1. Diagrammatic view of cross sections through the chorionic vesicle [of the garter snake] at various developmental stages to illustrate the formation of the omphaloplacenta. *A*, Zehr stage 15-20 *B*, Zehr stage 24. *C*, Zehr stage 27 and after.

folds indent the fetal placenta. The cuboidal uterine epithelial cells and ectodermal chorion cells are enlarged, suggesting increased activity.

Birds

All birds are oviparous, and the eggs need to be warmed and guarded. Parents or substitute parents incubate the eggs by sitting on them and transferring heat from their body to the eggs. In many species, such an exchange is enhanced by the formation of a so-called incubation patch of the ventral skin. This usually consists of a bare skin. The feathers are either permanently absent, as in doves, or are pulled out by the parent, as in water fowl, or fall out through hormonal action, as in most birds. In addition, there is usually a thickening of the epidermis, and an edema and increased vascularization of the skin.

Incubation Patterns

Parental Incubation. This may be carried out by one parent or by both. When both sexes incubate the eggs, each develops an incubation patch, as in the Podicipediformes. Procellariiformes, Columbiformes, Piciformes, some of the Charadriiformes, Gruiformes, Falconiformes, and Passeriformes. When the female incubates, she develops an incubation patch, as in the Galliformes, Strigiformes, most Falconiformes, Apiciloiformes, and some Passeiformes. When the male incubates, he develops an incubation patch, as in Phalaropes, Jacanas, some sandpipers, the Turmcidae, and the Tinamiformes. Not all species develop an incubation patch, however—for example, Cassin's auklet (*Ptychoramphus aleutica*) does not—and not all males develop an incubation patch when they are incubating, as, for example, bank swallows (*Riparia riparia*) (Drent, 1975).

Communal nesting. In this form of incubation, several females all lay their eggs in one nest (as in rheas, tinamous, anis and ostriches). In the groove-billed anis (*Crotophaga sulcirostus*), both males and females help incubate the eggs.

Nest parasitism. In this incubation pattern, a species other than the parents incubate the eggs. Cuckoos and cowbirds are such nest parasites. The eggs that are laid are strikingly similar in colour and shape to the egg of the host, but the eggs of the parasite are larger Wickler (1968) has excellent colour photographs showing these similarities. Different females within a species of nest parasites may lay different coloured eggs. Females within a species that lay similar-coloured eggs belong to the same gen (plural gentes). Intriguing questions are: How does the parasite female "know" the colour of the egg she is going to lay? One may propose that she may "know" which host species' nest to lay her eggs in because she is imprinted at the time of hatching on the nest and egg shell colours of the host's eggs and the egg from which she hatched. Then the question arises whether the male with which the mother mated influences the egg shell colour of his daughter. It is possible that males and females from the same gen mate with each other only and not with those of other genste.

Incubation without nest sitting. The following account is mainly taken from Welty (1962). The Megapodiidae of Indonesia and Australia use the heat of the sun, the heat from rotting leaves, and a combination of both to provide heat for incubation. The maleo (*Megacephalon maleo*) of Celebes lays her eggs (after a migration of as long as 32 km) in the black sand of the ocean beach or in warm soil next to hot springs or volcanic steam fissures. Females may lay eggs in communal nests.

The brush turkey (*Alectura lathami*) piles up rotting vegetation to provide heat. The mallee fowl (*Leipoa ocellata*) of the semi-desert region of Australia uses a combination of rotting vegetation and heat from the sun (which is maximized by the rearranging of the sand by the male) to incubate the eggs. The Egyptian plover (*Aegyptius pluvialis*) uses a combination of sitting on the eggs during the night and the heat of the sand in which the eggs are buried to provide the necessary heat.

Physiology and Endocrinology of Embryo Care

The physiology and especially the endocrinology of nest building or incubation activity have been studied in most detail in canaries (*Serinus canarius*), ring doves (*Streptopelia risoria*), chickens (*Gallus domesticus*), and turkeys (*Meleagris gallopavo*). In canaries, estrogen is the principal hormone that results in nest building. For defeathering and increased vascularity of the incubation patch, a synergistic interaction between prolactin and estrogen is necessary for optimal results. Estrogen, probably in conjunction with progesterone, increases the sensitivity of the skin receptors of the incubation patch to tactile stimuli; however, the building of a nest of normal size and composition or the lack of interest in nest building that the females normally show as egg laying approaches do not depend on the sensitivity of these receptors. In some other species there is evidence that estrogen and prolactin induce the development of the incubation patch, e.g., starlings (*Sturnus vulgaris*), red-winged blackbirds (*Agelaius phoeniceus*), English sparrows (*Passer domesticus*), and California quail (*Lophortyx californicus*), although the details of the action of these hormones may vary from species to species. Interestingly, in two species of phalaropes, *Phalaropus tricolor* and *P. lobatus*, in which the males are the only incubators, the incubation patch can be induced by a combination of testosterone and prolactin, but not by estrogen and prolactin. Induction of incubation behaviour has been accomplished in chickens by the injection of prolactin. The studies by Lehrman and his colleagues carried out with ring doves have shown that progesterone induces incubation behaviour and that incubation behaviour itself may then enhance prolactin secretion, which can maintain, but not initiate, incubation behaviour.

The incubation-inducing effect of progesterone is facilitated by previous experience with nest-building phases of behaviour. Early studies in the endocrine control of incubation behaviour of chickens indicated that prolactin was the principal hormone that controlled the onset of this behaviour. The evidence for this was the higher concentration of prolactin in the pituitaries of incubating or broody hens than in nonbroody hens. Furthermore, injections of mammalian prolactin preparations, which, in retrospect were probably not very pure induced incubation behaviour. With the use of a sensitive radioimmunoassay for prolactin, Sharp *et al.* (1979) found that plasma prolactin concentrations were about 23 per cent higher in incubating than in non-incubating bantam* hens, whereas plasma progesterone and LH concentrations were lower in the incubating than in non-incubating hens. The data available for the relationship between plasma prolactin concentrations and incubation behaviour in turkeys vary considerably. Etches et al. (1979) found no significant difference in plasma prolactin concentration between broody and non-broody hens in the same period after the start of egg laying. They observed that broodiness was always preceded by an increase in plasma prolactin concentration, but that such an increase was not always followed by broodiness. Burke and Dennison (1980) reported that in 6 of 7 hens that became broody (out of a total of 10 hens) the prolactin concentration increased and, on the average, reached a concentration of 500 ng/ml on the first day of broodiness and a peak concentration of about 1200 ng/ml. In the three hens that did not become broody, the peak concentrations of plasma prolactin, in the same period after the first ovulation as for the broody hens, were between about 500 ng/ml and 800 ng/ml for two hens and < 100 ng/ml for the third hen. For two of the three hens, as in the finding of Etches *et al.* (1979), an increase in plasma prolactin did not predict the onset of incubation behaviour. Burke and Papkoff (1980), El Halawani *et al.* (1980), and Proudman and Opel (1981) have reported four- to ninefold higher plasma prolactin concentrations in broody than in non-broody hens. Such a finding, of course, does not indicate causal relationship. El Halawani *et al.* (1980) showed that not making nests available to incubating turkey hens resulted in a precipitous decrease in plasma prolactin concentration. The concentration of this hormone increased again after the hens were returned to the nest, suggesting that nesting behaviour or exposure to a net results in prolactin secretion. It remains to be investigated whether administration of turkey prolactin induced incubation behaviour and, if so, under what circumstances.

MAMMALS

Prototheria

The duckbill platypus (*Ornithorhynchus anatinus*) incubates her eggs in a nest, and the echidna (*Tachyglossus ceuleata*) incubates her eggs (after a sojourn in the uterus estimated to be about 12-28 d) in a pouch. In both cases the young lives on the yolk provided at ovulation and the albumen laid down in the oviduct; the mother provides heat and protection.

Metatheria

The reproductive organs of metatherians undergo seasmal changes. Examples of the reproductive cycles of different marsupials are presented. The marsupial oocyte is surrounded by a cell membrane and has more yolk than the oocyte of placental mammals. The zygote gives rise to a laminar blastocyst and most of the dividing cells of the embryo are parts of the developing germ layers, whereas in placental animals an invasive trophoblast develops.

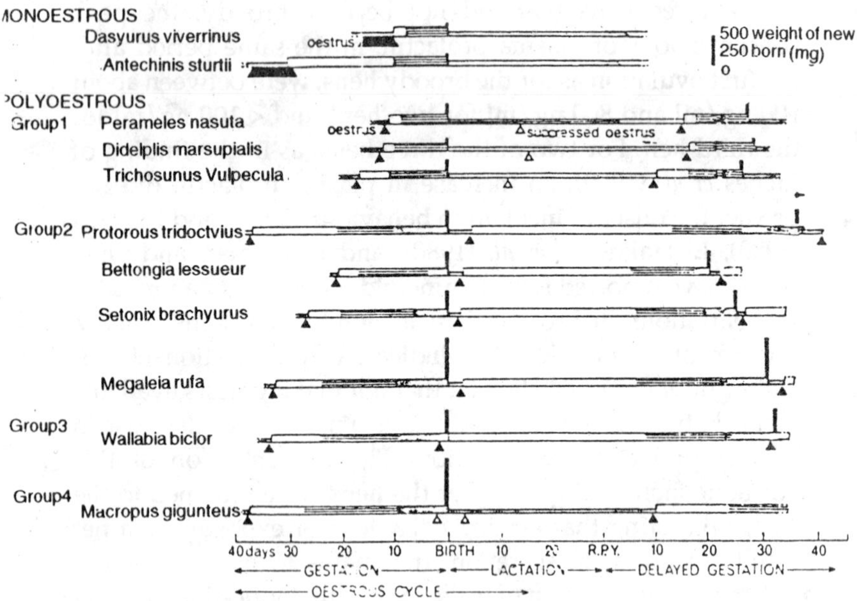

Fig. 9.2. The reproductive cycles (Δ-Δ) gestation and lengths (Δ-→) neonatal weights of 11 marsupials compared by reference to the day of born. Subsequent lactation suppresses the reproductive cycle at approximately the stage reached at parturition, and removal of the pouch young (RPY) releases it from inhibition, so that the suppressed events are completed. The sucessive stages of the cycle are represented in the horizontal bars as follows prelutea blank, luteal, horizontal hatch; postueal/proestrous, sucle.

The marsupial embryo takes up nutrients secreted by the endometrium during the preimplantation phase, which lasts more than half of the gestation period. There is no intimate contact between maternal and fetal tissues except in the bandicoot (*Isoodom macrourus*). This initimate contact between maternal and fetal tissues results in a more efficient transport of nutrients and waste products, and it is correlated with a shorter gestation than in related species that lack the intimate contact. Tynadale-Biscoe (1973) has speculated that shorter gestation is the result either of an allongraft rejection or of a response to avoid such a rejectino. In the Eurheria, a mechanism has developed that prevents the immunological alongraft rejection and so allows for a longer sustained pregnancy and more development prior to birth. Marsupials do not possess such a mechanism that prevents allograft rejection, and the more intimate contact of maternal and fetal tissues in the bandicoot than in other marsupials leads to earlier rejection and thus birth of the young.

In the quokka (*Setonix brachyurus*), the trammar wallaby (*Macropus eugenii*), and the red kangaroo (*M. rufus*), removal of the pouch young results in histological changes in the embryo followed by changes in the endometrium during the luteal phase, followed by division and enlargement of CL cells. Seven days after removal of the pouch young of the tammar wallaby, a reduction in prolactin secretion, a reactivation of the CL, an increase in endometrial secretion, and reactivation of the blastocyst have been observed to occur in the sequence listed. In the red-necked wallaby (*M. reforgriseus banksianus*), the same sequencem reactivation of the CL, increase in endomerial secretion, and reactivation of the blastocyst has been found.

Removal of the CL before day 2 of gestation results in a prolongation of the diapause, whereas removal of the CL between day 3 and 6 does not prevent the initiation of embryonic development but does result in the collapse of the embryo at the vesicle stage; removal of the CL after day 6 has no apparent effect on development, and the young are born at term.

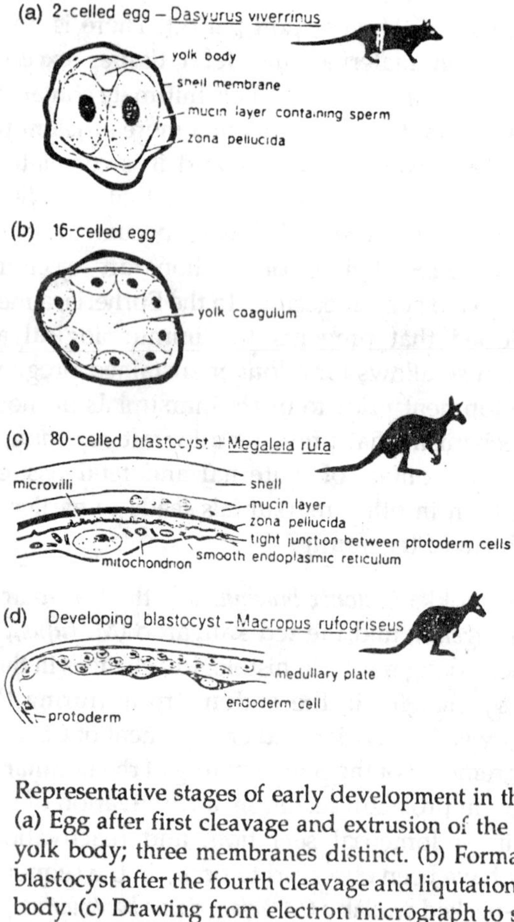

Fig. 9.3. Representative stages of early development in three species (a) Egg after first cleavage and extrusion of the noncellular yolk body; three membranes distinct. (b) Formation of the blastocyst after the fourth cleavage and liqutation of the yolk body. (c) Drawing from electron micrograph to show detail of undifferentiated protoderm cell. (d) Similar blastcyst after formation of endoderm cells and differentiation of medullary plate, desined to become the neural tube of the embryo.

Progesterone injections can initiate the development of the blastocyst in lactating kangaroos and tammar wallabies and in anestrous wallabies; continued progesterone injection can maintain such an embryo for as long as ten days. By blastocyst transfer, it has been shown that a blastocyst in diapause can resume development when transferred to the

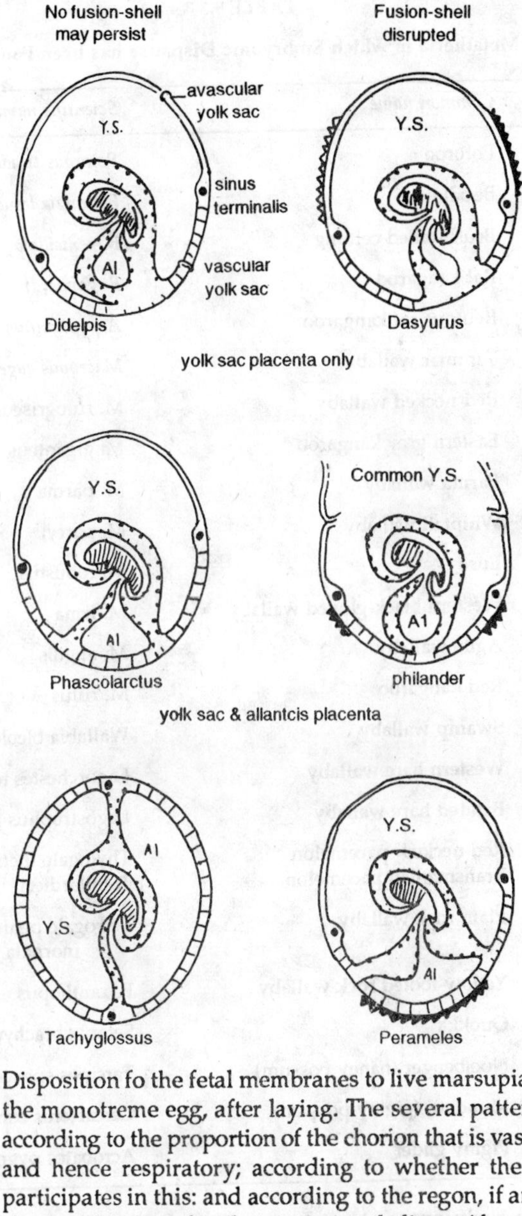

Fig. 9.4. Disposition fo the fetal membranes to live marsupials and in the monotreme egg, after laying. The several patterns differ according to the proportion of the chorion that is vascularized and hence respiratory; according to whether the allantics participates in this: and according to the regon, if any, of the chorion that invades the uterine euthelium. Al = Allantois, Am = Amnion, Y.S = Yk-sac.

TABLE 9.3

Metatheria in which Embryonic Dispause has been Found

Common name	*Scientific name*
Potoroo	*Potorous triadacytylus*
Boodle	*Bettongia lesueur*
Brush-tailed cettong	*B. penicillata*
Rat kangarod	*B. gaimardi*
Reufous rat kangaroo	*Aepyprymnus rufescens*
Tammar wallaby	*Macropus eugenii*
Red-necked wallaby	M. rufogriseus
Eastern grey kangaroo	M. giganteus
Parma wallaby	M. parma
Whiptail wallaby	M. parryi
Euro	M. robustus
Western black-gloved wallaby	M. Irma
Agile wallaby	M. angilis
Red kangaroo	M. rufus
Swamp wallaby	Wallabia bicolor
Western hare wallaby	Lagorchestes hirsutus
Banded hare wallaby	Lagostrophus fasciatus
Red-necked pacemelon	Thylogale thetis
Transmanian pacemelon	T. billardieri
Plain rock wallaby	Petrogale penicillata inornata
Yellow-footed rock wallaby	P. xanthopus
Quokka	Setonix brachyurus
Noolbenger (naney possum)	Tarsipes spencerae
Mundara (pigmy possum)	Cercartetus concinnus
Pigmy glider	Acrobates pygmaeus

uterus of a recipient up to 6 days more advanced in development than the recipient, but a blastocyst 2 days more advanced than the recipient will not develop. Transfer of tammer wallaby blastocysts into a recipient ovariectomized on day 8 showed that one-day-old as well as 8-day-old blastocysts developed in intact and ovariectomized females. This suggests that the blastocyst does not require a direct stimulation by the CL. Tyndale-Biscoe (1973) concluded, on the basis of various experiments, that "embryonic survival in these species beyond the blastocyst stage is completely dependent upon the onset of the luteal phase in the uterus, and that the corpus luteum provides a secretion which synchronizes the two phenomena." From these results it appears that the CL is necessary to induce uterine changes that are required for the development of the blastocyst, but the once these changes are induced, the CL is dispensable.

Eutheria

Ovum Transport. In most eutherian species, unfertilized eggs (oocytes) and fertilized eggs (zygotes) are transported, mainly by ciliary action, through the oviduct into the uterus after delays at the isthomoampullar and the tobo-uterine junctions. The duration of tubal transport for differnt species is summarized Table 9.4. Interestingly, for the three genera of bats in which the zygote remains in the oviduct for an extended time, there is a marked accentuation of secretory activity, after ovulation, in the oviduct containing the zygote or oocyte. This secretory activity may be a special adaptation of providing the zygote with a supportive environment. In the mare *(Equus caballus)*, the long-tongued bat *(Glossophaga soricina)* and possibly in the little bulldog bat *(Noctilio albiventris)*, zygotes are transported to the uterus, but unfertilized oocytes are retained in the oviduct. Why these three species differ from most other mammals and what mechanism is involved in the retention of the oocytes in the oviduct is unclear.

The incidence of oocytes that remain unfertilized is high in the American opossum *(Didelphis marsupialis)*, which ovulates up to 56 oocytes of which 30 per cent are not fertilized, and in the plains viscacha *(Lagostomus maximus)*, which may ovulate

TABLE 9.4

Chronology of preimplantation egg transport and development

Species	Tubal transport (days)	Development stage entering uterus	Implantation (days post fertilization)	Development stage at implantation
Marsupials				
Opossum (Didelphis)	1	1 cell, pronuclear	None, strictu sensu	
Native cat (Dasyurus)	1 (?)	1 cell, pronuclear	?	Probably late primitive streak stage
Insectivores				
Elephant shrew (Elephantulus myurus)	1-2 (?)	1-, 2- and 4-cell stages	?	Early blastocyst
Tree shrew (Urogale everetti)	?	Unilaminar blastocyst	?	Bilaminar blastocyst
Tree shrew (Tupaia longipes)	5-6 (est.)	Bilaminat blastocyst	?	Bilaminar blastocyst
Shrew (Sorex sp. Blarina brevicauda)	?	Small blastocyst	?	Bilaminar blastocyst
Chiroptera				
Vampire bat (Desmodus)	16+	Blastocyst, some entoderm differentiated	not less than 16 days	Blastocyst, bilamnar partially
Long tongued bat (Glossophaga)	12-14	Blastocyst	15	Blastocyst, bilaminar partially
Short-tailed bat (Carollia)	13-16	Blastocyst	?	?
Little brown bat (Myotis lucnugus)	?	Morula	not less than 10 days	Blastocyst, bilaminar
Free-tailed bat (Tadarida brasiliensis)	?	Advanced morula	?	Blastocyst, bilaminar

cont.....

Primates				
Rhesus monkey (Macaca)	4	16 cells	8-9	Blastocyst
Baboon (Papio sp.)	4-5	32 • cells	8-9	Blastocyst
Man (Homo)	3-4	?	6-7	Blastocyst
Rodents				
Rabbit (Oryctolangus curiculus)	3-3•	20+ cell morula	7	Bilaminar blastocyst
Mouse (Mus musculus)	3-3•	16+ cell morula	6	Blastocyst, some entoderm
Rat (Rattus norvegicus)	4	Morula	6	Blastocyst, some entoderm
Hamster (Cricetus auratus)	3•	4-8 cells	41/3	Blastocyst, some entoderm
Guinea pig (Cavia porcellus)	3•	8 cells	6	Unilaminar blastocyst
Mt. viscacha (Lagidium)	3-4	1-2 cells	7	Blastocyst
Carnivores				
Dog (Canis familiaris)	4-5	16 cell to young blastocyst	11-12'	Bilaminar blastocyst, 2500 •m diam.
Cat (Felis catus)	6-7	28-30 cell morula	13-14	Bilaminar blastocyst
Ferret (Mustela putorius)	5-6	32+ cells	12 •	Bilaminar blastocyst
Ungulates				
Horse (Equus caballus)	?	?	49-63	Elongaed chorionic vesicle with allantois
Pig (Sus scrofa)	2•-3 3/4	3-8+ cells	11-20(Progressivee)	Elongated chorionic vesicle with developed allantois
Sheep (Ovais aries)	2•2•	?	15-17 (progressive)	15-19 cm chorionic vesicle with allantois
Cow (Bos taurus)	4-4•	8-16 ceells	40+ (progressive)	Elongated chorionic vesicle with allantois

as many as 800 oocytes of which % to 99 per cent remain unfertilized. The mechanisms that account for these low rates of fertilization are unknown. Is there a deficit of capacitated sperm, or are the oocytes defective in some manner so that fertilization cannot occur?

Ovum transport through the oviduct can be affected by hormonal treatments, the effect depending on the species and the dose of hormone used. In the rabbit progestins seem to decrease the rate of oviduct transport during the first 2.5 days of pseudopenancy. In spite of experiments reported in Table 9.5, it is not clear whether the ovary is required for normal oviduct transport. For instance, ovariectomy of rats 7-27 h after presence of the vaginal plug or ovariectomy of guinea pigs 2 d after mating does not seem to affect ovum transport.

TABLE 9.5

Effect of Estradiolcyclopentyl Propionate on Ovum Transport in Different Laboratory Animals

Species	*Accelerated transport (μg)*	*Retention in oviduct (μg)*	*Dose Required for Interruption of pregnancy in 80% animals (μg)*	*Time eggs enter uter of controls (day post coitum)*
Guinea pig	50-100	250	10	4
Hamster	100	250	25	3 (afternoon)
Rat	10	No dose	10	5
Mouse	1	1	1	4
Rabbit	25	100	50	3*

*From literature

Development of the Zygote. During the sojourn in the oviduct, the oocyte may undergo changes; for example, the rabbit ovum loses its cumulus oophorus and corona radiata, but acquires an albumen layer. Furthermore, in most mammals, of course, fertilization takes place in the oviduct, with the exception of the Madagascar hedgehog (*Setifer setosus*), in which fertilization takes place in the ovarian follicle. The development of the zygote in most mammals thus starts in the oviduct. The extent to which the zygote develops in the oviduct is roughly proportional to the duration of its stay in the oviduct.

After reaching the uterus, oocytes and zygotes are transported by contradictions of the myometrium. In polyovular mammals, the zygotes are transported in such a way that they are equidistinct from each other at the time of implantation; this distance is thus inversely proportional to the number of zygotes that arrives in each uterine. The exact mechanism that makes such equidistant distribution possible is not known. In the rabbit, the stimulus for spacing is blastocyst expansion, which distends the uterus. Such blastocyst expansion does not occur in rats, mice, and other mammals, however; therefore, another mechanism must be involved, as Wimsatt (1975) has pointed out. The rabbit blastocyst eventually expands so much that it is not propelled further and the blastocyst is arrested. The expansion eventually causes muscular relaxation at the anti-mesometrial side and a ballooning of the wall to form an implantation dome. The role of steroid hormones in the spacing of embryos is not clear, although it may be expected to differ among species, as is the case with the effect of these hormones on transport of zygotes in the oviduct.

Zygotes in the uterus may not only be transported within the uterine horn in which they arrived, but they may also be transported to the other uterine horn. In some species, implantation is always in the right horn, although ovulation is either bilateral or sinistral, thus providing evidence that transuterine migration is common in these species. Transuterine migration has been found (1) in the bat (*Miniopterus natalensis*), which has a dominant left ovary, and in the little brown bat (*Myotis lucifugus*), with equal ovaries, and (2) in the following Artiodactyla with equal ovaries: the Uganda kob (*Adenota kob*), the common duiker (*Sylvicapra grimmia*), the dik dik (*Rhynotrachus kirki*), the impala (*Aepyceros melampus*), the muntjak (deer) (*Muntiacus muntjak*); and (3) in the water buck (*Kobus defassa*), which has a dominant left ovary. Transuterine migration has also been observed in species in which ovulation can occur from both ovaries, and implantation can occur in either horn or both horns of the uterus. Such migrations have been found among ungulates, carnivores, chiropteras and insectivores.

During the preimplantation period, the embryo takes up nutrients that are provided by oviductal secretions, by interstitial fluid and peritoneal fluid which enter the uterus, and by secretions from the endometrial glands and luminal epithelium of the uterus. In the rabbit, a protein, with a molecular weight of about 14,000-16,000, is secreted by uterine epithelial cells into the uterine lumen under the stimulation of progesterone. this protein, uteroglobulin, is secreted during the preimplantation phase and penetrates into the blastocyst; however, the function of this protein is still an enigma.

Site of Implantation. In most species, the entire endometrium of both horns of the uterus is receptive for implantation of the embryo. However, in some species the areas of receptiveness are restricted and the receptiveness may vary from only one horn to a small part of endometrium within a horn. The restricted areas in the uterus of ruminants, the so-called caruncles, are illustrative of such restricted sites.

Both blastocyst and uterus have to be ready for implantation, and the synchrony of development of each for implantation needs to be very close. In rabbits, the transfer of zygotes that are more than 2 or 3 days out of phase with the uterus of the recipient results in a high percentage of failures of implantation and embryonic mortality. Similar types of observations have been made for rats, mice and sheep. It is curious that the rather close synchrony required for implantation in the uterus does not apply to ectopic implantation. Mouse blastocysts transferred to such unlikely places as the anterior eye chamber, surface of the kidney, spleen or testis (1) can become implanted regardless of the endocrine status of the host.

Before discussing the sequence of the events involved in implantation, we should return to the embryo and consider the endocrinology of its survival. In the rat, the embryo can develop from the morula to the blastocyst stage after ovariectomy. In vitro two-celled rabbit embryos can develop into blastocysts in a defined medium without hormones. Progesterone can, however, increase the survival time of the blastocyst prior to implantation in ovariectomized rats and stimulate mitotic activity in the blastocysts. However, morulae transferred to ovariectomized rats treated with progesterone show a poor rate of survival. These data indicate that, although early embryos can survive and grow in an abormonal environment, their destiny is influenced by the hormones present.

The questions whether the blastocyst contains steroids and whether it synthesizes steroid hormones have not been answered unequivocally, as has been pointed out by Bullock (1977) and Sauer (1979). In rabbits, rats, and mice, steroid hormones may be present in the blastocyst, but even on this point there is disagreement. Even if it is accepted that the blastocysts contain steroids, then they may have their origin in uterine fluids and may not have been formed within the blastocyst.

As we shall see later in this chapter, in the rat, as little as 2-4 ng estradiol-17β injected locally into the uterus results in implantation of blastocysts in ovariectomized progesterone-treated rats. The question whether estrogen produced by or present on the blastocyst can induce implantation is thus an important one. After studying the concentration of receptors at the site of implantation and at interimplantation sites, in ovariectomized progesterone-treated rats injected systemically with 0.25 μg estradiol, Martel and Psychoyos (1981) concluded that the blastocyst did not secrete estrogens. Steroidogenesis has been established in the pig blastocyst, however. The estrone secreted by the pig blastocyst, as early as day 12 of gestation, is probably sulfated in the endometrium, then transported to the ovary, where it is hydrolyzed, and the resulting estrone then stimulates the maintenance of the CL.

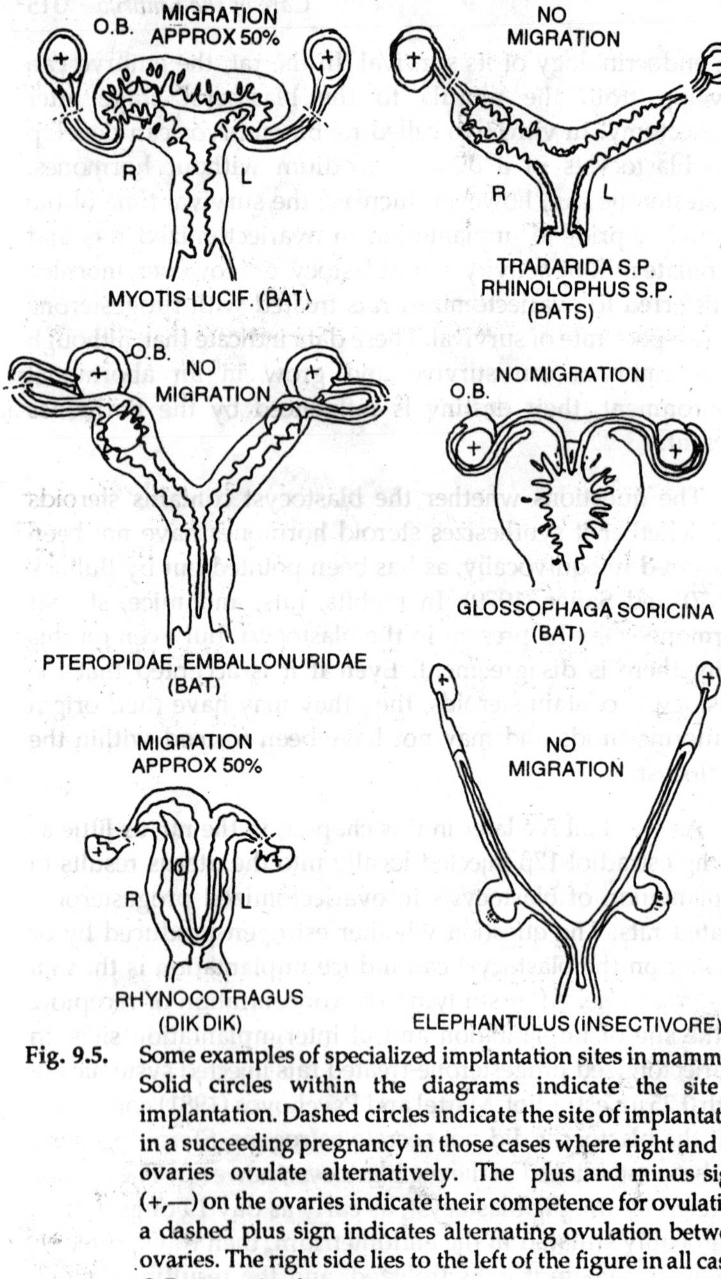

Fig. 9.5. Some examples of specialized implantation sites in mammals. Solid circles within the diagrams indicate the site of implantation. Dashed circles indicate the site of implantation in a succeeding pregnancy in those cases where right and left ovaries ovulate alternatively. The plus and minus signs (+,—) on the ovaries indicate their competence for ovulation; a dashed plus sign indicates alternating ovulation between ovaries. The right side lies to the left of the figure in all cases. O.B. = ovarian bursa.

[illegible]*on.* Implantation can be defined of the establis[illegible]ment of contract between the tophoblast of the differentiating blastocyst and the uterine wall, with the subsequent erosion of the uterine epithelium, and invasion and decidualization of the stroma by the blastocyst, so that it becomes embedded in the uterine wall. In swine, sheep, cude, and horses, no direct contact between trophoblast and endometrial stroma becomes established, and stromal decidualization does not occur, and in these cases the relation between embryo and uterus will, therefore, be called attachment and not implantation.

In the preparation of the uterus prior to implantation in the rat and mouse, the lumen takes on a slitlike appearance after ovulation, as the result of the loss of uterine fluid. Eventually, on day 5, the microvilli from opposing luminal surfaces are interlocked, so that the lumen is closed. On day 6 the luminal surface loses the regular microvilli and comes in close contact with the surface of the cells of the opposite wall. The lumen at this time is only about 1500 nm wide. The surface of the epithelial cells resembles that found during the attachment reaction of the trophoblast to the surface of the epithelium. Proliferation of the endometrial glands, which secrete a mucopolysaccharide may serve to nourish the embryo. There is little or no stromal differentiation prior to implantation, but after implantation the so-called decidual reaction, or decidualization, is principally the result of differentiation of stromal cells.

The preparation of the endometrium for implantation is under hormonal control and is dependent upon a complicated interaction between progesterone and estrogen. Progesterone is required for preparation of the uterus for implantation. In rats, estrogen made available endogenously or exogenously subsequent to this progesterone-induced preparation causes a short period of receptively for the blastocyst, so that implantation can occur. If implantation fails to take place, the uterus becomes either non-receptive or toxic to zygotes. This explains the precise synchrony between zygote and prepar[illegible]erus. The non-receptive state is maintained as long as p[illegible]sterone is available to the uterus. When secretion or administration of progesterone is terminated, the uterus recovers, and a neutral state prevails until the secretion of progesterone re-establishes the receptivity of the uterus.

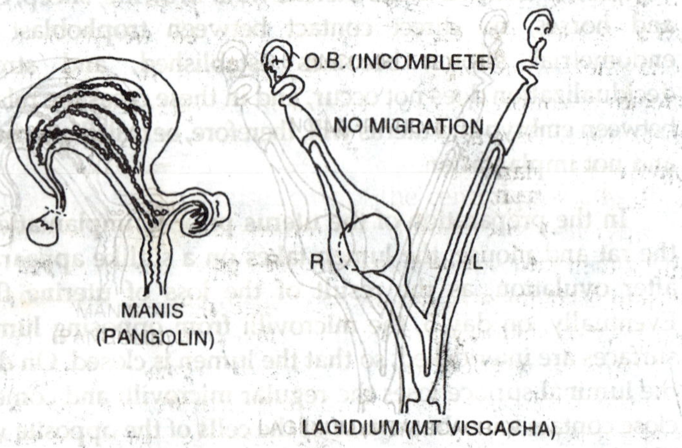

Fig. 9.6. Specialized implantation sites in the pangolin and the mountain viscacha. The beaded strips on the uterine wall in the pangolin indicate the location of the placental stripes of this species. The right horn lies to the left side in these figures. Note that the left ovary in the mountain viscacha is non-functional. O.B. = ovarian bursa.

During the reproductive cycle of the rat, coitus causes maintenance of the CL and progesterone secretion, and at day 2 estradiol concentrations in the plasma increase and remain elevated until about day 8. The details of the hormonal regulation have not been worked out in as great detail in the mouse, but the situation may be similar. It seems from an evolutionary point of view, however, most likely that the timing and the control of implantation in different species is correlated with the hormonal concentrations found in early pregnancy.

At the site of implantation there is a considerable increase in vascular permeability in the codometrium in response to the presence of the blastocyst. In the rat, there is an estrogen surge at the time of implantation. The surge and the ability of estrogen to induce implantation in the case of lactation-induced and experimentally, induced delayed implantation regests that estrogen induces changes in the blastocyst-endometrial relationship, although the nature of the blastocyst stimulus is still unknown. There is evidence that histance and/or prostaglandins (PGs) may be involved changes in vascular permeability in the uterus. The observation by Shemesh *et al.* (1979) that the bovine blastocyst can synthesize PGs. is suggestive for a possible role of these hormones in changing endometrial permeability. The estrone secreted by the pig embryo may serve as a luteotrophic signal, but it can also affect the uterine endometrium and thus affect the attachment of the blastocyst.

Preimplantation losses of about of 30 per cent of the embryos are found in the American opossum and the Australian native cat (*Dasyurus viverninus*), which ovulate on the average of 22 and 20 to 25 oocytes, respectively. In the pronghorn (*Antilocapra americana*), the preimplantation loss of embryos is about 50 per cent. The chorionic vesicles develop a necrotic tip at the oviductal end. The 4-5 embryos are initially about evenly distributed between a distal portion of the lumen nearer to the oviduct and a proximal portion near the uterine end. As the proximal vesicle expands, its necrotic tip eventually destroys the distal vesicles, so that only two embryos survive.

In most mammals, implantation of the embryo occurs when it has reached the blastocyst stage, except in ungulates, in which an elongated chorionic vesicle with allantois becomes attached to the endometrium. Implantation can be classified according to the degree of penetration of the uterine mucosa by the blastocyst, or according to the orientation of the blastocyst in the uterus, or according to the point of first attachment of the trophoblast. The degree of penetration of the uterine mucosa

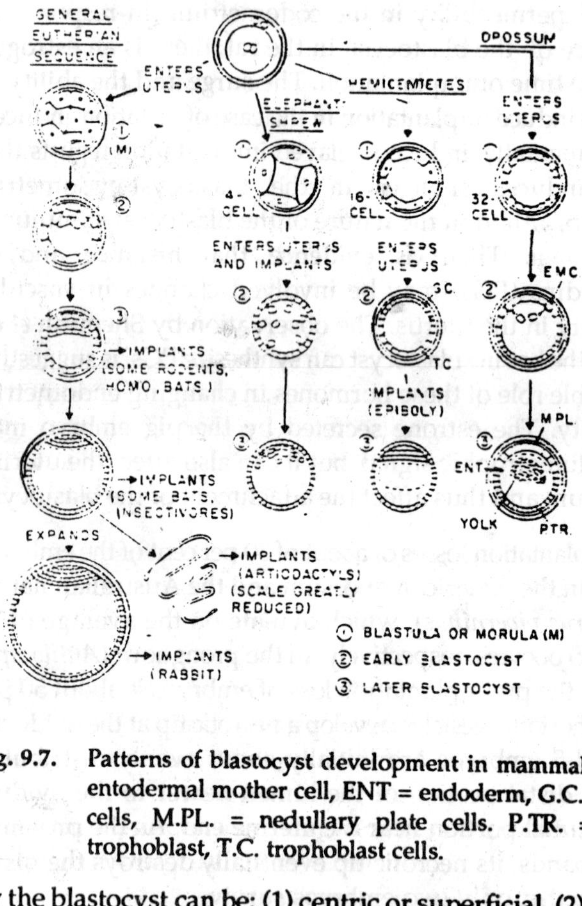

Fig. 9.7. **Patterns of blastocyst development in mammals. E.M.C. = entodermal mother cell, ENT = endoderm, G.C. = germinal cells, M.PL. = nedullary plate cells. P.TR. = primitive trophoblast, T.C. trophoblast cells.**

by the blastocyst can be: (1) centric or superficial, (2) eccentric, (3) partly interstitial, and (4) interstitial. Orientation of the blastocyst in the uterus can be such that the direction of the germ disc is: (1) mesometrial, with the disc facing toward to mesometrium, (2) antimesometrial with the disc facing away from the mesometrium; and (3) lateral, with the disc facing a position between 1 and 2. According to the point of first trophoblastic attachment, there are diffuse, antimesometrial, mesometrial, lateral, bilateral, circumferential (or equatorial), and cotyledonary attachments, with the cotyledonary type being a special case of diffuse attachment. The advantages of these various patterns of implantation are unknown. The reader is referred to Wimsatt's (1975) excellent critical review and interpretation of the occurrence of the different patterns of orientation of first trophoblastic attachment and disc attachment. In the Insectivora and Chiroptera, these show a great deal of variation among orders and suborders, in contrast to the lack of variation in rodents and primates.

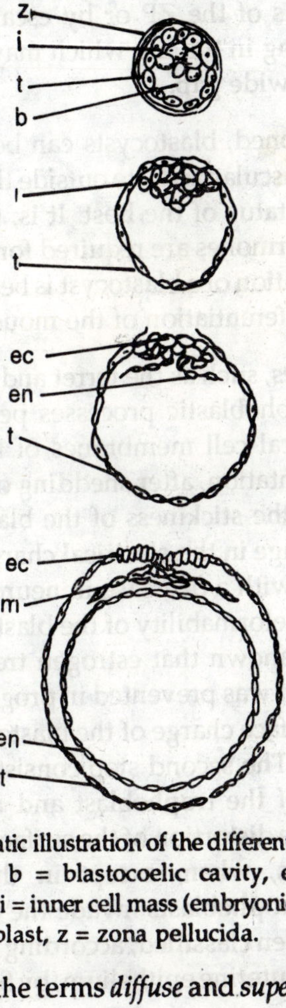

Fig. 9.8. Diagrammatic illustration of the differentiation of a mammalian blastocyst. b = blastocoelic cavity, ec = ectoderm, en = endoderm, i = inner cell mass (embryonic disc), m = mesoderm, t = trophoblast, z = zona pellucida.

We have used the terms *diffuse* and *superficial* implantation and pointed out the *cotyledonary* implantation is a special case of diffuse attachment.

The initial phase of implantation consists of adhesion of the blastocyst to the uterine epithelium. In some species, e.g., the rat and the armadillo, the earliest stages of implantation consist of interdigitation of the microvilli of the blastocyst and the uterine epithelium. In these species the blastocyst is free from the zona pellucida (ZP). This shedding of the ZP can take place by lysis of the ZP or by escape of the blastocyst through an opening in the ZP, which may be either a narrow escape route or a wide gap.

As we mentioned, blastocysts can become implanted in almost any well-vascularized site outside the uterus, regardless of the hormonal status of the host. It is, therefore, not likely that the uterine hormones are required for the shedding of the ZP. The differentiation of a blastocyst is beautiful micrographs illustrating the differentiation of the mouse blastocyst.

In some species, such as the ferret and the rabbit, the ZP is not shed, and trophoblastic processes penetrate the ZP and adhere to the apical cell membranes of the uterine luminal epithelium. Implantation, after shedding of the ZP, consists of three steps. First, the stickiness of the blastocyst changes, as the result of a change in the electrical charge of the blastocyst. This is correlated with a decrease in neuroaminic acid which may increase the deformability of the blastocyst. Nilsson *et al.* (1975) have since shown that estrogen treatment of mice, in which implantation was prevented in progesterone, induced a decrease in the surface charge of the blastocyst and decreased its occur genicity. The second step consists of interdignation of the microvilli of the trophoblast and uterine epithelium. The third step is the distortion of the surface of the trophoblast and of the uterus, when, except in the case of diffuse implantation, the trophoblasts invade the uterine epithelium. This process has been classified, according to the nature of the involvement of the uterine epithelium, by Schlafke and Enders

(1975) as intrusion, displacement, or fusion implantation. In intrusion implantation, as found, e.g., in the ferret (*Mustela putorius*), the trophoblast penetrates between the uterine luminal epithelium to the basal lamina of the uterine epithelium and extends beneath it. In displacement implantation, as found, e.g., in the rat and mouse, the uterine luminal epithelium is readily dislodged from the basal lamina, and the trophoblast comes to lie along areas previously occupied by these displaced uterine cells. In fusion implantation, as found, e.g., in the rabbit, the trophoblast penetrates by fusion of an area of the syncytial trophoblast to individual uterine luminal cells.

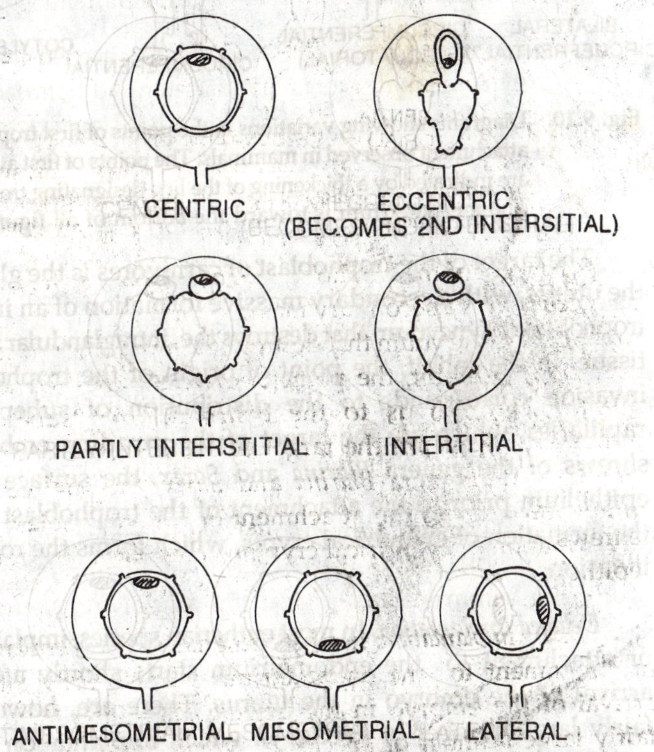

Fig. 9.9. Implantation patterns and orientations. The upper four drawings illustrate the common implantation patterns observed in mammals. The lower three drawings illustrate the orientation of the embryonic cell mass in relation to the mesometrium. The mesometrium is directed toward the bottom of all figures.

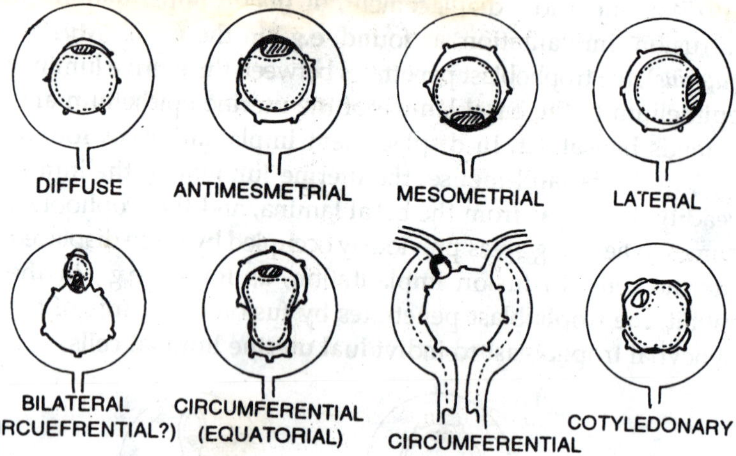

Fig. 9.10. Diagrams showing variations in the points of first trophoblastic attachment observed in mammals. The points of first attachment are indicated by a thickening of the line designating trophoblast. The mesometrium is toward the bottom of all figure.

The target of the trophoblast of carnivores is the glands of the uterus, with a secondary massive formation of an invasive trophoblastic syncytium that destroys the intraglandular stromal tissue. In the rabbit, the point of origin of the trophoblastic invasion corresponds to the distribution of subepithelial capillaries, which are the target of the invading embryo. In shrews of the genera *Blarina* and *Sorex,* the surface of the epithelium prior to the attachment of the trophoblast shows the formation of cylindrical crypts, which forms the routes of invasion.

Delayed Implantation. In most eutherian species, implantation or attachment to the endometrium starts shortly after the arrival of the embryo in the uterus. There are, however, a fairly large number of species in which implantation of the embryo is delayed. The species showing such delayed implantation can be divided into two groups, species with obligatory delayed implantation and species with facultative, usually lactation-induced, delayed implantation.

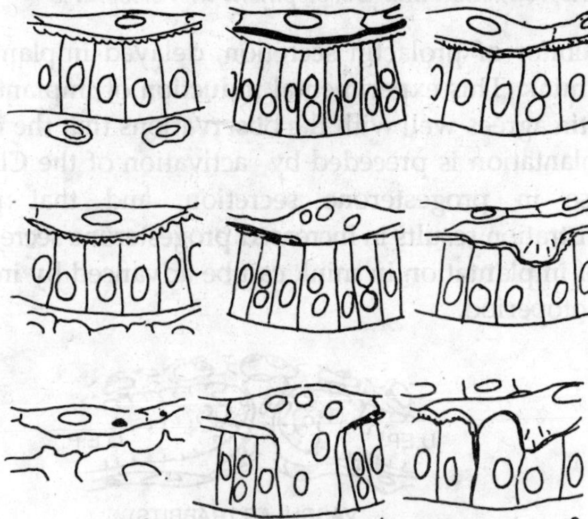

Fig. 9.11. Three types of interaction of trophoblast with uterine epithelium. epithelium during penetation of this. In all three types, apposition and adhesion precede penetration. However in (a), displacement penetration (rat, mouse) the uterine luminal epithelium is readily dislodged from the basal lamina, and the trophoblast come to lie along areas previously occupied by the placed uterine cells. In (b), fusion penetration the bit), the syncytial trophoblast fuses with a uterine luminal epithelial cell. In (c), intrusion penetration (ferret, others?), projections of the synthetic trophoblast penetrate between uterine epithelium cells.

Obligatory delayed implantation. found in the species. During the period of delay, the embryo is in embryonic is diapause, and little growth or expansion occurs during this period in most species; in others, such as the roe deer (*Capreolus capreolus*), the rate of growth of the embryo is slow.

The endocrinology of implantation in most species with obligatory delayed implantation, except the mink (*Mustela vison*), is still largely a mystery. In the mink, which has a short and variable period of delay, if mating is early, the delay is long, if mating is late, there may be no delay. Implantation can be induced by prolactin administration in intact and in hypophysectomized mink. Administration of bromocryptine, an inhibitor of prolactin secretion, delayed implantation in intact mink. This experimental induction of implantation by prolactin agrees well with the observations that the initiation of implantation is preceded by activation of the CL and an increase in progesterone secretion, and that prolactin administration results in increased progesterone secretion. The time of implantation in mink can be advanced by increasing the photoperiod.

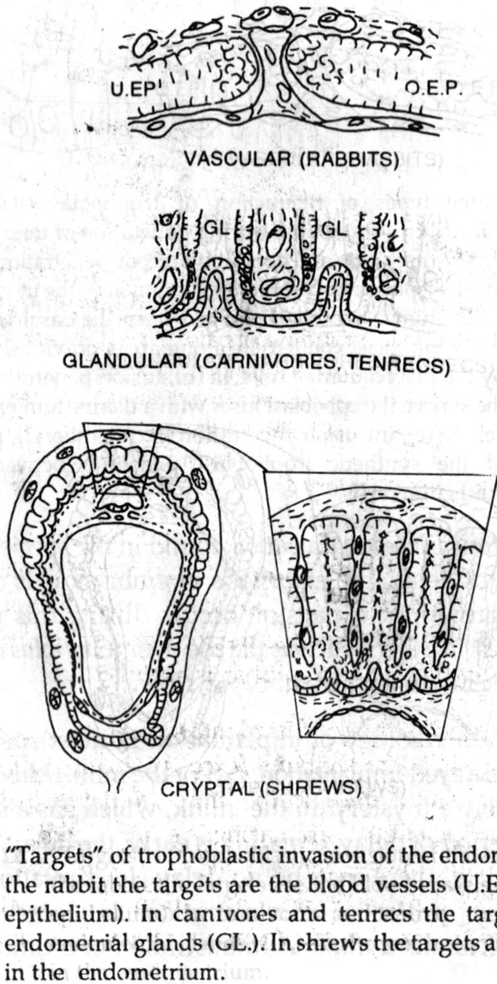

Fig. 9.12. "Targets" of trophoblastic invasion of the endometrium. [In the rabbit the targets are the blood vessels (U.EP. = uterine epithelium). In carnivores and tenrecs the targets are the endometrial glands (GL.). In shrews the targets are the crypts in the endometrium.

In the European badger (Canivenc and Bonnin, 1981), the western spotted skunk, and the short-tailed weasel implantation is preceded by an increase in progesterone secretion; however, administration of progesterone has failed to induce implantation in the European badger and the western spotted skunk. In the European badger, neither prolactin not LH administration has induced premature blastocyst implantation, although LH concentrations increase prior to spontaneous implantation.

In the nine-banded armadillo, plasma progesterone increased just prior to implantation, but implantation could be induced experimentally by ovariectomy, however, unilateral ovariectomy, independent of the presence of a CL on the removed ovary, did not have this effect. The findings that administration of progesterone (24 mg/d) or estrone (0.2 mg/d) inhibited ovariectomy-induced implantation makes it difficult to reconcile these findings with the increased progesterone concentration at the time of implantation.

In the roe deer, an increase in plasma estrogen and progesterone concentration is associated with implantation, but according to Aitken (1981), these changes are a consequence of implantation and not the cause of it. throughout the period of embryonic diapause, the ovaries look active histologically.

It has been possible to induce premature implantation of the blastocyst in the western spotted skunk, the American pine marten, the long-tailed weasel, and the sable by exposing the pregnant females to *increasing* day length and in the European badger and possibly the river otter (*Lutra canadensis*) by exposing the pregnant females to *decreasing* day length.

From the available data, it is not possible to formulate a generalization about the endocrine causes of implantation in species with obligatory delayed implamantation. It is not as difficult to understand why in the western subspecies of the spotted skunk (*Spilogale putorius latifrons*), which mates in the fall, delayed implantation is part of the reproductive cycle, but in the eastern subspecies (*Spilogale putorius ambarvalis*), which breeds in the spring, implantation is not delayed. Both subspecies give birth in the spring and, apparently for the western subspecies, mating in the fall was advantageous, whereas for the eastern subspecies, mating in the spring was advantageous. The ecological basis for the difference between these two subspecies with respect to the time of mating is probably quite complicated.

Facultative delayed implantation is induced by lactation and is found in the species listed in Table 9.4 The endocrinology of this type of delayed implantation has been studied in most detail in the rat and mouse. If the mouse or rat mates and conceives at postpartum estrus, the oocytes do not become implanted if the mother is using a critical number of pups. This critical number varies probably among strains within the two species. The lactation induced and maintained by the suckling young inhibits FSH release and thus prevents the surge of estrogen secretion, which normally occurs on day 4 post coitum and which is necessary for implantation. The length of the delay depends on (1) the number of young being nursed, (2) the parity of the mother, the delay becoming shorter with increased parity, and (3) the continued presence of the young, temporary removal reducing duration of the delay.

TABLE 9.6

Euthenan Species with Facultative Delayed Implantation

Rodentia
Lesser short-tailed gerbil, *Dipodillus simoni*
Mongolian gerbil, *Meriones ungulculatus*
Shaws jird, *M. shawi*
Fat-tailed gerbil, *M. crassus*
Bank vole, *Chethrionomys glareolus*
Red-tree mouse, *Phenacomys longicaudus*
Oldfield mouse, *Peromyscus polionotus*
Deer mouse, *P. maniculatus*
While footed mouse, *P. leucopus*
Cotton mouse, *P. gossypinus*
Pinon mouse, *P. truei*

Northern grasshoppermouse, *Onychomys leurogaster*
House mouse, *Mus musculus*
Laboratory rat, *Rattus norvegicus*
Australian brush rat, *R. fuscipes*
New Holland mouse, *Pseudomys novaehollandiae*
Fawn hopping-mouse, *Notomys cervinus*

Insectivores
Common shrew, *Sorex araneus*
Pygmy shrew, *S. minutus*
European water shrew, *Neomys fodiens*

Experimentally, delay can be induced by either hypophysectomy or ovariectomy, prior to the afternoon of day 4 post coitum thus preventing the estrogen release mentioned above, or by administration of a progesterone, such as medroxy-progesterone acetate. After ovariectomy, implantation can be induced by systemic administration of progesterone and estrogen in sequence, or by systemic administration of progesterone followed by local injection of as little as 2-4 ng estradiol-17β into the uterus. In the ovariectomized golden hamster (*Mesocricetus auratus*), progesterone administration alone is sufficient for the induction of implantation, but administration of progesterone and estrogen is followed by a higher incidence of implantations than progesterone administration.

In the mouse, administration of FSH is followed by implantation, whereas in the rat apparently FSH and LH are required for the induction of implantation, although MacDonald et al. (1967) obtained implantation with LH alone. Gidley-Baird (1981) has concluded that in the rat and the mouse, prolactin and LH are required for the progesterone secretion necessary for implantation, and that in the mouse FSH, and in the rat FSH and LH are required for the estrogen secretion.

In the common shrew (*Sorex araneus*) and the European water shrew (*Neomys fodiens*), the delayed implantation that occurs after a postpartum conception is in a sense an obligatory delayed implantation, because the gestation period is shorter than the (for the suckling young) indispensable lactation period, by four days for the common shrew and by eight days for the European water shrew. Implantation has to be deferred to prevent the necessity of having to nurse, simultaneously, two litters of different ages.

The endogenous or exogenous estrogen, which results in implantation in the rat and mouse, also increases the vascular permeability, as mentioned earlier in this chapter.

Immunological Aspects of Implantation and Gestation. An important fundamental question is why the embryo is not rejected by the uterus, because it is — except possibly for highly inbred lines — a foreign graft. We discuss briefly some of the possible mechanisms and the evidence for and against each proposed mechanism:

1. Since the fetus has 50 per cent of the genetic endowment of the mother, the rejection may not occur because of genetic similarity. The argument is invalid, because: (*a*) mothers do reject skin grafts of their offspring, and (*b*) eggs transferred from a totally unrelated donor become implanted and are carried to term normally.

2. One can consider that the mother lacks an immunological response, because the uterus is an immunologically privileged site, similar to the anterior eye chamber or the cheek pouch of the hamster. Arguments against this explanation are: (*a*) that most of the immunologically privileged sites lack lymph vessels, whereas the uterus has lymph vessels, and (*b*) that ectopic pregnancies, such as occur on the outside of the intestine or elsewhere in the abdominal cavity, are not rejected. Moreover, blastocysts placed under the kidney capsule experimentally are not rejected.

3. Pregnancy can be seen as reducing the maternal Immunological response. The main argument against this explanation is that skin grafts are equally rejected by pregnant and non-pregnant cows and mice, although there is some

evidence that skin grafts survive longer in pregnant than in non-pregnant rabbits.

4. The trophoblast, and later thee fetus, cannot induce an immune response. This explanation seems invalid because transplantation antigens can be detected early during embryonic development.

5. There is a non-cellular barrier with a high sialic acid concentration, which protects the embryo against the immune response of the mother however, trophoblast and other fetal antigens do reach the immunological system of the mother.

The Placenta. The placenta takes over the nutrition of the embryo, the exchange of gases between mother and embryo, and the removal of waste products from the embryo. In addition, the placenta serves as a mechanical barrier against shocks and, depending on the species, may secret gonadotrophic hormones (as in the horse and human), a lactogenic hormone (as in humans), and steroid hormones. The placenta develops either as a chorio-vitelline (yolk-sac) placenta or a chorio-allantoic placenta. The chorio-vitelline placenta is usually a transient structure in the Eutheria, but, as we have seen earlier in the chapter, is the definitive placenta of marsupials.

Placentas can be classified in different ways for instance, according to shape or according to the number of layers of tissue that separate fact and maternal blood streams.

Classification by Shape. Classification by shape gives the following groups:

1. *Diffuse Placenta.* Most of the outer surface of the chorion is covered by small villi or folds, which are in intimate contact with corresponding depressions of the endometrium. This type is found in swine, horses and donkeys.

2. *Cotyledonary or Multiplex Placenta.* The chorionic villi (cotyledons) are restricted to well-defined circular or oval areas on the chorion. The areas between these villi are relatively

smooth. The fetal cotyledons, which overlie the caruncles of the uterus, form a hand-in-glove arrangement with the caruncles. This type is found in cattle, sheep and goats.

3. *Zonary Placenta.* The chorionic villi are present as a band of placental tissue, which encircles the equatorial region of the chorionic sac. Complete girdles of villi are found in the dog, cat, and spotted hyaena (*Crocuta crocuta*); incomplete girdles of villi are found in the mink, the ferret, the polar bear (*Thalarctos maritinus*), and the brown bear (*Ursus arctos*). The placenta of carnivores shows the presence of so-called hemophagous organs. These consist of either marginal or central effusions of maternal blood, which is in close contact with cells of the chorion that apparently ingest the red blood cells. Marginal effusions are found in the cat, the dog, the coyote (*Canis latrans*) and the red fox (*Vulpes vulpes*); central effusions are found in the river otter (*Lutra canadensis*), the Eurasian otter (*L. lutra*), the badger (*Meles meles*), the raccoon (*Procyon lotor*), the wolverine (*Gulo gulo*), the ferret, and the mink.

4. *Discoid Placenta.* The placenta is discshaped and may be single, as in the human,or double, as in the primate families Ceropithecidae and Cebidae.

Classification by Number of Layers. For classification of the placenta based on the number of layers hat separate maternal and fetal circulations, because it agrees well with the data obtained in investigations of the fine structure of the placenta of different groups. The types of placenta and the components of the barrier between the maternal and fetal circulation are listed in Table 9.5 This table illustrates the considerable variation in the number of layers that form the barrier between maternal and fetal circulations. Steven (1975) emphasized that it should not be concluded that the efficiency of placental transfer increases as the number of layers decreases. In fact, he emphasized that the greater the number of placental layers, the greater the development of the animal at birth, suggesting a greater efficiency of nutrient transfer as the number of placental layers becomes larger.

TABLE 9.7

Classification of Eutherian Placentas, Based on Number of Layers of Tissue Separating Fetal and Maternal Circulation

Type of placenta	*Layers of tissue*	*Animals in which found*
Epitheliochorial	Endothelium of fetal capillaries, basement membrane of fetal capiliaries basement membrane of chorion, cells of chorion indented by fetal capillaries, zone of interdigitation of fetal and maternal microvilli, maternal epithelium, basement membrane of maternal epithelium, layer of scattered collagen fibers, basement membrane of maternal capillaries, endothelium of maternal capillaries	Artidactyla, Perissodactvla, Cetacea, Lemuridae
Epitheliochorial	Endothelium of fetal capillaries, basement membrane of cappillary (in late stages of gestation), basement membrane of trophoblast, layer of syncytial trophoblast, interstitial membrane, endothelium of maternal capillaries	Carnivora, Pinnepedia
Anmomonochorial	Endothelium of fetal capillaries, basement membrane of fetal capillaries, basement membrane of trophoblast, cytotrophoblast syncytiotrophoblast	Humans, armadillos, guinea pigs
Manodichorial	Endothelium of fetal capillaries, basement membrane of fetal capillaries, basement membrane of trophoblast, inner layer of trophoblast, outer layer of trophoblast	Rabbits, chipmunks
Memotrichorial	Endothelium of fetal capillaries, basement membrane of fetal capillaries, inner layer of trophoblast, middle layer of trophoblast, outer layer of trophoblast	Rats, mice, hamsters

Pregnancy Recognition. The secretion of gonadotrophins by the fetus may function as a signal to the ovary that the CL needs to be maintained for maintenance of the pregnancy; alternatively, steroid secretion by the placenta can serve to maintain pregnancy independent of the ovary. For some species, such as the dog, cat, and ferret, the secretion of progesterone for maintenance of pregnancy is largely independent of the presence of a fetus.

TABLE 9.8

Effect of Removal of the Ovaries of Pituitary, in First and Second Halves (Approximate) of a Pregnancy, on the Maintenance of Gestation in Various Species

Animal	*Length of gestation (days)*	*Ovariestomy*		*Hypophysectomy*	
		First half	*Second half*	*First half*	*Second half*
Sheep	148	-	+	-	+
Human	280	+	+	+	+
Monkey	165	+	+	+	+
"Guinea pig	68	±	+	+	+
Rat	22	-	±	—	+
Mouse	23	-	-	±	+
Hamster	16	-	-	-	±
Cat	63	-	±	n.d.	±
Cow	282	-	±	n.d.	n.d.
Ferret	42	-	±	—	±
Horse	350	-	+	n.d.	n.d.
Dog	61	-	n.d.	-	±
Goat	148	-	-	-	-
Sow	113	-	-	-	-
Rabbit	28	-	-	-	-
Opossum	16	-	-	n.d.	n.d.
Blue fox	52	-	-	n.d.	n.d.
Armadeo	150	Implantation may occur	-	n.d.	n.d.

+, Fetuses survive; =, some fetuses survive; -, fetus aborts; n.d. not determined.

We have discussed the possible mechanisms that may be involved in maintenance of the CL in general. In order to understand how gestation is maintained in various species, it is necessary to know whether the hypophysis and/or ovary are required for maintenance of pregnancy. Table 9.6 shows the species in which pregnancy is continued after ovariectomy or AP. To this should be added species in which the CL disappears during gestation, as in the African bats *Nycteris luteola* and *Traeniops after*, the black rhinoceros (*Diceros bicornis*), and the horse.

In the cases in which gestation is maintained after oophorectomy, it may be assumed that the placenta produces the hormones required for gestation. There are alternate possibilities for those cases in which pregnancy is maintained after hypophysectomy; i.e., the CL is independent of the pituitary or the placenta produces luteotrophic hormone, maintaining the CL, or the placenta produces the steroid hormones required for maintenance of gestation. On the basis of the available evidence, it appears that luteotrophins are produced by the placentas in monkeys, humans, horses, rats, mice, sheep and goats.

The endocrine relationships in the maintenance of pregnancy in (1) the rabbit, which is AP- and ovary-dependent; throughout pregnancy; (2) the rat, which during the first half of pregnancy is AP- and ovary-dependent and during the second half is AP-independent but ovary-dependent; (30 the sheep, which is ovary-and AP-dependent during the second half of pregnancy and, and (4) the human, who is early in gestation independent of AP and ovary for maintenance of pregnancy.

In the rabbit, the principal luteotrophin is probably LH. this maintains the estrogen secretion by the follicles, which, in turn, stimulates progestin secretion by the CL necessary for maintenance of gestation. In the rat, thee stimulus of coitus induces the release of prolactin (twice peer 24 h) and, together with LH, it maintains progestin secretion of the CL, whether the mating is fertile or not. By day 11, the placenta starts to secrete a luteotrophic complex which maintains the CL beyond the time of pseudopregnancy, and by day 18, the uterus secretes sufficient progesterone to maintain gestation, de Greef *et al.* (1977) have obtained evidence that the presence of decidual tissue in pseudopregnant rats increases the plasma progesterone concentrations in peripheral blood. The exact mechanism by which this occurs needs to be elucidated further, but the point of interest here is that progesterone can be secreted by the uterus in the absence of a placenta.

In the sheep, either LH alone or LH and prolactin maintain secretion of progestins, but the CL of pregnancy either is maintained because of an inhibition of the secretion of the uterine luteolytic factor or because the embryo secretes a luteotrophic hormone. After about 50 days, the placenta secretes sufficient progesterone itself to maintain pregnancy.

In the rhesus monkey (and possibly in the human), the regression of the CL, which would have occurred if no pregnancy had intervened, is prevented by the secretion of chorionic gonadotrophin. The "rescue" of the CL, which occurs at about day 9-12 of pregnancy, maintains the pregnancy until about day 25, when the CL starts to secrete enough progesterone for maintenance of pregnancy.

The human placenta, although it produces steroid hormones, lacks the complete set of enzymes required for synthesis of all the hormones it produces. This apparent problem is solved by a complicated exchange of precursors and hormones between fetus and placenta. The so-called feto-placental unit is responsible for the final hormone production. The human placenta secretes, in addition to hCG, another protein hormone, human placental lactogen (hPL), also called human chorionic somatomammotrophin (hCG). This hormone acts synergistically with hCG to regulate and promote progesterone and estrogen production by the placenta.

The equine placenta produces a gonadotrophic hormone, pregnant mare's serum gonadotrophin (PMSG), which has

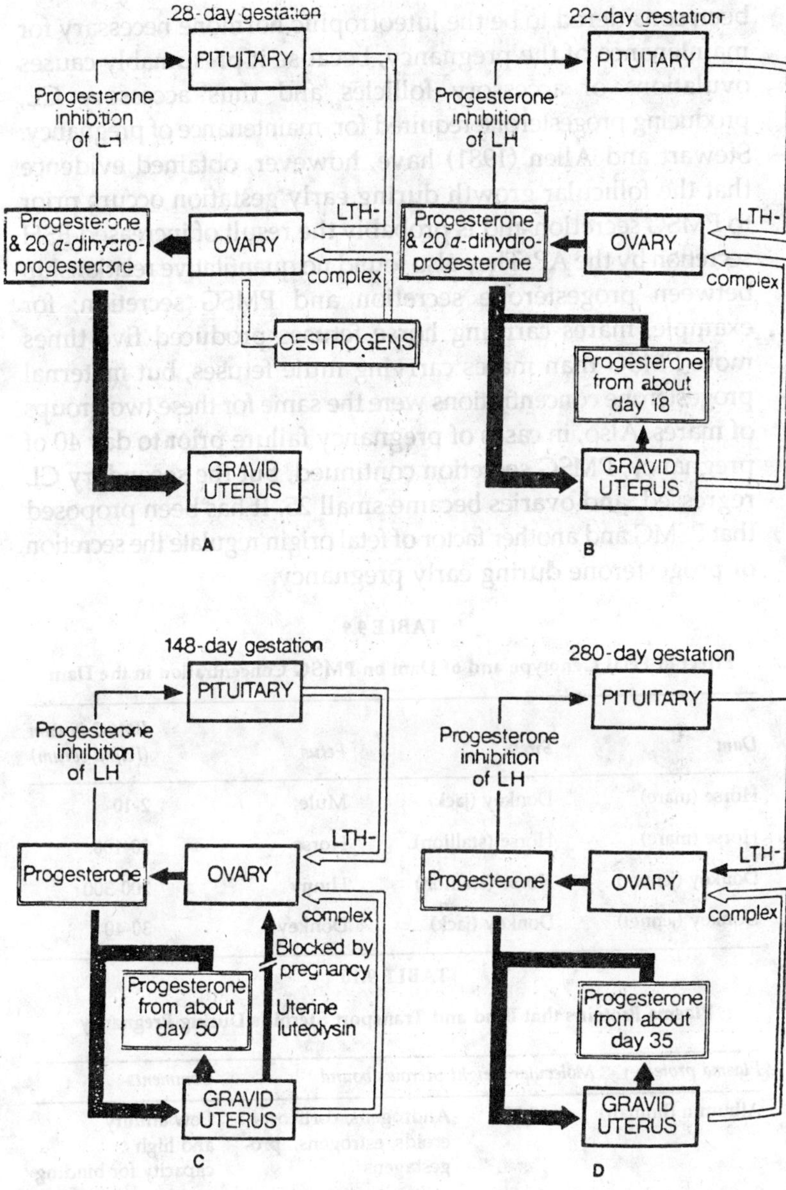

Fig. 9.13. Hormonal interrelationships in the maintenance of pregnancy in the rabbit (*A*), rat (*B*), sheet (*C*), and human (*D*).

been considered to be the luteotrophic hormone necessary for maintenance of the pregnancy, because it presumably causes ovulations of accessory follicles and thus accessory CL, producing progesterone required for maintenance of pregnancy. Stewart and Allen (1981) have, however, obtained evidence that the follicular growth during early gestation occurs prior to PMSG secretion and is probably the result of increased FSH secretion by the AP. They also found no quantitative relationship between progesterone secretion and PMSG secretion; for example, mares carrying horse fetuses produced five times more PMSG than mares carrying mule fetuses, but maternal progesterone concentrations were the same for these two groups of mares. Also, in cases of pregnancy failure prior to day 40 of pregnancy, PMSG secretion continued, but the secondary CL regressed, and ovaries became small 25. It has been proposed that PSMG and another factor of fetal origin regulate the secretion of progesterone during early pregnancy.

TABLE 9.9

Effect of Fetal Genotype and of Dam on PMSG Concentration in the Dam

Dam	*Sire*	*Fetus*	*PMSG in dam (IU/ml serum)*
Horse (mare)	Donkey (jack)	Mule	2-10
Horse (mare)	Horse (stallion)	Horse	20-100
Donkey (jennet)	Horse (stallion)	Hinny	200-300
Donkey (jennet)	Donkey (jack)	Donkey	30-40

TABLE 9.10

Plasma Proteins that Bind and Transport Steroids During Pregnancy

Plasma protein	*Molecular weight*	*Steroids bound*	*Comments*
Albumin (human)	69,000	Androgens, corticosteroids, estrogens, progestagens	Low affinity and high capacity for binding
a_1-Acid glycoprotein, AAG (human)	41,000	Progesterone, testosterone	Medium affinity and medium capacity for binding

Transcortin, CBG (human, monkey, guinea pig, rat and many other species)	52,000	Cortisol, corticosterone, progesterone	High affinity, low capacity for binding
Sex steroid-binding protein, SBP (human)	52,000 to 100,000	Estradiol, testosterone	High affinity, low capacity; SBP concentration increases in pregnancy
Progesterone-binding protein, PBP (guinea pig, coypu)	100,000	Progesterone	High affinity for binding; PBP concentration increases in pregnancy

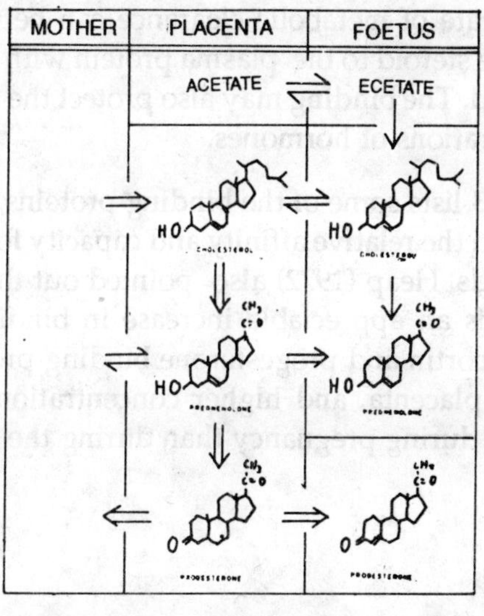

Fig. 9.14. Present concept of the metabolism of cholesterol, pregnenolone, and progesterone in the human feto-placement unit at midgestation.

The origin of the PMSG was originally thought to be the mare, because the hormone is produced by the endometrial cups, but these cups have been shown to be of fetal origin, and Stewart and Allen (1981) have proposed that the term *PMSG* be replace with *chorionic gonadotrophin* (CG). Not surprisingly, in view of the fetal origin of the hormone, the secretion of PMSG is strongly affected by the genotype of of the fetus.

Table shows the effect of intraspecific crosses and or pure matings on the secretion of PMSG.

It should be kept in mind in any consideration of plasma concentrations of hormones that higher concentrations in pregnant than in non-pregnant animals can be the result of either a higher production of such hormones or a slower metabolic clearance rate, or both Heap (1972) pointed out that the higher progesterone concentration during pregnancy in women is the result of higher rate of production, but that in guinea pigs it is the result of a slower metabolic clearance rate. This slower rate of metabolic clearance can be the result of binding of the steroid to the plasma protein with high affinity for this steroid. The binding may also protect the fetus against high concentrations of hormones.

Table 9.13 lists some of the binding proteins, the steroids they bind, and the relative affinity and capacity for binding of certain steroids. Heap (1972) also pointed out that species in which there is an appreciable increase in binding proteins, such as transcortin and progesterone binding protein, have a hemochorial placenta, and higher concentrations of plasma progesterone during pregnancy than during the cycle.

10

Reproduction and Immunology

The interrelationships between the immune and the reproductive system are too many and too complex to be given extensive treatment in this book. Here we attempt only to mention briefly some of the fundamental questions, some medical and biological phenomena, and some applications of immunological techniques to reproductive physiology. Much of the discussion will be limited to mammals, because the overwhelming amount of research has been done on this class of animals.

We will discuss the use of immunological techniques that have advanced knowledge about reproductive physiology, the relationship between immunology and male reproduction, and the relationship between immunology and female reproduction.

Immunological Techniques

Three immunological techniques, radioimmunoassay (RIA), passive and active immunization, and localization of sites of hormone production will be discussed.

Radioimmunoassay

The account of the development of this technique should be read in the Nobel Prize lecture by Yallow (1978), who was one of the developers of this valuable technique. Radioimmunoassay can be defined as "Assay techniques based on competitive binding of a known amount of isotope labeled hormone and variable amounts of unlabeled hormone ('unknown') to anti-hormone antibody molecules." For protein hormones, antibodies are raised usually in rabbits or larger laboratory or domestic animals. For the production of steroid hormone antisera, the steroid needs to be conjugated as a hapten to a protein. The position of the steroid molecule which the protein is conjugated to a large extent determines the specifically of the antisera. Details about the techniques used for the RIA of different hormones can be found in Jaffe and Behrman (1974).

The great sensitivity of RIA has made it possible to make great strides in the understanding of reproductive endocrinology. The technique has made it possible to evaluate the dynamic changes in hormone concentrations in plasma and organs during so-called spontaneous physiological events, and before, during, and after experimental treatments. Much of the evidence used in previous chapters has been based on RIAs of various hormones.

Passive and Active Immunization against Hormones

Passive immunization (i.e., with the injection of antisera), has made it possible to study the effect of short-term inactivations of the recipient's own hormones, thus allowing the study of the requirement of an endogenous hormone for a specific physiologic function or event. We have mentioned an example of this in the study of the function of LH and FSH in the reproduction of the male and the female rat.

Active immunization, in which the animals to be studied are induced to produce antibodies against the injected antigens, makes it possible to study the long-term effects of selective hormone deprivation. Information on the effects of antibody production against gonadal steroids (conjugated to a protein) in mammals is provided in the excellent review by Nieschlag and Wickings (1978). According to these authors, active immunization against testosterone has the following repercussions: (1) plasma testosterone concentrations become very high (up to 100 times the normal concentration); (2) testosterone binding in serum is increased, as is testosterone production; and (3) testosterone clearance is decreased. Plasma luteinizing hormone concentrations are also higher in testosterone-immunized animals than in controls, but plasma FSH concentrations are similar in experimental and control animals. Several functions that are testosterone-dependent are adversely affected in testosterone-immunized animals; e.g., the accessory reproductive organs are smaller and show histologic signs of atrophy; male offspring of female rabbits actively immunized against testosterone are feminized; and male rabbits immunized against testosterone show less sexual activity than controls. The male pheromone produced by male boars in fatty tissue is reduced in testosterone-immunized boars. We thus find the apparently paradoxical situation of high testosterone titers and diminished activity of testosterone-dependent functions. This can be explained by the lack of negative feedback as a result of the antibodies inactivating androgen activity.

Localization of Sites of Hormone Production

Fluorochromes coupled to an antigen or its antibody make it possible to lacalize a particular hormone in specific cells. This technique can be used for qualitative and quantitative procedures, but an unlabeled antibody-enzyme technique can be used instead. These techniques can be used for investigations with both light and electron microscopes.

IMMUNOLOGY AND MALE REPRODUCTION MATING PREFERENCES AND MAJOR HISTOCOMPATIBILITY GENES

In mice, there are strain differences among males in their preferences for females, which are similar or dissimilar with respect to the male in the major histocompatibility (*H-2* complex) genes. For example, Andrews and Boyse (1978) have shown that *B-6* and *B6-T1a*[a] males mate preferentially with B6-T1a[a] females when given a choice of *B6* or *B6-T1a*[a] females. Further investigations have shown that in female mice, a recognition-of-identity gene *(Ri-1)* influences mating preference of males, and that in males there is another gene *(Ri-2)* that influences their mating preference. The approximate location of these genes within the major histocompatibility complex has been established.

Antigens of the Male Reproductive System

Bishop (1968) has used the enzyme sorbitol dehydrogenase (SDH), which is associated with spermatogenesis throughout the animal kingdom, both for monitoring spermatogenetic activity and as an antigen to induce aspermatogenesis in guinea pigs. He found that antigen to be effective when used for active immunization, but antibodies to SDH raised in rabbits did not affect spermatogenesis in guinea pigs.

The secondary reproductive organs of various mammals (e.g., dog, rabbit, rat, mouse, hamster and gerbil) have highly organ-specific antigens. In some cases, such antigens are shared by related genera, e.g., mice, rats, hamsters, and gerbils but they are not shared with rabbits and guinea pigs. In rats, the presence of these antigens has been shown to be androgen-dependent. Such antigens can be used to determine the embryologic relationships among accessory organs and the phylogenetic relationships among species.

Seminal plasma, as expected, contains antigens also found in accessory reproductive organs. An example of such an antigen is the sperm-coating protein antigen (SCA) found in the seminal vesicles, seminal plasma, and on the ejaculated sperm in humans and rabbits, but not on either epididymal or testicular sperm. Neither this nor any other antigen secreted by the accessory organs is essential for fertilization, as is evident from the fact that epididymal sperm are capable of fertilizing oocytes. According to Beer and Billingham (1976), no autoimmune lesions have been produced in any of the accessory organs of reproduction except possibly in the seminal vesicles of the guinea pig.

The testis is an immunologically privileged site, because

the blood-testis barrier prevents self-tolerance. If the sperm penetrate the blood-testis barrier, autoimmune responses may be expected. They do indeed occur and may, in turn, lead to destruction of the germ cells, azoospermia, or permanent sterility.

An important practical aspect of autoimmune sterility involves the consequences of vasectomy. There is evidence in guinea pigs rabbits hesus monkeys and humans that after vasectomy there is an increase in sperm antibodies in the blood. This increase may be associated with orchitis, e.g., in rabbits, but in guinea pigs. Tung and Alexander (1977) found that the histo-pathological changes that followed vasectomy were not correlated with a humoral immune response to sperm. Tung (1978) reported that peritoneal exudate cells from long-term vasoligated guinea pigs to syngeneic recipients induced allergic orchitis. It appears, therefore, that there is both a specific and a non-specific immune response in the testicular damage after vasectomy.

These findings have clear implications for the practice of vasectomy in humans, especially if future restoration of fertility is contemplated. Alexander (1975) has found, however, that in rhesus monkeys, the sperm count after vasovasotomy (rejoining of the cut ends of the ductus deferens) was initially-lower than prior to vasectomy, but that after about eight months, sperm counts were normal. A more serious potential side effect that orchitis after vasectomy is the finding by Alexander and Clarkson (1978) that the severity of atherosclerosis induced by diet is higher in vasectomized than in either sham-vasectomized or control cynomolgus monkeys (*Macaca fascicularis*).

Sperm Surface Antingens

It is known that ABO blood group antigens are expressed on sperm, but it is not known whether there is selection in the woman's genital system for sperm with a specific antigen. The possibility of such selection is indicated by the observation that antibodies to ABO antigens are found in the cervical mucus. So far there is no evidence that ABO incompatibility is a common cause of infertility.

There is suggestive evidence in the rat that sperm bearing determinants different from those on the oocyte may have a selective advantage over sperm bearing determinants identical to those in the oocyte. Such a phenomenon would result in a greater heterozygosity than in the absence of such selective fertilization.

We have discussed in peculiar inheritance of the gene for short tail (brachyury) in the mouse. Recently, it has been found that the genes for this characteristic can be detected be serological methods as surface antigens on the sperm. In view of the possibility of selective fertilization, this might of a partial explanation for the unique inheritance of the T-locus genes; it would not, however, explain the effects of late insemination.

IMMUNOLOGY AND FEMALE REPRODUCTION

Antigens of the Male Reproductive System and the Female

The question has arisen whether the antigens in the seminal plasma and on the sperm might not cause immunization in the female. This aspect of the immunology of reproduction has been recently reviewed by Jones (1976), and the major part of the following account is taken from this review. Acute anaphylactic reactions to semen in women are apparently extremely rare, but they have been reported to occur. The evidence that isoimmunization against semen antigens is important in explaining long-term infertility in women without major or minor organic causes of infertility is not convincing for mammals. In chickens, Went-worth and Mellen (1964) found that after repeated artificial inseminations, the antibody titer to rooster sperm became higher than that in virgin females and that the duration of fertility was negatively correlated (r= -0.46) with sperm antibody titer. It is, of course, possible to raise antibody titers to semen products by active immunization, using classical immunological methods, such as injection of antigen with or without Freund's adjuvant.

After such immunization, fertility in turkey hens was reduced below that in controls, the extent of the reduced fertility being dependent to some extent on the male line from which semen was used for artificial insemination. In laboratory mammals and in cattle (lottes, 1976), active immunization against sperm has resulted in a decrease in fertility, mainly, at least in rabbits, through interference with fertilization.

Immunization with either sperm or semen sometimes introduces many antigens, and thus the antibodies produced are not very specific. Instead of sperm and semen, an isozyme specific for male germ cells, lactate dehydrogenase-C_4 (LDH-C_4, also called LDH-X), has been isolated and used to immunize mammals in attempts to affect their fertility. This isozyme may function as an autoantigen, and an antiserum against LDH-C_4 shows considerable cross-reactivity to the C_4 isozyme found in several species.

Lerum and Goldberg (1974) have reported that after active immunization with LDH-X on day 1-4 post coitus, mice showed a decrease in incidence of pregnancy. Their observations are in accord with observations on rabbits, that such immunization significantly decreased the percentage of embryos to corpora lutea to 29.3 per cent from the 88.3 per cent found in controls. These investigators also found that, in mice, passive immunization with rabbit antiserum to the mouse LDH-X antigen on days 1-4, or days 4-7, or days 7-13 post coitus, reduced the incidence of pregnancy. An antiserum to a differently prepared isozyme diminished the incidence of fertilization *in vitro*, but neither diminished the *in vivo* fertilization nor the survival of embryos. At first thought, it seems difficult to explain the effect of antiserum to a sperm-specific antigen, when such an antiserum is injected as late as 7 d post coitus. According to Lerum and Goldberg (1974) and Goldberg (1974), the effectiveness of the immunization at 7 d post coitus may be the result of supernumerary sperm within the perivitelline space of fertilized oocytes (zygotes), which provide specific recognition sites for the LDH-X antibody. They speculate that the induced heteroantigenicity leads to immunological rejection of the blastocyst analogous to the rejection of blastocysts of another species by the uterus.

Active immunization against LDH-X in male rabbits and guinea pigs leads to infertility. In the guinea pig, the aspermatogenesis so induced in transient: in 50 per cent of the affected males, fertility is eventually restored and normal offspring are born. Goldberg (1977) did not find aspermatogenesis after active immunization of guinea pigs with LDH-X and explained the results obtained by Wellerson *et al.* (1974) as due to a protein impurity in their preparation of LDH-X.

Antigens of Female Germ Cells

Mature mammalian oocytes are surrounded by a gelatinous-like layer, the zona pellucida (ZP). This layer has been assigned a large number of functions, e.g., sperm recognition, block to polyspermy, osmotic regulation and others. The ZP has antigenic properties, and after treatment of oocytes with ZP antibodies, several functions of the ZP are altered, so that sperm fail to attach themselves to the oocyte surface and fertilization fails to occur.

The question whether women produce autoantibodies against the ZP and, if so, whether this affects fertility has been investigated by Shivers and Dunbar (1977). They found that the sera from 22 infertile women produced a strong antibody reaction (measured by an immunofluorescence technique) in 6 cases and a moderate response in 9 cases. These numbers are too small to drawn definite conclusions about the biological significance of such autoantibodies, but the results suggest that, at least in some cases, infertility may be explained on this basis. If is unfortunate that no measurements were made on the sera of fertile women.

Blood Groups and Reproduction

It humans, the best-known instance of interaction between blood groups and reproduction is the rhesus factor. About 85 per cent of Caucasians, 93 per cent of all black people, and virtually all Mongolian people possess a blood antigen that is also found in the rhesus monkey. The antigen is inherited as a simple dominant factor. In a woman who lacks the rhesus factor (i.e., is Rh negative), problems may arise if her fetus is Rh positive and the woman has been sensitized to the Rh factor and has produced antibodies to it. The antibodies may attack the antigens on the fetus' red blood cells, destroying the red blood cell, resulting in erythroblastosts foetalis. If this occurs late enough during pregnancy, the fetus may be given intra-uterine blood transfusions, or after either normal or prematurely induced delivery, it may be given blood transfusions immediately after birth. With the idea that prevention is better than cure, women at risk of immunization with the Rh factor are passively immunized by administration of anti-Rh antibody. The antibody should be administered prior to antibody formation by the mother, so that no sensitization occurs.

Some interesting complications have arisen with the Rh factor. Sensitization can occur by either a previous blood transfusion with Rh-positive blood or by the passage of Rh-positive blood from fetus to mother at the time of delivery. Usually during the first pregnancy (if the mother has not received blood transfusion with Rh-positive blood), the chances of sensitization are small, and the chance of damage to the fetus is small. It has been reported, however, that if immunization during the first pregnancy occurs, the fetus is almost always male. The sensitization of the mother occurs either as the result of fetal bleeding, which causes red blood cells to enter the material circulation, or from the mother's own sensitization during *her* fetal life in the uterus of an Rh-positive mother. Of 25 Rh-negative women who were carrying their first Rh-positive fetus and who developed antibodies prior to delivery, 21 were daughters of Rh-positive mothers. In 23 of 25 cases, the sensitization was produced by an ABO-compatible, Rh-positive male fetus. The sensitization by the male fetus may be either the result of a synergistic effect of the Y-antigen on the immunogenicity of the erythrocytes or the result of a greater traffic of fetal and maternal red blood cells in the placentas of male fetuses as compared to the placentas of female fetuses.

ABO blood groups play a role in determining the effect of Rh sensitization. The risk of immunization by an ABO-incompatible, Rh-positive pregnancy is about 12.5-20 per cent of that of an ABO compatible. Rh-positive pregnancy (Beer and Billingham, (1976). One explanation for this interaction is as follows: If an Rh-negative mother with blood group 0 carries an Rh-positive fetus with blood group A, then fetal red blood cells that enter the maternal circulation will be destroyed by the anti-A antibodies of the mother. The cells will be removal in the liver, probably before sensitization of the mother can occur.

Evidence that ABO incompatibility between mother and fetus (when not complicated by differences in Rh factor) affects the success of pregnancy is inconsistent. There may, however, be an effect of blood groups on sex ratio. Allen (1975), combining data from the literature, found that the sex ratio of babies born to mothers who had as least one A gene (AO, AA, AB) was higher (1.15) when they had a different blood group than their mothers, and lower (1.01) when they had the same blood group as their mothers. The sex ratio of babies that lacked the A gene (i.e., blood groups B and O) was high (1.11) when their blood group was the same as their mothers and low (1.04) when their blood group was different from their mothers. They physiological and immunological differences for these differences in sex ratio have not been determined and require further investigations.

Immunology of Gestation

Probably one of the most intriguing and possibly one of

the most fundamental questions in mammalian reproduction is: Why is the fetus, which is an allograft, not rejected by the mother? We will discuss some of the explanations that have been proposed.

Immune Reponse Early during Gestation. Immediately (6-24 h) after insemination and fertilization, an early pregnancy factor can be detected by a rosette-inhibition test in the serum of mice, women and sheep. This early pregnancy factor, which in mice is present from 6 h after fertilization until 4-6 d before birth has, as the above test suggests, immunorepressive properties and may thus influence or be responsible for prevention of the allograft injection. The early pregnancy factor can be inactivated at 72°C, but remains stable at 56°C; it is not dialyzable and not species specific. Cerini *et al.* (1976), using a hemagglutination test, were able to detect antigens of 14-day-old sheep embryos in the maternal blood of pregnant ewes from day 6 after mating. It is not clear, however, whether this antigen and the early pregnancy factor are identical.

The early pregnancy factor is present prior to implantation, and the question arises whether implantation involves immunological response mechanisms. As Johnson (1976) has pointed out, in highly inbred lines, fetus and mother may be considered to be not immunologically different. This argues against the requirement for an immunological mechanism in implantation. This and other evidence, reviewed critically by Johnson (1976) led him to conclude: "There is therefore no good evidence to suggest that immunological interactions between mother and early embryo are either essential for facilitatory in the peri-im-plantation period in any specific manner."

Immune Response after Implantation. After implantation proceeds, the question why the fetus is not rejected becomes considerably more interesting. Some of the explanations that have been proposed for the non-rejection of the fetus are, for example: (1) inability of the fetus to produce antigens because of immaturity, (2) implantation in an immunologically privileged site, (3) suppression of the mother's immune response by non-specific factors, and (4) special properties of the trophoblast, among which are special immunological properties and the possession of a special immunological barrier.

1. The idea that the fetus is not capable of producing antigens is not an appropriate explanation for the non-rejection of the fetus, because it has been shown that major transplantation antigens are present as early as the two-cell stage of the embryo.

2. There are several immunologically privileged sites in which tissue from an immunologically incompatible donor can be transplanted without evoking a rejection, as, for example the anterior eye chamber and the cheek pouch of the hamster, which are characterized by the absence of lymph vessels. The uterus, however, is not such a privileged site, because it does have lymph vessels, and more importantly, tissue grafts made in the uterus are rejected. However, the uterus after formation of decidual tissue, does respond differently from other sites under special conditions; for example, experimentally immunized female mice reject alien embryos transplanted to sites other than the uterus, but do not reject such embryos transplanted to the uterus. The decidual tissue that develops under the implanting trophoblast may interfere with lymphatic drainage away from the zygote.

3. There are several factors secreted only during pregnancy, and these interfere with lymphocyte functions that are important in the immune response. Examples of such factors are pig uterine protein secretions, human chorionic gondotrophin (hCG) (Contractor and Davies, 1973), human placental lactogen (HPL) (Billington, 1979), progesterone, estrogens, corticosteroids, and alpha feto-protein (AFP). The AFP binds estrogens and this AFP-estrogen complex becomes bound to lymphocytes and makes them less responsive to exogenous antigens. Billington (1979) also mentions the possibility that the developing embryo may provide a source of prolonged exposure of the mother's

immune system to relatively weak antigens and that these antigens themselves or as antigen-antibody complexes may block either lymphocyte receptors or target antigens and thus diminish the immune response of the mother.

4. The trophoblast has some special properties, which suggest that it may be an immunologically privileged tissue. For example, a choriocarcinoma, a malignant derivative of trophoblast cells, is refractory to transplantation immunity directed against the paternal antigens. Beer and Billingham (1978) state, "Small grafts of trophoblastic cells obtained from ectoplacental cones behave as if they are invulnerable to transplantation immunity." Thus non-malignant trophoblastic tissue shows some resistance to immunity.

In virgin rats, removal of the para-aortic lymph nodes, several weeks before pregnancy, results in smaller F_1 hybrid fetoplacental units than in controls. It has also been found that the para-aortic lymph nodes that drain the uterus show a striking hypertrophy during pregnancy. Beer and Billingham (1978) proposed the hypothesis that the lymphocytes within these stimulated para-aortic lymph nodes generate a non-cytopathogenic humoral immunity, which contributes to the immunologically privileged status of the trophoblast. These lines of evidence support the idea that the trophoblast may be a privileged immunological tissue. The immunology of the hydatiform mole, however, suggests otherwise. The hydatiform mole is a curious example of a trophoblast abnormality. It is the product of an abnormal pregnancy and has grossly swollen villi, but no embryo, cord, or amniotic membrane. The hydatiform mole consists of trophoblast cells only and frequently becomes an invasive malignant chorionic carcinoma. Kajii and Ohama (1977) determined that hydatiform moles were androgenetic in origin and might be either the result of fertilization of an oocyte by a sperm that was diploid as the result of non-division of meiosis or the result of non-division of a normally fertilized oocyte at blastomere mitosis. Human major histocompatibility antigens of paternal origin are present on such hydatiform moles, and the mother produces antibodies against these antigens. This evidence argues against the hypothesis that the trophoblast is an immunologically privileged tissue.

The trophoblast has paternally inherited alloantigens, and the uterus is immunocompetent; therefore, either the trophoblast must have some intrinsic immunological properties or the maternal immune response must be different from its response to other tissues.

In humans, the trophoblast is invested by a fibrinoid layer, which is rarely if ever complete, however. In mice, there is a peritrophoblastic sialomucin coating, which may make the trophoblast immunologically inert. This inertness may be the result of free carboxyl groups of sialic acid, which cause a net negative charge that repels negatively charged lymphocytes. Human chorionic gonadotrophin (hCG), which decreases lymphocyte activity, may further increase the negative charge of lymphocytes. Uterine protein secretions in pigs, human chorionic somatomammotrophin and hCG in humans suppress the response of lymphocytes to phytohemagglutinins.

Intergeneric hybrids between goat ♀ and sheep ♂ show earlier embryonic mortality if the goat carries such a hybrid fetus for the second time than if she carries it for the first time. This finding and the ultra-structural damage to maternal uterine blood vessels suggest that there is an immunological basis for this *particular* rejection. It should be kept in mind that interspecific hybrids, e.g., between horse and donkey, are carried to term normally.

The placenta provides immunological protection for the fetus possibly by soaking up immune complexes and autoantibodies. The placenta also forms a selective barrier for immune globulins and antigens between mother and fetus. For instance, immunoglobulin G (IgG) (molecular weight 150,000) crosses the human placenta, but some do not, e.g.,IgA (MW monomer = 160,000 and MW polymer = 400,000 in humans), IgD (MW = 150,000), IgE (MW = 200,000) and IgM (MW =900,000). Maternal white and red blood cells may pass to the fetus, and fetal red blood cells may pass to the mother.

Rejection of Allograft and Short Gestation in Marsupials. We mentioned that the short gestation period of marsupials might be related to peculiariteis of the immunological response of the mother to the fetal allograft. Beer and Billing ham (1976 reject the hypothesis that the onset of parturition may be the result of an immunological sensitization of the mother to the fetoplacental unit in placental mammals. Their principal arguments are that the hypothesis would predict that successive pregnancies would become shorter as the mother became sensitized to the antigens of the fetus, and initial pregnancies should be abnormally long in cases where the mother had an immunological deficiency disease. Neither of these predictions has been fulfilled.

At the end of gestation the decreased secretion of protein hormones such as hCG by the placenta removes the suppression of the immune system so that rejection of the fetus and subsequent parturition occur. Recently Siiteri and Stites (1982) have presented persuasive evidence that progesterone that is present in high concentrations in pregnant Eutherians may be the immunosuppressive agent that prevents rejection of the fetus and that a decrease in progesterone secretion induces rejection and subsequent parturition.

After birth, limited passive immunity is transferred from the mother to the young via the immunoglobulins in the colostrum and milk. In birds, antibodies are transferred to the yolk, which is absorbed into the gut of the hatchling during the first few days after hatching.

It is beyond the scope of this book to discuss the pathological, clinical, and contraceptive or abortifacient aspects of immunological defects and applications. The interested reader is referred to the book by Beer and Billingham (1976) and to reviews in the book edited by Scott and Jones (1976).

11

Intersexes

We have mentioned earlier that intersexuality is found more frequently among fishes than among other vertebrates. The phylogenetic relationships among these groups in which functional hermaphroditism has been found to occur regularly in one form or another, e.g., synchronous, protogynous, or protandrous.

Fishes

Among the lower fishes functional hermaphroditism does not seem to be common. Here we briefly discuss a few cases that have been found.

Cyclostomes

The male and female Atlantic hagfish (*Myxine glutinosa*) each function as male and female respectively, although the gonad is often hermaphroditic. In other hagfishes, *Polistotrema stouti* (*Bdellostoma stouti*) and *Eptatretus burgeri* (*B. burgeri*), hermaphrodites are apparently rare. Lampreys are also not hermophroditic as adults.

Elasmobranchs

In this group hermaphroditism is rare. In *Squalus acanthias*, the spiny dogfish, it has apparently never been found, despite the thousands of animals used in comparative anatomy classes. In the lesser spotted dogfish (*Scyliorhinus canicula*), a few hermaphrodites have been described by Atz (1964), who also cites evidence that in the sharks *Notorynchus* and *Hexanchus*, the ovary is associated with a rudimentary testis.

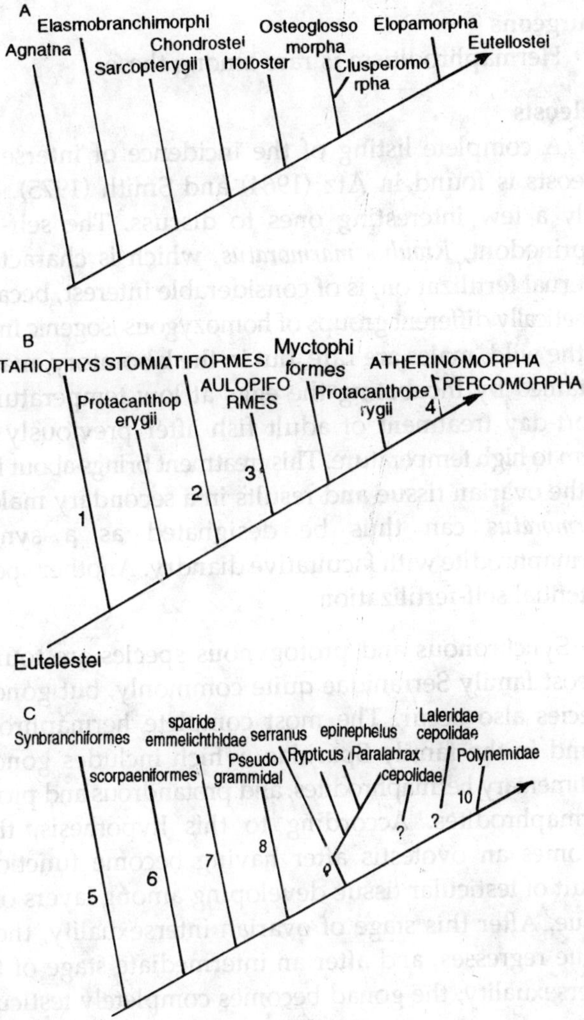

Fig. 11.1. Phylogenetic relationships of hermaphroditic fishes. Compiled from various sources. A, Lower fishes, Agnatha to Euteleostei. B, Qstariophysi to Percomorpha. C, Percomorpha Groups with hermaphroditic representatives are listed in capitals. Numbers indicate lineages in which hermaphroditism has arisen independently.

Sturgeons

Hermaphroditism is rare among these.

Teleosts

A complete listing of the incidence of intersexuality in teleosts is found in Atz (1964) and Smith (1975). We select only a few interesting ones to discuss. The self-fertilizing cyprinodont, *Rivulus marmoratus*, which is characterized by internal fertilization, is of considerable interest, because it has genetically different groups of homozygous isogenic individuals. In the wild, males are rare, but in the laboratory, males can be obtained by incubating the eggs at low temperatures, or by short-day treatment of adult fish after previously exposing them to high temperature. This treatment brings about involution of the ovarian tissue and results in a secondary male. *Rivulus marmoratus* can thus be designated as a synchronous hermaphrodite with facultative diandry. Another species with potential self-fertilization.

Synchronous and protogynous species are found in the teleost family Serranidae quite commonly, but gonochoristic species also occur. The most complete hermaphroditism is found in the family Sparidae, which includes gonochorists, rudimentary hermaphrodites, and protandrous and protogynous hermaphrodites. According to this hypothesis, the ovary becomes an ovotestis after having become functional as a result of testicular tissue developing among layers of ovarian tissue. After this stage of ovarian intersexuality, the ovarian tissue regresses, and after an intermediate stage of testicular intersexuality, the gonad becomes completely testicular.

One of the best-studied cases of protogynous hermaphroditism is that of the swamp eel or rice-field eel, *Monopterus albus*, of the order Synbranchiformes. In this species, female function and male function are completely separated in time, both gametogenically and hormonally. The animal first spawns oocytes and secretes estrogens and later releases sperm and secretes androgens, but at no time are functional

TABLE 11.1

Distinguishing Characteristics of Intersexes of Intersexual Condition (Presence of Male and Female Characteristics in the Same Individual)

Type of intersex intersexual Condition	*Characteristics*	*Example*
Hermaphrodite	Same individual has ovarian and testicular tissue.	
Synchronous	Ripe spermatozoa and oocytes are present simultaneously the same individual. In some cases self-fertilization occurs.	*Rivulus marmoratus*
Protandrous	Individual functions first as a male (testis produces sperm) and subsequently as a female (ovary produces oocytes).	Teleost family of Sparidae
Protogynous	Individual functions first as a female (ovary pro-duces oocytes) and subsequently as a male (testis produces sperm).	Teleost family of Labridae
Rudimentary	Individual functions as a male (testis produces sperm), but ovarian tissue is present, on the individual functions as a female (ovary produces oocytes), but tresticular tissue is present.	*Amphiprion bicinctus*
Pseudohermaphrodite	Individual has gonads of one sex, but secondary sex organs and or secondary characteristics of the other sex.	Testicular feminization syndrome in humans and mice
Sex reversal	Individual functions as a male, but acquires ovarian function and then function as a female, or individual functions as a female and acquires testicular function and then functions as a male.	Protandry and protogyny
Sex inversion	Secondary sex characteristics of the individual change without a change in the gonadal tissue, therefore without sex reversal.	Adrenogenital syndrome in women
Diandry	Some individuals differentiate as males and remain males; others first pass through a functional female stage before transforming into males.	Undifferentiated races of frogs

cont...

Gonochorism		
Primary	Genetic sex agrees with phenotypic sex.	Most classes of vertebrates except teleosts
Secondary	Genetic sex and prenotype do not agree	*Myopus schisticolor*

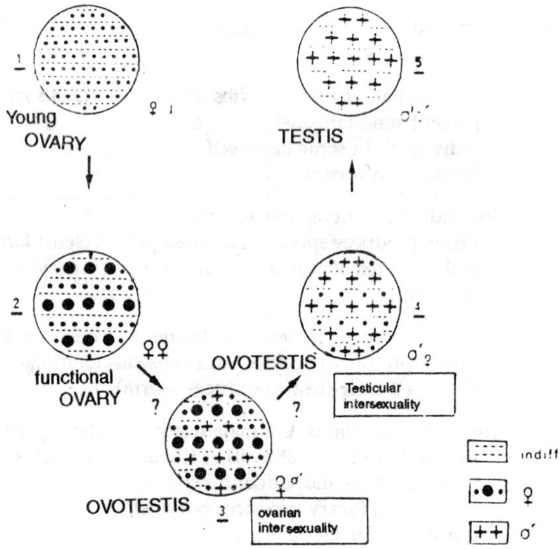

Fig. 11.2. Hypothetica scheme of gonadal evolution of the fish *Epinephelus*.

ovarian and testicular tissue present simultaneously in the same individual, so that the rice-field eel does not follow the hypothetical scheme of Figure 3.2. The increase in androgen secretion is highly correlated with the observation that 3b-hydroxysteroid dehydrogenase (3b-HSD), an enzyme that plays a key role in steroidogenesis, shows a weak reaction during the female phase and a strong reaction during the intersexual and male phase. During the female phase, enzyme activity is found is the granulosa cells of some large follicles and some scattered interstitial cells; during the intersexual and male phase, the enzyme activity is localized mainly in interstitial cells. The decrease in estradiol secretion and the increase in testosterone secretion associated with the sex reversal of the rice-field eel might be suspected to be causative factors of the structural changes in the germinal tissue. Experimentally, however, neither the administration of various androgens (testosterone, methyltestosterone, 11-ketosterone) nor the administration of estrone causes changes in the sex differentiation, although androgens stimulate spermatogenesis, and estrone increases the incidence of small oocytes and causes the destruction of Leydig cells and spermatogenetic cells. Androgens do not affect either the cryptic spermatogonia or the ovarian follicles of females. It seems unlikely, in view of these experiments and others, in which estrogen synthesis was blocked by cyano-ketone without affecting sex reversal, that sex steroids cause sex reversal.

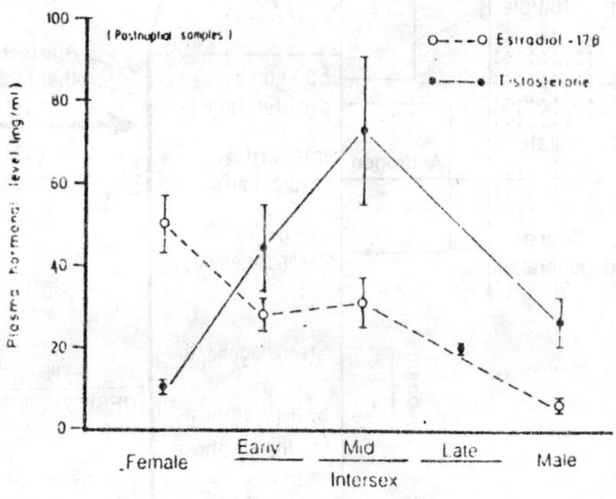

Fig. 11.3. Plasma levels of testosterone and estradiol-17β in *Monopterus* at various sexual phases.

Injections of mammalian LH cause precocious sex reversal in these eels by stimulating testicular lobule formation and by stimulating the 3β-HSD activity of the interstitial cells.

In some protogynous genera, *Coris, Thalassoma,* and *Halichoeres,* precocious sex reversal can be induced by administration of androgen. Ovariectomy of the fighting fish, *Betta splendens,* and the paradise fish, *Macropodus opercularis,* causes a high incidence of sex reversals, but the reasons for this sex reversal are not understood.

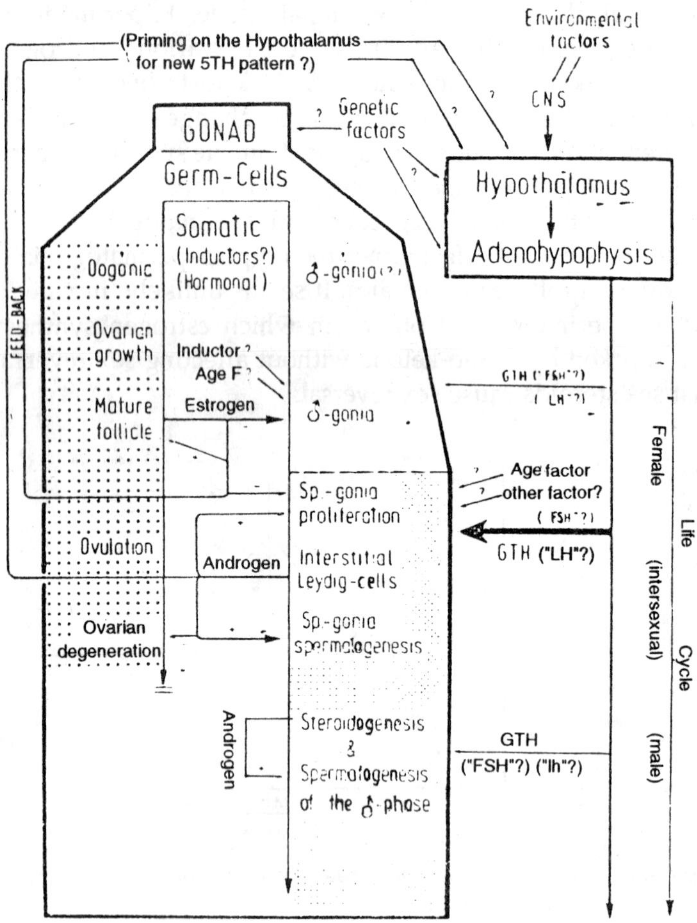

Fig. 11.4. A proposed scheme for the endocrine events of natural sex reversal in *Monopterus*, showing the various possible controls in the life-cycle of the process, which involves the nature of the germ cells, the gonadal endocrine interaction and the hypothalamo-adenohypophysiogonada axis, with genetic and age factor as the ultimate basis of control.

Hermaphroditism has arisen several times independently in teleost fishes. Smith (1975) cites the following evidence:

1. *Separation of testicular and ovarian tissue.*This varies, from testes and ovaries each encapsulated in their own connective tissue (Aulopiformes) to male and female cells scattered throughout the germinal epithelium (*Epinephelus*).

2. *Location of testes with respect to the ovary.* In the Aulopiformes, testes lie dorsal to the ovaries, but in the Sparidae the testes are on the ventral wall of the gonad.

3. *Arrangement of gonoducts.* In the Aulopiformes, each testis has its own duct, which either leaves the body separately or after it has joined the contralateral duct. In *Epinephelus*, the sperm run through lacunae that develop in the outer wall as the gonad changes from ovary into testes. The lack of correlation among these features in different groups of fishes supports the idea of the independent origin of hermaphrodites in teleosts.

Warren (1975) has discussed the advantages of sequential hermaphroditism. The advantages most obvious when there is an age-specific fecundity, which is not distributed in the same way for males and females. An animal can then function as that sex which has, at that age, the highest fecundity, and so gain an advantage over gonochorists. Protandry may have the greatest advantages when female fecundity increases with age and mating occurs at random. Protogyny would be most advantageous when female fecundity decreases with age or when male fecundity is low at early ages (as a result of either inexperience, territoriality, in which a smaller male has less chance to mate than a larger, older male, or female selection of mates).

Choat and Robertson (1975) have discussed protogynous hermaphroditism in detail for the family Scaridae. In 20 of 23 species, the larger individuals are more brightly coloured than the smaller ones, with an overwhelming probability that a brightly coloured individual is a male. Functionally, males that are drab coloured are generally primary males. Bright colouration is generally associated with pair spawning and may function for species and sex recognition. In group-spawning species, bright colours might be disadvantageous, because they might provoke attacks, whereas drab colour would be advantageous because the animals might be considered females and not be attacked. It may thus be an advantage to an individual to start as a female, especially if few males are required for fertilization, and reproduce. When the individual becomes older and thus better able to defend itself, it may be advantageous for it to become a male and fertilize the eggs of many females.

Fricke and Fricke (1977) have described the sex reversal in coral-reef fishes with a restricted distribution. In two species, *Amphiprion bicinctus* and *A. akallopisus*, which live on anemones, males are born as much and females as females, but after birth both undergo first a male ten a female phase. The testes of functional males always contain immature oocytes, but the ovaries of functional females contain no testicular tissue. The social units of these species consist of a large female, a small male, and a number of subadults and juveniles that are offspring of the adult pair. They are close, monogamous groups, with pair bonds and individual recognition. The female controls the sex and the growth of the subordinates. After removal of the dominant. (α) female of *A. akallopisus*, the β male takes the α position and changes to a female in less than 63 days. In *A. bicinctus*, an original β male laid eggs 26 days after removal of the _ female. The subadults are stunted in their growth, but develop rapidly after the adult dies. This system of socially controlled sex change and a large-sized female but a small male is advantageous because it allows for a constant supply of potential males and females from the subadult population; it reduces the necessity for migration and thus the risk of predation. The advantage of a large female is that it allows for greater oocyte production (body weight and ovarian weight are highly correlated), whereas the smaller male is able to fertilize and guard offspring, but requires less food than a larger individual.

Amphibians

In certain toads, e.g., *Bufo bufo*, the normally inactive Bidder's organ becomes a functional ovary after castration of normal males. This is, therefore, a form a protandry, although it does not normally occur in nature except possibly when the testes are destroyed by a disease that does not also kill the animal. The undifferentiated races of frogs, which during their development go through a hermaphroditic stage, but they do not function as sequential hermaphrodites. Foote (1964) provides information about aberrant manifestations of hermaphroditism and experimentally induced hermaphrodites in amphibia.

Reptiles

Among reptiles, there are several unisexual parthenogenetic species (Cole, 1975), but no hermaphroditic species have been found. Intersexuality is an abnormal occurrence, although, intersexes will occur in some species as the result of incubation at non-cyclic temperature variations. Forbes (1964) has summarized much of the information on non-functional hermaphroditism among reptiles.

Birds

In birds, intersexes may be the result of genetic selection, embryological conditions, or unnatural treatment.

Genetic Selection

In pigeons, *Columba livia*, selection has resulted in a line of birds with a hermaphroditic left gonad in males. In crosses of *Cairina moschata* o x *Anas platyrhynchos* o, the left gonad of offspring consists of connective tissue, cords of undifferentiated cells, and large cells with 3β-HSD activity, the right gonad is rudimentary, consisting of mesonephric tubules. In the reciprocal cross, the left ovary is normal and eggs are produced in the first year, but the ovarian tissue regresses, medullary tissue prolife-rates, and testicular tubules develop with or without gonocytes. The right gonad initially contains oocytes, but late tubules that may be sterile or may contain spermatocytes develop.

Freemartins

It cattle whenever fraternal twins are developed the female cell becomes sterile and is called the *freemartin*. Thus, the "freemartins" that are found among embryos developing in double-yolked eggs with embryos of different sex. Some feminization of the testes occurs but little or no masculinization of the ovaries.

Removal of the Functional Left Ovary

This operation, in certain species, leads to the proliferation of the rudimentary right gonad into a testis or ovotestis. This change has been induced in chickens, *Gallus domesticus*; muscovy ducks, *Cairina moschata*; and ring doves, *Streptopelia risoria*; however, it does not occur in mallard ducks, *Anas platyrhynchos*; turkeys, *Meleagris gallopavo*, pheasants, *Phasianus colchicus*; bobwhite quail, *Colinus virginianus*; or starlings, *Sturnus vulgaris*. In hybrids (*Anas platyrhynchos* o x *Cairina moschata* o), ovariectomy results in the development of a small right gonad with germ cells present in 3/14 of the cases only (i.e., testis, 2 atretic follicles); whereas in the reciprocal cross, after ovariectomy one finds a small nodule, which consists mostly of mesonephric nodules and which is very similar to the right gonad found in non-ovariectomized females of this hybrid.

In chickens, the available evidence indicates that the older the bird at ovariectomy, the greater the probability that the right gonad will become an ovotestis, which may contain follicles of ovulatory size and tubules with spermatozoa; the younger the bird at ovariectomy, the greater the probability that the right gonad will develop into a testis (van Tienhoven, 1968). The development of the right gonad in an ovariectomized hen can be inhibited by the injections of physiological doses of estrogens (van Tienhoven, 1968, for review) and also by a steroid-free extract of ovarian tissue).

Ovariectomy of 13 *Cairina moschata* at 2-15 days of age resulted in the development of the right gonad with atretic follicles in 11 cases, testicular tissue in 1 case, and a structure with clear cells in 1 case. When the ovary was removed between 40 days and 6 months, in 1 of the 5 birds the right gonad was an ovary, 3 had spermatocytes, and 1 had clear cells. It thus seems that in the chicken and muscovy duck, age at ovariectomy has opposite effects on the development of the right gonad. In the Japanese quail (*Coturnix coturnix japonica*), the right gonad proliferates after ovariectomy into a structure consisting of medullary tubules with fat-laden cells.

Mention should be made of the often misquoted paper by Crew (1923). Crew found a bird that laid eggs and late sired offspring. At autopsy, the animal was found to have *two* testes *and* a diseases ovary. This is thus a case of a true hermaphrodite, and it should not be confused with the *sex reversal* that occurs after loss of the ovary (experimentally or by disease). As the ductus deferens is lacking in such birds, even after development of the rudimentary gonad into a testis, such animals cannot be used as sires. Frankenhuis (1974) tried to obtain self-fertilization in chickens by grafting. He transplanted the right ovary of a female embryo 10 days old into a gonadectomized cockerel 10 days old and the left ovary into an ovariectomized pullet 10 days old. In theory, mating of the two animals having the transplants would amount to self-fertilization, if fertilization occurred. Unfortunately, the implantation of the ovary disturbed the oviduct, so that oocytes were not caught. Frankenhuis is pursing this novel idea.

Pesticides and Herbicides

Feminization of the testes occurs in some species as the result of administration of the following compounds.

1. *Organophosphates:* Parathion (chicken Japanese quail); azinphos (pheasant, gray partridge, red-legged partridge).

2. *Organo-chlorinated pesticides:* Aldrine (chicken, Japanese quail, pheasant); endosulfan or thodian (chicken, Japanese quail); DDT (chicken, Japanese quail); Lindane (Japanese quail).

3. *Herbicides:*2, 4-D (pheasant, gray partridge, red-legged partridge, Japanese quail); 2, 4, 5-T (chicken, Japanese quail); simazine (chicken, Japanese quail).

4. *Synergized pyrethrines:* (Chickens, Japanese quail); no examples given by Lutz and Lutz-Ostertag (1975).

Mammals

The intersexuality in mammals has already been discussed. It needs to be added that freemartins have occasionally been observed in pigs and sheep and that the XXY condition has been observed in species other than cats. The calico cat is, however, easily recognizable, and the fact that it is usually sterile, if it is a male, makes the coat colour a diagnostic character.

We want to discuss the freemartin in some detail, because the explanation for its occurrence has evaded many investigators for a long time, although bit by bit, parts of the problem have been solved. In 1911, Tandler and Keller suggested that the freemartin was the result of the anastomosis of the central blood vessels of the bull and heifer fetus. Lillie (1917), who was apparently unaware of Tandler and Keller's work, came to the same conclusion and suggested that male sex hormone secreted by the bull fetus might be the cause of the modification of the ovary of the heifer calf. Subsequently, Jost *et al.* (1963) demonstrated that injections of androgens started as early as day 40-42 of gestation caused modification of the external genitalia of the heifer fetus (thus demonstrating that the androgens reached the embryo), but failed to modify the ovaries. It was late discovered that freemartins as well as their bull twins were chimeras with respect to the somatic cells, XX/XY for the heifer and XY/XX for the bull. That the chimerism is not the cause of the freemartin, however, is shown by the following evidence; (1) The germ cells of the female do not carry XY chromosomes and those of the male do not carry XX chromosomes; at least, the evidence that they do is either indirect or weak. (2) There is no correlation between the incidence

TABLE 11.2

Genotype, Gonadal Phenotype, and Other Characteristics of Mammalian Intersex, and Explanation for the Condition.

Species	*Genotype*	*Phenotype*	*Explanation of condition*
Goat (*Capra hircus*)	PP; XX (P=polled)	Testes; female appearance. enlarged clitoris	*P* sex-reverses gonads but not germ cells; primorial germ cells do not survive in absence of follicle cells
Pig (*Sus scrofa*)	XX	Testes	Unknown
Mouse (*Mus musculus*)	XX; *Sxr*	Testes; sterile, H-Y antigen present	*Sxr* causes progressive necrosis of germ cells
	XO; *Sxr*	Testes; germ cells present but few sperm, mostly abnormal	
	XO	Normal fertile female	
	XY; *Sxr*	Testes; low sperm count, high incidence abnormal sperm	
Cattle (*Bos taurus*)	XX/XY (twin to bull with anastomosis)	Testicular gonads; no Müllerian duct derivatives	H-Y antigen coating of PGC in genital ridge causes testicular development
Cat (*Felis catus*)	XXY (calico cat)	Testes small and usually sterile, analogous with Klinefelter's syndrome (Table 11.3)	XX chromosomes cause gonia to be "female," they then degenerate in medullary environment
Dog (cocker spaniel)(*Canis familiaris*)	XX	Fertile female with ovotestes, with permanent testicular tubules, H-Y antigen positive	H-Y gene transmitted ancmalously
Rat (*Rattus norvegicus*)	XY	Testes in inguinal canal; germical arrest; no Wolffian ducts; no Müllerian duct	Insensitivity to androgen testicular termniation syndrome

of somatic XY cells and the degree of freemartinism. (3) By experimentally separating the blood vessels of cattle twins *after* the exchange of XX-XY cells, but *prior* to differentiation of the gonads, XX/XY chimeras were produced with normal ovarian tissue.

It now appears that the male twin the male may secrete H-Y antigen, which coats the germ cells (provided they have receptors for this antigen), and that this H-Y antigen then subverts the development of the gonad of the female into an ovary. The "problem" of the marmoset, which does not produce freemartins, although the placental blood vessels anastomose and the twins are chimeras may be explained by assuming that either the H-Y antigen is not secreted and distributed by the male fetus or that the germ cells lack the receptors for this antigen, or both. Intersexes in humans are listed in Table 11.3, together with a possible explanation of the observed phenomena.

The question may be asked why functional hermaphroditism occurs as part of the normal reproductive pattern of some species. Charnov *et al.* (1976) have, on theoretical grounds, concluded that low mobility favours hermaphroditism, because males will not need to develop structures or behaviours to hold females. This implies that one individual can serve as both a male and a female. Since low mobility also implies that only a few females can be fertilized, and thus there are few chances for spreading genes, it would be advantageous to function as a female simultaneously or sequentially.

TABLE 11.3

Intersexes in Humans, Genotype, Characteristics and Gonadal Phenotype, and Explanation for the Condition

Genotype				
Total Nos. of chromosomes	*Sex chromosomes*	*Characteristics and gonadal phenotype*	*Explanation*	*Reference*
45	XO	Turner's syndrome; streak gonads	Lack of the second X chromosome affects development of the ovary	German et al., 1978

46	XX	Turner's syndrome, gonads	One X chromosomes usually abnormal	Armstrong, 1964; Jacobs, 1966
46	XX	Resembles Klinefelter's syndrome (XXY); azospermia; male sexual behaviour	Mosaic XX/XY with loss of Y; Translation to either to X or to autosomae; autosoma mutant with same effect as Y	Vague et al., 1977
46	XY	Testes in inguinal canal; no Müllerian ducts or Wolffians ducts; female external genitalia, breasts developed, female sexual behaviour; testicular feminization syndrome	Somatic tissues are insensitive to androgens	Armstrong, 1964
46	XY	Streak gonads	Sex reversal on x-chromosomes	German et al., 1978
46	XX or XX/XY	Ovary on one side; testes on other side	Unknown	Armstrong, 1964
47	XXX	Normal ovaries; sometimes menstrual problems	No effect on ovaries by the extra X chromosome	Armstrong, 1964; Jacobs, 1966
47	XXY	Klinefelter's syndrome, usually small testes, usually sterile; breast development	Regression of germ cells in of medulla because XX	Armstrong, 1964; Jacobs, 1966
47	XYY	Testes normal or small	No effect from extra Y	Armstrong, 1964 Jacobs, 1966
48	XXXX	Normal ovaries; normal menstrual history	No harmful effect of extra X chromosomes in ovary	Jacobs, 1966
48	XXXY	Testes small; sterile; breast development greater than in 47-XXY	Regression of germ cells in medulla because of the extra X chromosomes	Armstrong, 1964; Jacobs, 1966
48	XXYY	Testes small; secondary sex characters poorly development	Unknown	Armstrong, 1964; Jacobs, 1966

12

Patterns of Seasonality

Having considered some exogenous factors that appear to control gonadal growth and activity, we can better appreciate the rhythmicity of reproductive seasonality in nature. This chapter reviews some reproductive patterns of free-living vertebrates and attempts to identify features of the environment to which seasonality is adjusted. The proximate factors or cues hat provide signals to the individual's hypothalamic pituitary-gonadal axis initiate development of reproductive potential. Other specific environmental signals are ultimate cues and may precipitate overt reproductive behaviour after gonadal maturation, but they may not necessarily have a direct effect on the gonads.

Most wild vertebrates have an annual cycle of sexual activity and circulating hormones, but the cycle has been determined for only a few species. In the little brown bat (*Myotis lucifugus*) the baseline levels of testosterone obtain during most of the year and range from 2.5 to 11.7 ng/ml, but in midsummer, before fall mating, there is a surge to 59.1 ± 9.2 ng/ml. Comparable premating increases are known for some other species that have been closely monitored. The Asiatic elephant (*Elephas maximus*) experiences a non-breeding low of 0.7 ng/ml but a reproductive high of 44.7 during musth, the state of maximal sexual aggressiveness in male proboscideans.

In females, gonadal steroids increase concurrently with seasonal activity. In fish, steroid hormones stimulate reproductive activity in those species with internal fertilization, but in some species with external fertilization, prostaglandins govern breeding.

This chapter is most descriptive. Some of the observed patterns in nature may suggest the factors that serve as *Zeitgebers*. The patterns observed in nature can be divided into *strategies* and *tactics*. These terms have been widely and loosely applied by field biologists. Here, a reproductive strategy is the totality of traits that maintain a population in its natural range. A tactic is a modification of the strategy so as to adjust it to environmental changes from year to year or to minor or local differences.

Discussions of the adaptive significance of life history strategies and of fitness frequently assume that observed events strictly follow genetic programming. These assumptions usually neglect the highly variable Pleistocene and Holocene environments from which modern species have emerged. If, indeed, reproductive strategies were rigidly fixed and represented "finely tuned" adaptations, these modern species probably would not have survived the Pleistocene. We shall see may reproductive patterns that are remarkably well adapted to existing conditions, but are also variable and apparently better described as prenotypic manifestations of a genotype that is flexible in its expression.

Ectomerms are generally temperature sensitive and frequently response to combinations of factors that include temperature. Among most of the trouts, charrs, and salmon (Salmonidae), spawning is frequently cued by low or declining temperatures, and these fish tend to be autumnal breeders. Most minnows (Cyprinidae), on the other hand, are triggered by rising temperatures, and they usually spawn in the spring and summer. Almost all amphibians require a humid atmosphere, if not standing water, before they spawn. Although most species are found in permanently humid habitats, the occurrence of water itself may not necessarily constitute a distinct variable about which seasonality is scheduled. In many narrative accounts of amphibian reproductive, however, there is the implication that water or rain functions as a cue. Also, amphibians in temperate latitudes spawn in a broad spectrum of ambient temperatures, and clearly there is not only one season for amphibian reproduction. Instead, in some regions different species of anurans spawn in waves, so that a minimal number of species occupy a given pond at any given period. Also, totally terrestrial amphibians usually differ in reproductive seasonality from those that spawn in water.

Reptiles are diverse in their seasonality. Within a restricted region, some mate in the spring and others in the autumn. We have already seen that gametogenesis occurs in the spring in some reptiles and at the end of summer in others, and this pattern is complicated by sperm storage, in males and/or females. In addition, among related species oviparous forms may have a reproductive seasonality very different from closely related live-bearing forms in the same region.

Among birds there is the strongest tendency for uniformity in seasonability. In the mid- and high latitudes most species adhere to a pattern of vernal nesting, and photoperiodic stimuli initiate gonadal recrudescence. At lower latitudes important cues are less rigidly coupled to the annual photocycle, for nesting in some tropical birds in entirely dissociated from the Gregorian calendar. In tropical regions factors other than light seem to cue the nesting of many species.

In midlatitudes larger species of mammals (which usually have a long gestation) tend to reflect more clearly environmental changes in their breeding seasons, but many small species sometimes depart from a strict dependence on meteorological cues. Marine mammals and some rats exhibit the greatest regularity in mating and parturition, for many of them breed in synchrony with migration. Tropical mammals follow a diversity of cues.

Aquatic Reproduction

Fish and many amphibians deposit eggs in water. The

aqueous environment provides protection from sudden temperature changes, and water temperature is a critical factor in the reproduction in many aquatic vertebrates. Light, especially at the red end of the spectrum, does not penetrate far through water, which limits the growth of green plants and the production of oxygen. The mass of water, approximately 840 times that of air at sea level, accounts for stratification of dissolved oxygen, minerals, and temperature, but it provides momentum to marine currents and provides force to wave and tidal action. Oxygen is increasingly soluble at lower temperatures but is nevertheless scarce in deep waters, whether temperatures can be very low. Oxygen content may be highest in polar waters, where surface ocean waters may be -1.9°C. Thus, there are marked oxygen gradients correlated with both depth and temperature, and spawning of fish is adjusted to these variations.

Because of the mass of water, environmental changes are much slower than in aerial habitats. Daily and seasonal variations in the physical features (oxygen, mineral content, and heat) may be rather slight in a given area but are conspicuously different from one region to another. The maximal amount of dissolved oxygen in water is approximately one-thirtieth the oxygen content of air at sea level, but carbon dioxide is approximately 28 times as soluble in water as in oxygen. In shallow, sunlit waters photosynthesis releases oxygen in the daytime, and there can be substantial daily cycles of dissolved oxygen; but where light penetration is weak, oxygen is sparse. This is important in spawning, for developing eggs and larvae have high oxygen demands.

It is ironic that the breeding seasons of the most abundant and diverse vertebrates, the fish, are the poorest known. Although many fish are major sources of food and have been carefully studied and some sport species and ornamental tropicals are virtually domestic, these examples represent a small fraction of the known species. Knowledge of the timing of reproduction of most other species accumulates from the efforts of enthusiastic students, and their attention is concentrated on freshwater and littoral fish.

Fish of Marine Waters

The tremendous mass of water in the oceans accounts for extensive environmental stability in a given area as well as great diversity from place to place. The high latitudes are continuously very cold but experience drastic variations in day length, which largely account for the polar summer productivity and seasonality of reproduction.

Latitudinal Trends

Sharks and rays occupy a specific position apart from bony fish, because of either their unique aspects of viviparity or their hard egg cases. The egg cases allow for a prolonged fetal growth, and their presence in the sea over an extended period does not necessarily indicate a prolonged period of oviposition. Nevertheless, oviposition does appear to occur throughout much of the year in many egg-laying species. Among live-bearing elasmobranchs, some are very clearly seasonal and other lack seasonality.

At high latitudes spawning is cued by temperature and tends to be compressed into a rather brief period. Cold-adapted fish at the warmest borders of their geographic ranges breed in the coldest months. In the Pacific waters off the coast of Canada, the sablefish (*Anoplopoma fimbria*) spawns at the coldest time of the year; eggs are more abundant in January and February, and the peak for larvae is in April. Warm-adapted species at the coldest parts of their geographic ranges spawn in the warmest months. As part of this general pattern, warm-adapted fish tend to have follicles in various stages of development, and they may spawn repeatedly over extended periods, in contrast to cold-adapted species, which spawn only once a year. In general these patterns seem to be adapted to seasons of maximal of plankton. At high latitudes plankton is abundant for a relatively short season, and in warmer water plankton occurs for much longer seasons but at lower levels of abundance.

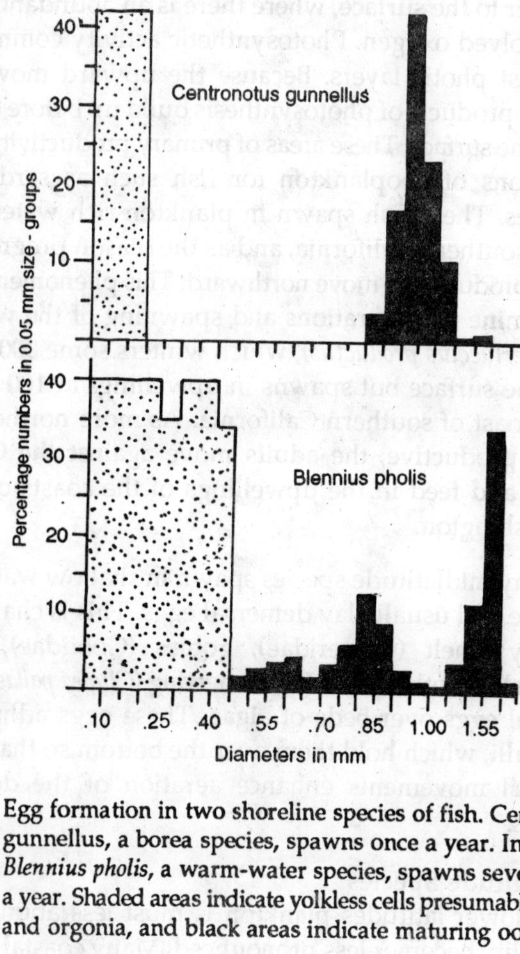

Fig. 12.1. Egg formation in two shoreline species of fish. Centronotus gunnellus, a borea species, spawns once a year. In contrast, *Blennius pholis*, a warm-water species, spawns several times a year. Shaded areas indicate yolkless cells presumably follicles and orgonia, and black areas indicate maturing oocytes.

Midlatitude Species

In the open ocean, successful spawning frequently depends upon adequate plankton for pelagic larvae. Plankton levels vary in a fairly regular fashion with seasonal changes in upwelling and solar radiation, but they are subject to the more variable factor of wind intensity.

Upwellings occur along the western coasts of North and South America, and they vary seasonally. They bring nutrient-rich water to the surface, where there is an abundance of light and dissolved oxygen. Photosynthetic activity commences in the lowest photic layers. Because the upward movement is slow, the products of photosynthesis build up before the water reaches the surface. These areas of primary productivity support populations of zooplankton for fish such as sardines and anchovies. These fish spawn in plankton-rich waters off the coast of southern California, and as the season progresses, the areas of productivity move northward. This phenomenon seems to determine the migrations and spawning of the whiting or hake (*Merluccius productus*), which winters some 300 to 400 m below the surface but spawns in upwellings at 100 m or less off the coast of southern California. As more northerly areas become productive, the adults move against the California Current and feed in the upwellings of the coasts of Oregon and Washington.

Many midlatitude species spawn in shallow waters along the shore and usually lay demersal eggs. This is characteristic of many smelt (Osmeridae), gobies (Gobiidae), sculpins (Cottidae) and others. The Pacific herring (*Clupea pallasi*) spawns demersal eggs over beds of algae. These eggs adhere to the algal thalli, which hold them over the bottom so that currents and tidal movements enhance aeration of the developing embryos.

Low-Latitude Species

At lower latitudes plankton is must less abundant, and seasonality becomes less pronounced. Many coastal species of fish tend to produce fewer ova at one time, but they spawn over protracted periods, frequently at specific times. This seems to be true not only of coastal species but of coral reef fish as well. In the Gulf of Guinea, between 6 and 7°N, the clupeid fish *Hisha africa* breeds throughout the year, and among snappers (Lutianidae) in marine waters of the African tropics, reproduction is essentially aseasonal. A number of subtropical gobies (Gobiidae) appear to have rather protracted spawning periods, which have locally variably peaks.

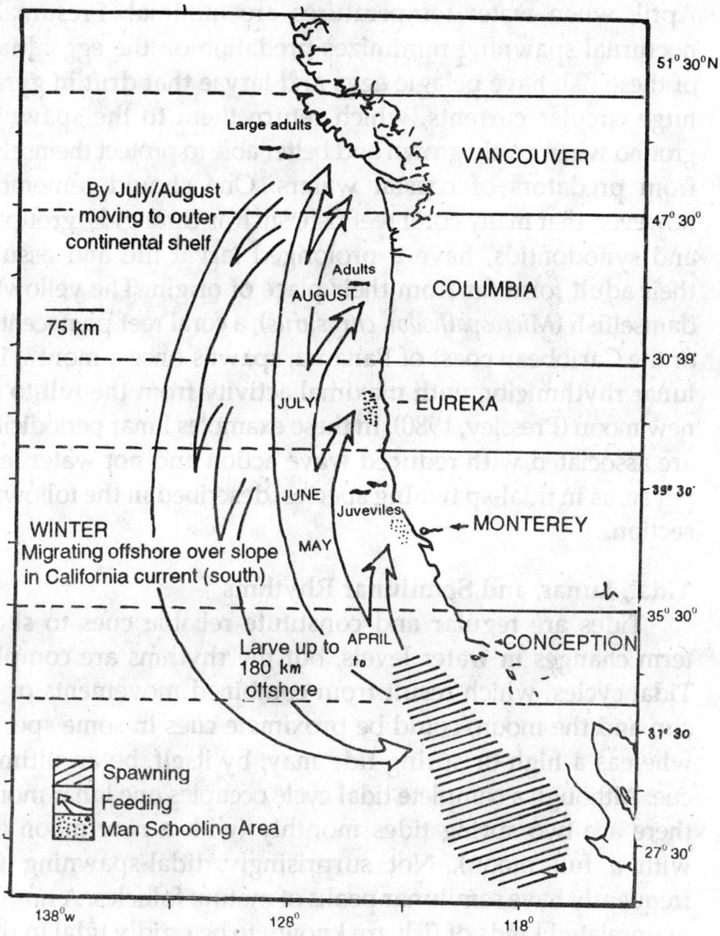

Fig. 12.2. Migratory and spawning patterns of the Pacific hake (Mercullius productus).

Among some 25 species of coastal tropical marine fish of Palau are many that spawn in spring and fall, independent of rain and temperature. Spawning aggregations are nocturnal and frequently occur when the moon is new or full and when the wind is minimal, thus providing a quiet environment for the eggs and larvae. Among 35 species of reef fish in the Caribbean off Jamaica, most spawn between February and April, when water temperatures are minimal. Presumably nocturnal spawning minimizes predation on the eggs. Many of these fish have pelagic eggs and larvae that drift in gyrals, huge circular currents, which return them to the spawning ground when partly grown and better able to protect themselves from predators of coastal waters. One should remember, however, that many coral reef fish, such as carangids, groupers, and synodontids, have a prolonged larval life and assume their adult form far from their place of origin. The yellowtail damselfish (*Microspathodon chrysurus*), a coral reef pomacentrid of the Caribbean coast of Panama, spawns once a month, in a lunar rhythmicity, with maximal activity from the full to the new moon (Pressley, 1980). In these examples lunar periodicities are associated with reduced wave action and not water level per se, as in tidal-spawning species, described in the following section.

Tidal, Lunar, and Semilunar Rhythms

Tides are regular and constitute reliable cues to short-term changes in water levels, but the rhythms are complex. Tidal cycles, which result from combined movements of the sun and the moon, could be proximate cues in some species, whereas a high or spring tide may, by itself, be an ultimate cue. Although a complete tidal cycle occupies one lunar month, there are two spring tides monthly (with a new moon and with a full moon). Not surprisingly, tidal-spawning fish frequently have semilunar peaks of mature follicles. A number of unrelated kinds of fish are known to be rigidly tidal in their spawning periodicity, and this pattern has clearly evolved many times.

Tidal spawning can be regarded as an extreme effort to place eggs in an oxygen-rich environment and also to escape the populations of predators of the intertidal zone. The associated rhythm of oocyte development indicates that this pattern of oviposition is probably an old one.

Several species of coastal fish spawn in synchrony with the near-spring tidal cycle: spawning occurs at high tides, and two weeks later, at the next high tide, the larvae are washed to the sea. Grunion (*Leuresthes*) of the Pacific Coast of North American spawn on sandy beaches at high tide. A female partly buries herself in a vertical position in the sand. As she releases her eggs, one or several males release milt next to her, and this fluid runs down her body to carry sperm to the eggs. The fertilized eggs remain in the wet sand for two weeks. If the fry are not carried out to sea on the next high tide two weeks later, they are capable of surviving an additional two weeks. The Japanese puffer or *fugu* (*Fugu niphobles*), on the east coast of Japan, spawns at the high tide of either the full or new moon. Eggs remain in sand from 3 to 10 cm below the surface, but mortality is 100%. Reproduction depends upon those eggs that are promptly washed to sea. The significance of tidal spawning in the *fugu* is equivocal. On the west coast of Japan (Japan Sea), moreover, the spring tides and daily high tides are minor, and there the puffer does not exhibit a tidal rhythm of spawning.

Fig. 12.3. The synchrony of high rides (light line) and spawning (wide lines) of the fugu, or Japanese puffer (*Fugu niphobles*).

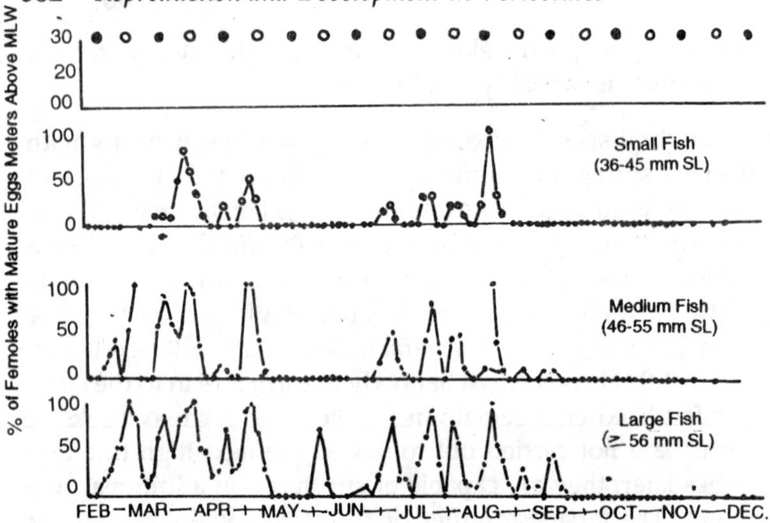

Fig. 12.4. Incidence of different-sized groups (SL = standard length) of the mummichog (*Fundus heteroclitus*) with mature ovarian eggs. Open circles at the top show the occurrence of full moons, and solid circles indicate new moons. The dotted line represents the predicted tidal level above the mean low water (MLW) of the two daily high tides.

The mummichog (*Fundulus heteroclitus*), a topminnow, spawns at spring tides on the Atlantic Coast of North America. Synchronized with the tidal rhythm is a rhythm of ovarian activity, with eggs maturing at 14-day intervals. This ovarian cycle, although essentially lunar, seems to result from successive semilunar cycles, in which alternate peaks exceed the intervening maxima. This little fish spawns when high tides flood grassy flatlands, and the eggs lodge in the axils of a salt march grass (*Spartina alternifolia*) or in the empty shell of a mussel (*Geukensia demissa*). As in some other tidal-spawning species, the eggs develop in air (Taylor et al., 1979). These eggs normally hatch when submerged in water, either two or four weeks later. When experimentally submerged, the eggs begin to swell, the embryo turns within its egg some 5 minutes later, and hatching occurs 15 to 20 minutes after submergence. *F. heteroclitus* has two distinct reproductive periods — one from late winter to early spring and another from summer to early fall — between which there is no spawning. Breeding adults differ in age. The first spawning ends at the same time for all size groups, but smaller fish resume reproductive activity after the larger cohorts, perhaps because of their greater energy demand for food. Two other cyprinodonts (*Fundulus pulvereus* and *Adinia xenica*) also have biweekly cycles of ovarian activity, with preovulatory follicles maturing synchronously with the approach of semilunar or biweekly spring tides.

Tidal reproduction of the Atlantic silverside (*Menidia menidia*) has been carefully documented. These little fish move to the intertidal zone biweekly at the height of the high tides, when current velocities are minimal, perhaps so as to release sperm when there is little water movement. The eggs lodge in the axils of the leaves of cordgrass (*Spartina alternifolia*) and other protective sites and are exposed to air about 10 hours daily, between high tides. Fine threads that develop on the spawned eggs aid in holding them to the substrate. Hatching occurs at the next spring tide two weeks later and is restricted to the hours of darkness. Embryo survival is greatest in the highest intertidal zone, apparently because of minimal predation at that level.

Several species of "southern trout" (Galaxiidae) of New Zealand are well-known to have tidal spawning rhythms. *Galaxias attenuatus* spawns in shallow water at the peak of a spring tide, and the eggs become lodged about the bases of temporarily submerged grasses. Thus, the eggs are protected from desiccation until the next spring tide two weeks later, at which time the larvae are carried out to sea. The inshore spawning migrations of this fish are synchronized with the tidal cycle. *Galaxias maculatus* also spawns on flooded grassy flats, and the eggs develop in the axils of grasses. The eggs are ready to hatch within two weeks, and the larvae drift to open water with the next spring tide.

Benthic Species

Deep waters are said to be virtually devoid of marked

seasonal fluctuations of temperature, food supply, and oxygen, and there is no light. Among a group of benthic fish there is slight evidence of seasonal spawning. Snailfish, small marine species of the family Liparidae, occur from tide pools to 3500 m or deeper. Truly abyssal species living at depths of 2100 m or more spawn throughout the year in contrast to marked seasonality of shallow-water species. *Osteodiscus cascadiae*, which lives between 1900 and 2997 m, spawns aseasonally, but *Careproctus melanurus*, of shallow waters, spawns only in the summer months.

Benthic environments, however, may not be totally aseasonal. Down to depths of 2000 m some species of rattails appear to have seasonality of growth and spawning. In the northeast Atlantic Ocean (50°N, 13°W) there is a rapid fall of phytoplankton, which forms a sedimentary accretion in the spring and early summer. This detritus is derived from diatoms and dinoflagellates. Based on the date of phytoplankton bloom, the rate of fall was 100 to 150 m a day. Such fluctuations in the sea floor might account for synchronized reproductive cycles in four species of benthic echinoderms. There is some evidence that benthic echinoderms and crustaceans are productively seasonal. At depths of from 1330 to 3834 m the brittlestar (*Ophiura ljungmani*) has maximal gonadal mass in the winter months, and most young appear in the summer. There are substantial seasonal variations in community oxygen consumption in the northern Pacific Ocean: north of Hawaii (31°N, 159°W) at a depth of 5900 m sentiment oxygen consumption in June is almost four times higher than in November (Sumith and Boldwin, 1984). Clearly, the benthos is not without seasonal environment charges and it appears possible that seasonality may also be found in some benthic fish.

Some species of flatnosed cod (Moridae) and the grenadiers or rattails live at 2000 m. It growth rates and spawning they seem to reflect seasonal charges at the surface, presumably descent of energy from plankton. *Corypeaenoides* rupestris (also called *Macrourus rupestris*) off the coast of Scomtland has an autumnal spawning.

Fish of Freshwaters

Seasonality of reproduction in freshwater fish usually follows one of several patterns: (1) some species develop a single generation of oocytes and spawn once a year, (2) some breed over an extended season with more than one generation of oocytes, and (3) still others have successive generations of oocytes and spawn sporadically. This classification covers most reproductive strategies of freshwater fish, but it does not correlate these patterns with environmental events. The complex of patterns described in the following sections will usually fall into one of the three categories of Qasim and Qayyum, but there are many diverse environmental conditions to which these patterns are adapted.

Quiet Waters

Lakes and ponds very in size but have distinctive features in common not found in rapidly moving waters. A lake has convection currents that result from seasonal temperature changes, prevailing winds, and streams entering and leaving. A deep lake will have some features of a marine sea with respect to stratification of light, temperature, and dissolved oxygen. Shallow lakes and ponds may light penetrating to the bottom, with a rapid turnover of oxygen at all depths.

Tropical Lakes. In tropical lakes environmental cues depend not only on the cycle of precipitation but also on the depth and breadth of the lake. Rains increase nutrition introduced from inflowing streams, and changes in plankton populations are also caused by fluctuations in water depths and shorelines. This suggests that reproductive seasonality varies with available nutrition and perhaps with the chemical composition of the water. Temperature and day length changes, so important in medlatitudes, are much less pronounced in the tropics and of minor importance to fish reproduction. For those species with pelagic larvae, plankton may be a major factor in their survival. In tropical lakes the growth of plankton is promoted not only by rain, but also by increased productivity from upwellings, and lack of wave action.

In Lake Tanganyika, species of herring (Clupeidae) have spawning peaks coinciding with high populations of plankton. By way of contrast. African Cichlidae, especially those that carry the eggs in their buccal cavity, are rather aseasonal. Among a group of mouthbrooding cichlids in Lake Malawi, nine breed throughout the year, with bimodal peaks at times of plankton bloom. Food is important not only for energy in the production of eggs but also to sustain metabolism during the period of care of the young. By carrying the freshly spawned eggs in the buccal cavity, a fish may create its own habitat for the eggs and small young. Some sympatric species of *Tilapia* have discrete and separate spawning seasons, timed to different phases of the annual cycle of rains. Similarly, in Lake Victoria three of the five major species of catfish breed more or less continuously, but increases in spawning tend to occur between rainy seasons. Like some of the cichlids, these catfish provide parental care. Likewise an Indian catfish (*Glyptosternum pecinopterum*) is aseasonal and spawns several times a year, and the ovaries in one fish may contain ova in several stages of maturation. In contrast the cichlid fish *Entroplus maculatus* and *E. suratensis* spawn twice annually in Sri Lanka, during periods of maximal water clarity, and do not breed when seasonal rains render the water murky.

Many oviparous cichlid fish of Africa extended breeding seasons with local peaks in spawning, and these peaks may coincide with times of high water. Some species spawn over land that is temporarily inundated. Because hatching occurs soon after spawning and because growth is rapid, the young move to the main watercourses before the floodwaters recede. This same pattern also exists in tropical South America and southeast Asia.

A unique tropical pattern occurs in the cyprinodont "instant fish," which live in temporary ponds in East Africa and arid parts of South America. These fish have a rigidly defined annual life-cycle, for all the adults die when ponds dry up after the rainy season. Their eggs, which remain in the dry soil, enter diapause and thus survive the seasonal drought. The eggs resume development when the ponds are flooded almost a year later. Final embryonic development is cued by submergence. These fish (species of *Nothobranchius, Aphyosemion,* and *Cynolebias*) have a rather restricted reproductive season and are called instant fish because of their rapid development when submerged in water. A through review of this group is presented by Simpson (1979).

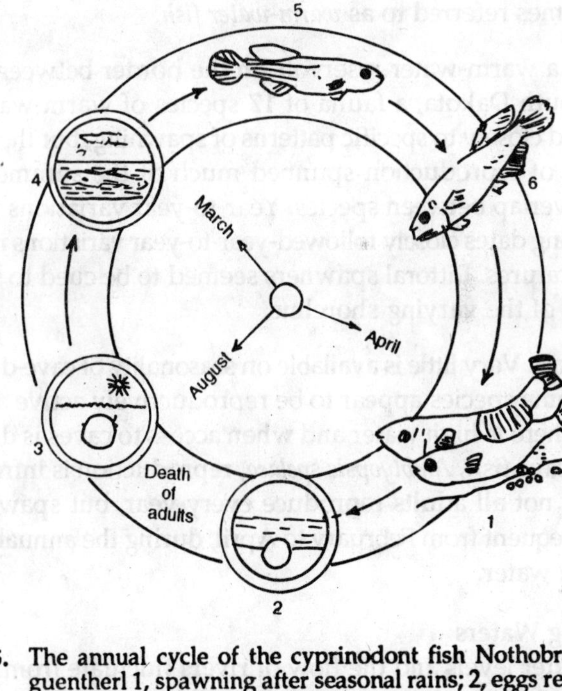

Fig. 12.5. The annual cycle of the cyprinodont fish Nothobranchius guentherl 1, spawning after seasonal rains; 2, eggs remain in the bottom of temporary pools; 3, water evaporates during the rainless periods during which time the adults perish; 4, the return of the rains again floods the pools causing some eggs to hatch; 5, the young grow rapidly; 6, spawning follows some four to eight weeks after hatching.

Although the tropical Asiatic anabantid fish *Trichogaster pectoralis* breeds throughout the year, peaks of breeding coincide with seasonal rains, and there is some suggestion that rains are followed by an increase in food supply for the fry.

Temperate Lakes. In lakes and ponds sufficiently far from the equator so that winter ice forms, there is a kaleidoscopic array of vernal changes, any or all of which might constitute proximate cues and precipitate gonadal recrudescence. In midlatitude lakes spawning is usually a spring or summer event, and there is much evidence that both light and temperature play major roles in timing of gonadal development. This is especially true if the water is rather shallow, for such ponds and lakes become warm by summer and their fish are sometimes referred to as *warm-water fish.*

In a warm-water reservoir on the border between North and South Dakota, a fauna of 17 species of warm-water fish adhered closely to specific patterns of spawning, but the overall season of reproduction spunned much of the summer, with little overlap between species. Year-to-year variations in peak spawning dates closely followed-year-to-year variations in water temperatures. Littoral spawners seemed to be cued to specific depths of the varying shoreline.

Caves. Very little is available on seasonality of cave-dwelling fish. Some species appear to be reproductively active at times when there is high water and when access to caves is difficult. In the cave fish, *Amblyopsis spelaea,* reproduction is infrequent; that is, not all adults reproduce every year, but spawning is most frequent from February to April, during the annual period of high water.

Moving Waters

Water levels and the flow of rivers fluctuate from season to season, depending upon patterns of precipitation and snow melt. When waters are low, surrounding banks are dry; but as rainfall saturates the land, seepage into rives basins enriches the mineral and oxygen content of the rivers. In time the water levels rise and create optimal conditions for spawning of many species of fish as well as for growth of food supplies.

In tropical rivers and floodplains, there is heavy pressure on fish populations. Both predation and oxygen deprivation account for large losses, and in these environments fish either spawn large numbers of eggs or provide some sort of parental care. The spawning of small species is concentrated in small pools and pounds that become isolated by receding waters during dry seasons. Shallow waters are, however, vulnerable to oxygen depletion when temperatures rise, and high mortality of eggs and fry can result from overheating and suffocation.

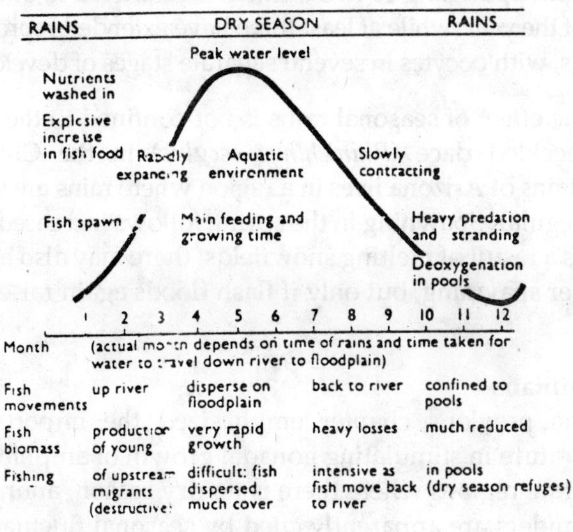

Fig. 12.6. The seasonal relationship between rainfall and spawning of teleosts in a tropical river system.

Among many tropical riverine fish, spawning seems to be stimulated by seasonal rains. In many fish in the Amazon Basin, spawning commences soon after flooding, indicating that gonadal ripening had occurred prior to the earliest heavy rains. Wild populations of the platyfish (*Xiphophorus variatus*) breed at the onset of seasonal rains, which result in an improvement in their streams, and reproduction ceases near the start of the dry period. This cycle seems to synchronize gonadal maturation and mating within a given stream system. In coastal rivers of tropical Australia the spangled perch (*Therapon unicolor*) spawns when water levels rise and turbidity increases as a result of annual rains, and gonadal preparation seems to be induced by rising temperatures and increasing day lengths.

If there is a lack of great seasonality of flow in a tropical stream spawning periods then resemble seasonality in temperate regions. Among six species of characoid fish in Panama (at 10°N), the spawning of two of them is restricted to one or two days of the year, while at least three have extended reproductive periods, with oocytes in several separate stages of development.

The effect of seasonal rains is not confined to the tropics. the speckled dace (*Rhinichthys osculus*) in the Chiricahua Mountains of Arizona lives in a region where rains are seasonal but irregular. Spawning in the spring follows increased stream flow as a result of melting snowfields; there may also be a late-summer spawning, but only if flash floods again raise stream levels.

Amphibians

The previous chapter emphasized the importance of temperature in stimulating gonadal growth of amphibians. In temperate regions where there is no dry season, anurans and salamanders are apparently cued by seasonal fluctuations in temperature. At any latitude, however, in regions with marked wet and dry periods, amphibians seem to adapt more to soil moisture and standing water for proper reproductive conditions than to ambient temperature, and this is true whether rainfall is irregular or seasonal. Anuran reproductive seasonality in midlatitudes is suppressed by low winter temperatures and resumes as temperatures rise and snow changes to rain. In arid regions, where standing water is temporary, rains are required before amphibians can spawn. In low latitudes with continuous high humidity, mating may be more or less aseasonal, but oviposition may occur within a restricted period. Seasonality can be quite different for terrestrially ovipositing species.

At any latitude, terrestrial and arboreal amphibians may have cycle that diverge from those of species that spawn in water. For example, most species of the Nearctic genus *Ambystoma* move to water in late winter or spring, but those that lay eggs on land do so in autumn. There is a great range of spawning times of salamanders in warm climates or the tropics.

Although most amphibians in variable environments breed when there is a combination of rising temperatures, warm rains, and increasing day length, little experimental work has been done to separate and identify environmental factors. As pointed out in the previous chapter, day length, the sound produced by falling rain, and atmospheric humidity may function as cues in amphibian reproduction. Additionally, rain, as opposed to falling snow, would raise the temperature of pond water.

In regions with a fairly regular annual cycle of temperature change, there are corresponding cycles of gonadal recrudescence. In lower latitudes, where temperature changes only slightly and precipitation may be sporadic, the gonads must be in a state of readiness for prolonged periods. In these regions gonadal recrudescence may not be cyclic, although spawning usually is.

Temperate Species

Zoologists are concentrated in the midlatitudes, and most field studies are made in such regions. Therefore until recent years our knowledge of amphibian reproductive cycle in the wild has been largely concerned with species of the temperate regions. In such environments gonadal recrudescence often starts in late summer, and mating occurs on rising temperatures in the spring. In toads (*Bufo bufo*), for example, the ovulatory cycle begins a few weeks after egg laying, in late spring or summer, and oocytes reach near-maximal size by November, prior to hibernation. Thus the females are replete with mature eggs in late winter and early spring, when both sexes migrate to the spawning ponds.

Some amphibians inhabit springs or lakes, in which obviously no land-to-water migration takes place. In these environments there is frequently no annual cycle of temperature change, and gametogenesis, which is heat sensitive, is continuous. In spring-inhabiting frogs of the genus *Telmatobius*, spermatogenesis is continuous throughout the year at constant water temperatures near 0°C, and ovaries also show no seasonal changes.

Northeastern United States has a rich and diverse fauna of salamanders, and the marked climatic seasonality must account for the annual regularity of reproduction in these amphibians. The seasonality of reproduction is nevertheless diverse. Almost all of these species have internal fertilization involving the transfer of a spermatophore, with the possibility of sperm storage. As a result of this situation, there can be seasons of mating and also of oviposition, and these two periods may be separated by months.

Tropical Species

Anurans that breed in the water, whether in the tropics or at higher latitudes, are most active during periods of high atmospheric humidity. Because tropical rains may be irregular, pond-breeding frogs and toads spawn sporadically. In an equatorial region in the upper Amazon Valley in Ecuador, Duellman (1978) recognized 11 reproductive modes, based on sites of oviposition, habitat, and type of larval development. Aseasonality is characteristic of those species that lay eggs on land, and most of the terrestrially egg laying species have direct development, with the larva stage being passed within the egg. In pond egg. In pond-spawning species oviposition increases after heavy rains. Thus although there is anuran reproduction throughout the year, specific groups have their own seasonality.

Pond-spawning anurans in deserts appear when sudden rains create temporary ponds or only puddles, and such precipitation is commonly irregular. Consequently, those anurans adapted to anexerit existence must be prepared to oviposit during periods in which standing water may briefly occur. The spadefoot toads (*Scaphioupus* spp.) if the New World breed it temporary ponds and are well-known for their rigid adherence to rainfall as an environmental signal. They move from their subterranean retreats to ponds during spring rains. The rapid growth and early metamorphosis of their tadpoles are acuptations to such a spawning site.

Cyrorana platycephalus, a lepudactylid frog of Australia, burrows into the soil during droughts. Females preserve ripe ova through the nonreproductive season and thus are prepared to oviposit whenever there is adequate rain. Ripe ova may be retained from one year to the next. In the Durban region of South Africa the microhylid frog *Brevipes adspersus* burrows deeply in the dry season and its emergence coincides with the appearance of the alate stage of termites on which they feed. The termites do not likely constitute a seasonal cue for the frogs, but it is possible that both amphibian and insect respondent to the same cues.

In the subtropical rice frog (*Rana limnocharis*) of Taiwan, there are two rather distinct spawning periods: vermal spawning throughout. March and April and a second spawning in July and August. Each time, adults assemble in the shallow water of freshly cultivated fields when early evening temperatures reach 20°C. This movement occurs when the fields are plowed and flooded, which together seem to constitute an immediate attractant for the rice frog; odor from the freshly disturbed soil is suspected to guide the frogs to these fields.

In several well-studied species of both live-bearing and egg-laying caecilians, mating may apparently occur cover in extended period or throughout the year, but the young appear near the onset of the rainy season. In French Guyana *Typhlorectes compressicaudus*, a vaziparous caecilian, has an annual reproductive season, with spermatogenesis in June and July, during the rainy season. Young are born in me dry season, while water is still high.

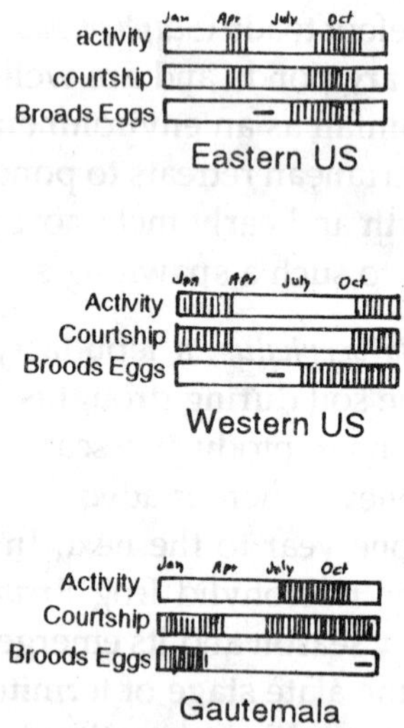

Fig. 12.7. General patterns of reproductivity of three geographic groups of plethodontid salamanders. The horizontal bar indicates oviposition. Vertical lines indicate periods of surface activity. For Guatemalan species, only terrestrial taxa were studied.

Geographic and environmental vacations in seasonality of plethodontid salamanders suggest the relative important of ambient temperature and humidity. In eastern United States low temperatures restrict general activity to the warm months. In the western states, where winter precipitation (in low elevations) falls as rain and where summer rain is spurse, activity is seen form autumn to spring, when the soil is moist. In both regions oviposition is mostly in the spring. In Guatemal, where there are continuously mild temperatures and a discrete dry season, activity continues throughout the year.

Spermatogenesis, moreover, is un-interrupted by any seasonal quiescence. Oviposition, which is either terrestrial or arboreal is confined to a brief person in autumn, and breeding continues through the dry period. The eggs hatch near the onset of the annual rains. Thus reproductive activity, and activity in general, is initially suppressed by low temperatures. When temperatures are sufficiently warm to permit locomotor activity as well as gametogenesis, humidity appears to define reproductive seasonality in these amphibians.

Biennial Reproduction

In some amphibians, either at high latitudes or high altitudes, females tend to develop mature ova every second or third year. This characteristic is not confined to amphibians having short seasons of activity. In *Typhlonectes compressidcudus*, a Neotropical caecilian, individual females are gravid biennially, with only half the population giving birth to young in any given year. All caecilians appear to be biennial in their reproduction (Wake, 1970). Regularity of spawning in the toad (*Bufo bufo*) seems to depend upon the individual's ability to accumulate energy stores sufficient to form mature oocytes. In Switzerland (Zurich) this limitation may account for a pattern of biennial spawning in most females. In montane populations of the fire salamander (*Salamandra salamandra quadri-virgata*), breeding is biennial. Biennial reproduction is probably more widespread among the amphibia than is generally recognized.

Reptiles

Most reptiles exhibit reproductive seasonality, whether in the temperate regions, where climatic conditions change conspicuously from month to month or near the equator, where photoperiod and temperature are rather constant. Although not all species in a given region follow a uniform schedule of steroidogenesis, gametogenesis, mating, and birth or oviposition, most taxa have a rather regular annual pattern that want only moderately from year to year.

Timing of reproductive cycles is probably more variable

for reptiles than for other terrestrial vertebrates. Not only are there distinct patterns of gonadal recrudescence (refining different patterns of seasonality of gametogenesis, steroidogenesis, and curcuaning levels of gonadotrophins), but reptilian cycles are also complicated by the occasional occurrence of viviparity and sperm storage. Unfortunately, the hormonal background of these variations has been studied in only a few species.

Temperate Species

Although vernal mating characterizes most species, the timing of gonadal activity varies some reptiles undergo spermatogenesis in the spring and mate shortly thereafter, and others experience spermatogenesis in the autumn and store spermatozoa in the ecidyrides for mating the following spring. In such reptiles oviposition occurs shortly after mating, and gonadal regression follows shortly thereafter.

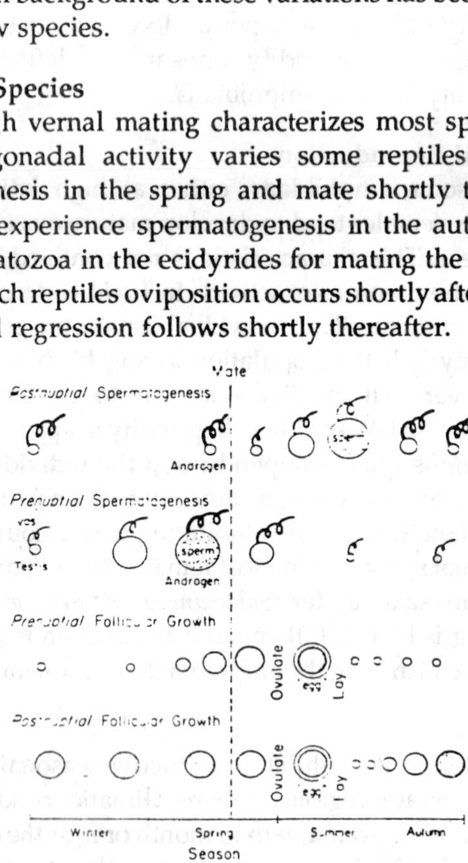

Fig. 12.8. Two major types of reptillian gonadal activity relative to season of mating. Examples represent oviparous species mating in the spring, with gametogenesis postnuptial.

Two contrasting species in one area will illustrate the

extent of divergence of gomada activity. Both the cobra (*Naja naja*) and the soft-shelled turtle (*Trionyx sinensis*) hibernate in the Hong Kong area, and both mate soon after emergence from dormancy, in the spring. Although the ovulatory cycles of the two reptiles are very similar, the calendar of testicular activity differs. Spermatogenesis is prenuptial in the cobra and postnuptial in the turtle. After the production of spermatozoa by the turtle in late summer, the testes regress and the spermatozoa move to the cauda epididymides, where they remain over the winter. In the cobra, steroidogenesis is maximal in late winter, but steroidogenesis is essentially a late-summer or autumnal activity in the turtle. In some vipers, such as the copperhead (*Agkistrodon contortrix*), spermatogenesis is postnuptial, and mating is both in the autumn and the spring. In these vipers, oogenesis is prenuptial.

In the oviparous *Sceloporus undulatus*, testicular recrudescence begins in mid-summer, after mating. Spermatid formation follows in the autumn; spermatozoa are stored in the epididymides over the winter, and oogenesis takes place in the spring. Many midlatitude oviparous iguanid lizards have this pattern, in which the two sexes appear to be following different cues; spermatogenesis is initiated in the late summer, and oogenesis commences the following spring. In those species in which gametogenesis of males and females is not simultaneous, the hypothalamus would appear differentially responsive to environmental stimuli. In the majority of oviparous lizards the young emerge from their eggs at the end of the summer or in autumn, shortly before entrance into hibernation. This season would seem to be the least appropriate and is in marked contrast to the season of birth of viviparous lizards, which are born in the spring.

In some temperate reptiles the initial breeding of young adult females is slightly delayed. Emergence from hibernation may be concurrent in both young adults and those that have reproduced the previous year, but the young females tend to mate a week or more after the older females mate. In the viviparous *Lacerta vivipara*, young adult females (breeding for the first time) mate somewhat after the time at which older adult females mate.

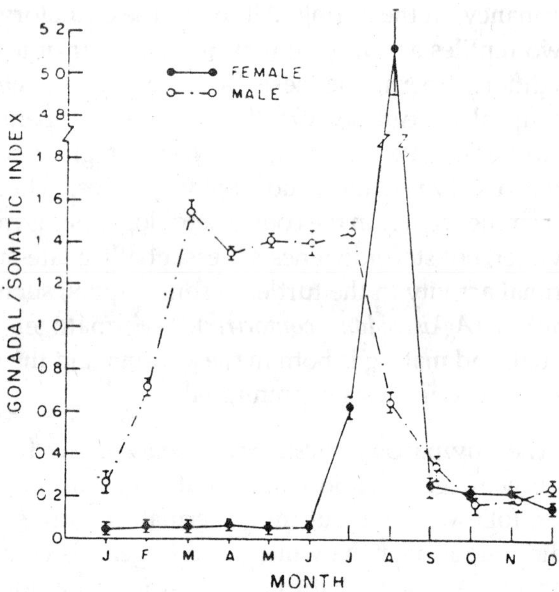

Fig. 12.9. Monthly changes in gonadosomatic index of Sceloporus grammicus microlepidotus, a live-bearing iguanid lizard, illustrating vernal spermatogenesis and autumnal folliculogenesis. Values are means ± SE. Gonadosomatic index = gonadal mass/body mass.

Live-bearing in reptiles generally alters the pattern of seasonality. In temperate regions at least, viviparity seems to be accompanied by winter activity, in contrast to sympatric egg-laying species. In live-bearing iguanid lizards testicular activity commences in the spring and continues through the summer, but ovarian growth commences in mid-or late summer, at the time of testicular regression. In *Sceloporus grammicus*, sperm are stored in the epididymides and mating occurs in September (Guillette and Casas-Andreu, 1980); in *S. jarrovi*, mating is in November. Similarly, in the live-bearing *Sceloporus formosus* of southern Mexico, testicular growth occurs in the early spring and regression by late spring, and the development of oocytes takes place anytime from late spring until early December. In *Sceloporus formosus* the time of mating is unknown, but it clearly follows testicular regression and probably involves storage of sperm. These species are active in the winter and bask in sunny weather, and parturition is in the spring. In the elapid genus *Austrelaps* of Australia, however mating occurs in the spring and young are born in the following summer.

Tropical Species

The concept of tropical aseasonality is based on few studies, but several tropical reptiles are known to have a discrete hiatus in reproductive activity. Clear-cut evidence for acyclic breeding is known for only a few tropical squamates. In those species in which males may have spermatozoa throughout the year, cyclicity may be apparent in the females. In tropical rain forests reptiles are likely to breed throughout the year, but in grassland species reproductive cycles may be synchronized by seasonal rains. Among several species of snakes in Cambodia, each has a distinct seasonality, but there is no uniformity across specific lines.

Among seven species of anoles (*Anolis* spp.) studied in tropical environments on Caribbean islands, reproductive activity persisted throughout the year, judging from the state of both male and female gonads. Minimal gonadal activity characterized the "winter" months, from October to January, and reproduction appears to depend upon a combination of day length and temperature cycles. In Caribbean anoles ovarian activity is closely tied to rainfall. Folliculogenesis declines with the arrival of the dry season, but rainfall has no effect on testicular activity.

There are sexually dissimilar gonadal cycles in the viviparous boid snake, *Candoia carinata*, of the southwest Pacific. In the males testicular activity, as measured by seminiferous tubule diameter, is greatest from April to June, after which epididymal spermatozoa are preserved. There appears to be no cyclicity to ovarian activity, however. The opposite situation obtains in the anguid lizard, *Barisia monticola*, of Central America. The males are continuously spermatogenic, but there is a tendency for reproductive seasonality in females.

In the absence of marked seasonal changes in whether, there may be prolonged reproductive seasons in tropical squamates, which has been documented for several species. The teiid lizards, *Ameiva quadrilineata* and *A. festiva*, of Costa Rica are reproductively active at all seasons. Females with ripe follicles or oviductal eggs have small fat bodies, whereas reproductively inactive females have large fat bodies (Smith, 1968). Among a group of four species of geckos in Brazil there is a marked seasonal cycle in testis volume, with a reduction in the dry period of the year; but oogenesis and ovulation continue throughout the year. Possibly sperm storage allows this lack of synchrony between the sexes.

In Tanzania the live-bearing skink *Mabuya striata* breeds throughout the year, but in the same area the oviparous agamid lizard *Agama cyanogaster* lays eggs only in the wet season, with young hatching only in the dry season. Similarly, in Zambia, where there are marked seasonal variations in rainfall, the oviparous skink *Mabuya quinquetaeniata* lays eggs at the onset of the rainy period, but *Mabuya striata* breeds throughout the year.

Chelonians

Among midlatitude freshwater turtles, egg laying is usually a vernal event, following the formation of sperm the previous year. In at least four families of New World turtles, mating takes place in the spring, spermatogenesis is postnuptial, and sperm are stored over the winter in the cauda epididymides. Nesting of the highly aquatic snapping turtle (*Chelydra serpentina*) is closely correlated with rising temperatures in shallow water, most especially with accumulated heat units, or degree-days, which are needed for final follicle maturation.

In lower latitudes, where temperatures are moderate and annual variation is reduced, reproduction may be correlated with rainfall, either positively or negatively. Some species lay only in the dry season, and others lay only in the rainy period. Although testicular recrudescence occurs in late summer in the Caspian terrapin (*Mauremys caspica rivulata*), most mating activity occurs in the winter, during months of maximal precipitation and minimal temperatures, day length, and testicular mass. At this time the epididymides are swollen with spermatozoa.

The timing of chelonian reproduction appears to accommodate maximal survival of the eggs rather than to provide for hatching at the most favuorable time for survival of the newly hatched young. Many species of freshwater and terrestrial turtles deposit one or more clutches in the summer, and the young overwinter within the egg and emerge the following spring. Delayed emergence is independent of any particular sequence of suppressive seasonal climatic changes in temperature and/or humidity, but it allows hatchlings to emerge when environmental conditions are most favourable for their survival. In many chelonians the eggs are far more vulnerable than the young are, but in some turtles, such as the marine-turtles, there is great predation on the young.

Some of the large marine turtles tend to lay eggs during irregularly recurring visits to isolated sandy beaches. This irregularity has been confirmed by the identification of marked individuals over a period of years. The green turtle (*Chelonia mydas*) remains at sea for several years, and females lay several times within weeks, every second or third year. On ascension Island, green turtles experience difficulty in digging nests in the hard dry sand, which may account for nesting during the season of heaviest precipitation. On the Comoro Islands green turtles lay eggs sporadically, without a discrete nesting season. The loggerhead (*Caretta caretta*) appears to be irregular in its spawning periodicity. It also oviposits at intervals of from two to several years and shows little rhythmicity in its return to its nesting beaches. The reproductive periodicity of marine turtles is probably determined partly by the time required to make a complete migration and also to accumulate sufficient fat stores for a clutch of eggs.

Some species of chelonians divide a season's production of eggs among two or several clutches. Because the nest is not subsequently guarded, it is quite vulnerable to predation, and prehatching loss may be high. Multiple nesting is of obvious advantage in reducing loss to predators. The loggerhead turtle (*Caretta caretta*) has been known to lay up to seven clutches a season (Frazer and Richardson, 1986). The leatherback turtle (*Dermochelys coriacea*) lays a mean of four clutches annually, with a maximum of ten.

Biennial Reproduction

In temperate regions, with a reduced period of annual activity and lower mean temperatures, reptiles require prolonged periods to accumulate fat for embryonic growth. This environment introduces another variable in reproductive periodicity: in a number of reptiles in temperate regions, reproduction is biennial, and alternate years are devoted to accumulation of adipose tissue, most of which supports development of a subsequent generation of young. The accumulation of fat for reproduction is most critical for females, and biennial reproduction does not affect males.

In several species of vipers (Viperidae) females breed in alternate years. In northern Idaho the rattlesnake (*Crotalus viridis*) shows a strong tendency to follow a biennial pattern. The mean percentage of reproductive females in any given year varies from 54 to 77, and reproductive activity is positively correlated with fat stores. Similarly, in northwestern Texas *Crotalus atrox* produces young every other year, and biennial breeding seems to be predicated on the amount of body fat. The common adder (*Vipera berus*), in the Swiss Alps, reproduces every second or third year, and birth takes place just before its entrance into hibernation. Species of *Austrelaps,* an Australian elapid snake, appear to breed in two years out of three, perhaps depending on prey availability.

In Japan the skink *Eumeces okadae* reproduces biennially, and approximately half of the females lay eggs in any year. In this little lizard, females seldom eat while they are brooding their eggs, allowing little time to restore their fat supplies. The anguid lizard. *Barisia monticola,* is a tropical live-bearer and lives at 2000 m in Costa Rica, where the temperature is constant. It produces young every other year. On the Isle of Portland (Dorset) in the British Isles (50°N), the live-bearing *Anguis fragilis* breeds every other year. The young are born just before the adults enter hibernation, and there is little time for the adults to accumulate sufficient energy for reproduction the following spring.

Significance of Reptilian Seasonality

With the preceding live-bearing iguanid lizards the adjustment in seasonality of mating results in birth of young in the spring, when there is an abundance of small arthropods upon which they can feed. In sympatric oviparous species, however, the young emerge from eggs in the late summer or autumn. In the tropics the association of precipitation and reproduction is apparently far more important for egg-laying species than it is for live-bearers. Food may not constitute a limiting factor for young lizards in the tropics (Stamps and Tanaka, 1981), and the lack of seasonality among the viviparous squamates in the tropics suggests that seasonality among egg-laying species is an adaptation to humidity. Oviposition in *Anolis sagrei* is correlated with atmospheric humidity, suggesting that it may not require rainfall per se.

Reptilian reproductive appears to be scheduled to be scheduled so that eggs are laid at the most propitious time, which may be determined at least partly by soil and/or atmospheric moisture. Oviparous species usually have clearly apparent seasons of oviposition. Live-bearing species appear to be free of this constraint, and tropical reptiles produce young over more prolonged periods.

Birds

Avian seasonality is far more uniform than seasonality for other classes. All birds are oviparous. Sperm storage is brief or absent, and embryonic diapause, when it occurs, is brief and postovipositional. In most birds, especially the smaller, altricial species, incubation may occupy only two weeks, and growth of prefledging young is very rapid. It is thus possible to concentrate as avian reproductive cycle into relatively discrete short periods.

Reproduction, migration, and molt are the three major activities demanding energy, above that required for growth and maintenance in birds, and avian life cycles are adjusted so as to minimize concurrent demands of these activities. For example, a bird that regularly makes a long migration may have a relatively restricted reproductive period with a single brood, in contrast to a sedentary species that has a more prolonged nesting season and may produce several successive broods. Several broods do not place an unmanageable demand upon the energy of a species that does not need to store energy for a migration. The timing of the molt is adjusted so as not to conflict with nesting or migration. The schedule for these three events is a relatively rigid budget of the species' accumulation and expenditure of energy.

Temperate Latitude Species

The reproduction of midlatitude birds is almost always vernal. At the highest latitudes the breeding season of most species is restricted to a period when days are long and environmental productivity is intense. All birds in temperate and polar regions are stimulated by day length. The critical day length varies: species that are stimulated by a relatively short day length breed earlier than do those that require a longer stimulatory period. The duration of reproduction is arrested by a refractory period in almost all birds.

In birds it is apparent that nesting begins after the most favourable date for nestling survival. This suggests that there is some physiological imbalance between the readiness to lay and the nutritional demands of the young. When provided with supplemental food (fish), magpies (*Pica pica*) laid their eggs somewhat earlier than did a control group that foraged for all its own food. The provision of supplementary food has been shown to advance the time of nesting for many species.

Many species show year-to-year variations in the time of nesting and egg laying. The great tit (*Parus major*) varies in the date of first laying, and approximately 40% of this variation seems to be genetically determined, so that the date of laying can be readily adjusted to changing environments.

The allocation of time and energy to these events can be seen among some Old World warblers. The chiffchaff (*Phylloscopus collybita*) regularly rears two broods and may winter from southern Britain and Europe. Some other Eurasian species (e.g., *Phylloscopus trochilus* and *P. sibilatrix*) are usually single brooded and make extensive migrations to the southern half of Africa. The chiffchaff has a postnuptial molt, in contrast to the willow warbler (*P. trochilus*), which molts after arriving on the wintering ground in southern Africa. The chiffchaff clearly extends more time and energy in reproduction than the willow postnuptial molt is delayed until after the autumn migration.

In the far north, where the reproductive season is extremely abbreviated, there is little elasticity in the nesting calendar. In the white wagtail (*Motacilla albai* in central Finland, older females, which are nesting for the second year, arrive on the nesting ground approximately a week before the younger females, which are easily distinguished in the field. Dates of egg laying do not differ between the two age groups.

Many species of birds cross the equator as part of their annual migration and are then exposed to another set of vernal cues, including resurgence of plant growth and a stimulatory photoperiod. Although most species of birds are refractory to these stimuli, a few species may sometimes nest on their "wintering" grounds, where it is summer. The white stork (*Ciconia ciconia*) and the house martin (*Deltcon urbica*) have both been known to breed on their winter range in Africa, and the bee eater regularly breeds twice a year, once in each hemisphere.

Low-Latitude Species

In tropical regions where there is pronounced seasonality of precipitation, many birds appear to use rainfall as an ultimate cue. Most species nest at the onset of the rains, and rainfall is frequently followed by a resurgence of vegetation and an increase in insect populations.

Among 117 species of insectivorous birds in the Serengeti, none breeds during the dry season, between July and October. Insect abundance coincides rather closely with seasonal rainfall, and peaks in nesting follow those of rainfall and insect numbers by about two months. In *Acacia* woodlands in Tanzania, insectivorous birds breed before the rains commence, possibly because the fresh leafy growth of forest trees precedes the seasonal rains and forms the basis for the development of insect populations.

Many species of African and Australian estrildine finches occupy open areas, and they nest well after the onset of the rainy season. They build nests as grasses mature, and they not only construct their nests from fresh grass but also feed their young largely on developing seeds in the milk stage. In the soft condition, grass seeds consists of a rich endosperm that lacks cytoplasmic separations of cellulose. Such soft seeds are not only physically easy for the young finches to eat but probably almost totally digestible.

In the Galapagos there is a marked alternation of season, from hot and wet to cool and dry. Galapagos finches

(Geospininae) nest during the rainy season, closely following local and year-to-year variations in precipitation. At this time food is relatively abundant, and the feeding patterns of the different finches tend to converge.

The role of rainfall in arid regions is not restricted to low latitudes. In Arizona (32-33°N), Abert's towhee (*Pipilio aberti*) nests regularly in the spring but nests a second time in mid- or late summer in years in which there is a substantial late-summer rain.

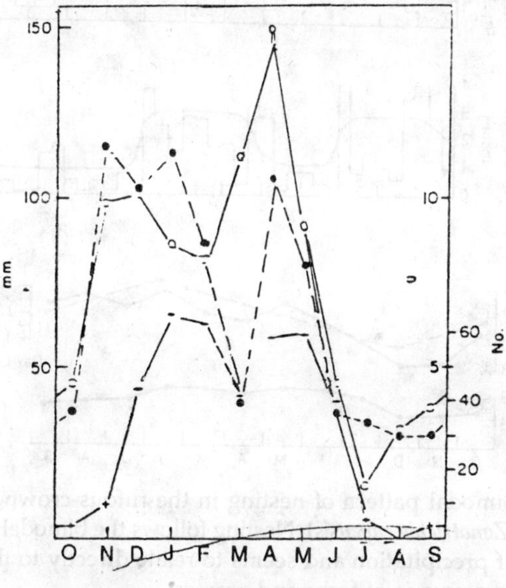

Fig. 12.10. The distributor of nesting insect birds (+), showing two peaks, which lag behind those of mean (sold circles) and mean daily catch of insects (open circles). Breeding are mean rainfall are for the decade 1964 to 1973.

There are many more anecdotal references repeating the preceding correlation of nesting and rainfall in arid regions. Such patterns have clear advantages, but the underlying ultimate cues are not always apparent. Most experimental studies on reproductive cues or factors have been conducted with temperate species, which are exposed to marked seasonal changes in day length. In the previous chapter it was pointed out that atmospheric humidity may constitute an ultimate cue for birds.

Some species, in contrast, avoid nesting in the wet season of the year. The jungle fowl (*Gallus gallus*) in the Greater and Lesser Sunda Islands has a variable nesting season that is controlled by local rainfall patterns. Probably because of its ground-nesting habit, reproduction is confined to dry seasons.

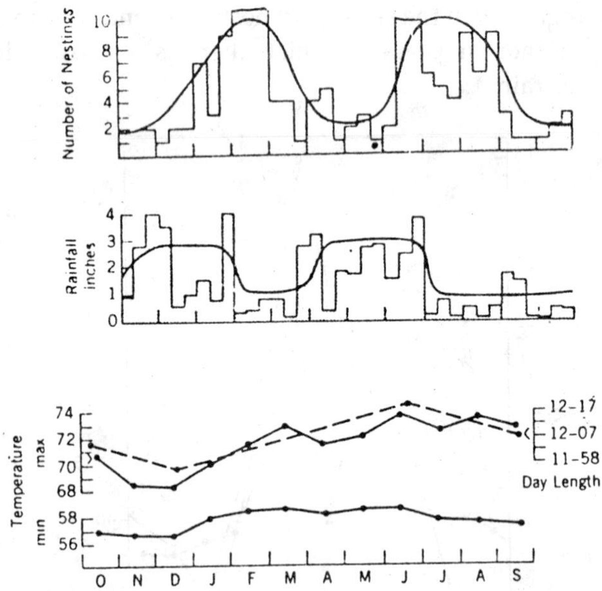

Fig. 12.11. Bimodal pattern of nesting in the rufous-crowned sparrow (*Zonotrichia capensis*). Nesting follows the bimodal seasonality of precipitation and seems to relate directly to the seasonal appearance of forbs and grasses.

At the equator the rufous-crowned sparrow (*Zonotrichia capensis*) has a bimodal nesting pattern, apparently cued by the seasonal appearance of some plants, which sprout after seasonal rains. Also, in three contiguous but contrasting habitats in Argentina (27°S), this species shows a similarity of gonadal development but nevertheless differs in nesting times. Locally, favourable nesting conditions, especially growth of ground cover, appear to stimulate nesting. Experimentally, this species displays gonadal growth when kept under midlatitude photocycles, indicating that the stimulatory effect of long days has not been lost by its residence in the tropics. It is likely, then, that equatorial photocycles are more or less continuously stimulatory, but that nesting is triggered by local non-photoperiodic signals.

Non-seasonal Species

Several non-migratory species of birds may have semiannual breeding seasons while occupying relatively uniform environments. In the presence of an adequate food supply and nearly constant schedule of rainfall, temperature, and day length, some species breed twice a year or on a schedule not in phase with the calendar. The rufous-crowned sparrow (*Zonotrichia capensis*) of Colombia has two separate breeding periods annually, and swiftlets (*Collocalia* spp.) of southeast Asia may have been annual nesting periods. The subtropical wideawake or sooty tem (*Sterna fuscata*) has discrete nesting periods that are not necessarily correlated with environmental cues. Its breeding times are interrupted by rather brief non-reproductive phases. In colonies near the latitudinal limits of its geographic range, nesting is clearly annual; but close to the equator, nesting recurs at intervals of less than 12 months. In the West Indies (20°N) this term nests regularly in the spring. On Ascension Island (7°S), however, the nesting cycle recurs every 9.6 months. On the Hawaiian Islands (20°N) nesting may be semiannual, with separate colonies being out of phase with one another.

In the absence of any *Zeitgeber*, reproduction in the sooty term might be initiated and synchronized by mutual sexual stimulation that affects the population as a unit. In such a situation, if this suggestion is valid, a refractory period would end reproduction more or less at the same time for the entire colony. Here, rhythmicity is presumably achieved by social stimulation.

Feral pigeons (*Columbia livia*), which are semidomestic,

may have irregular patterns of nesting. This pigeon breeds in the winter in lofts of vacant buildings near Tampere (Finland). Egg laying and rearing of young continue from December to mid-August, with a refractory period from September. This population molds from August to October. It has been suggested that pigeons (Columbidae) do not become refractory.

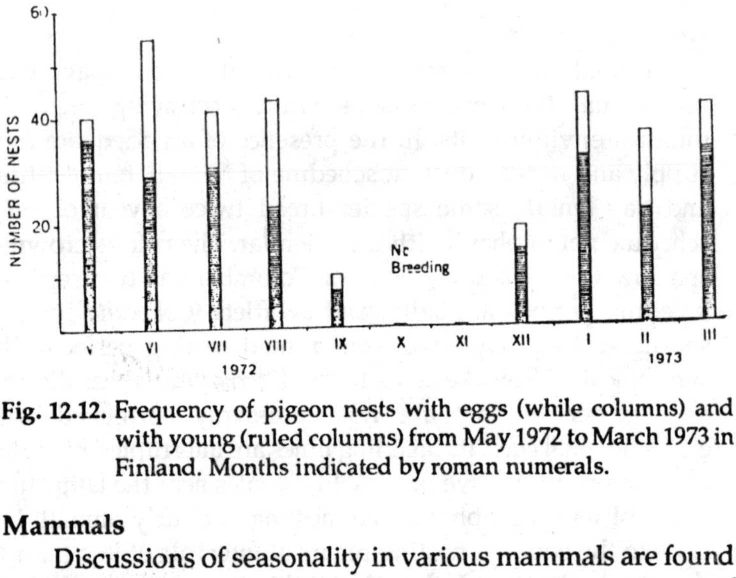

Fig. 12.12. Frequency of pigeon nests with eggs (while columns) and with young (ruled columns) from May 1972 to March 1973 in Finland. Months indicated by roman numerals.

Mammals

Discussions of seasonality in various mammals are found in a large number of widely scattered reports. Here patterns of seasonality are reviewed primarily as adaptations to habitat, climate, migration and hibernation. Additionally, the general reproductive cycles are discussed for some of the more common, widespread, or representative taxa.

In many respects mammalian reproductive seasonality resembles that of birds. In spite of the avian pattern of diurnal activity and the mammalian pattern of nocturnality, both groups contain many photosensitive species. The majority of mid- and high-latitude mammals are markedly seasonal and usually breed in the spring. Some mammals are very sensitive to variations in food supply. Forest-dwelling and meadow-dwelling rodents, for example, may have an abbreviated reproductive season in years of sparse food production, whereas in years of plenty they breed through the summer or sometimes into the following winter. A few species, especially among the artiodactyls, mate in late summer or autumn, under the influence of short days. Mammalian seasonality, although free in most groups from the complication of migration, is frequently modified by hibernation, embryonic diapause (or delayed implantation), postpartum estrus, and sperm storage.

Tropical Environments

A tropical climate suggests a warm, lush, and constant milieu, free from fluctuations. Although true with respect to photoperiod and more or less true for temperature, tropical environments are not equable for rainfall, and many low-latitude mammals are conspicuously seasonal in their gonadal activity and timing of parturition.

In the New World the genus *Didelphis* is represented by three species, one or more of which may be found from southern Canada to cool temperate South America. Breeding seems to commence with the annual increase in day length; reproductive seasonality is conspicuous but highly variable with latitude and altitude. In low latitudes, where the photocycle does not offer pronounced changes in day length, reproduction typically occurs in rainy periods.

Tropical crocidurine shrews have been little studied, but most appear to have rather extended reproductive seasons. The subtropical house shrew or musk shrew (*Suncus murinus*) appears to breed for very prolonged periods and may well be aseasonal. It is commensal about human dwellings and may be somewhat removed from natural cues.

Although one would expect that tropical bats, being free from the constraints of hibernation, would be aseasonal in their reproductive cycles, this frequency is not the case. A few examples will serve to illustrate the breadth of variation breeding of low-latitude bats.

Among tropical bats there is a tendency for reproduction to be associated with seasonal precipitation in regions where rainfall is regularly seasonal. Among a group of insectivorous bats in northeast Brazil, most timed their reproduction so that the young became independent during times of maximal rainfall. The flat-headed bat (*Neoplatymops mattogrossensis*) breeds near the close of the annual wet season, and lactation continues through the dry period. There are a few exceptions to this pattern. *Glossophaga soricina* produces one brood in the dry season and a second during the rainy season. The yellow-winged bat (*Lavia frons*), of East Africa, lives in a region where the rains are sporadic but concentrated. It is closely associated with *Acacia tortilis* and seems to subsist largely on insects that abound near that tree. Development of leaves and or blossoms follows the rains and is, in turn, followed by an increase of insects, which provide a positive energy flow to the economy of the bats. Parturition occurs at the onset of the more intense rainy period, from April to May.

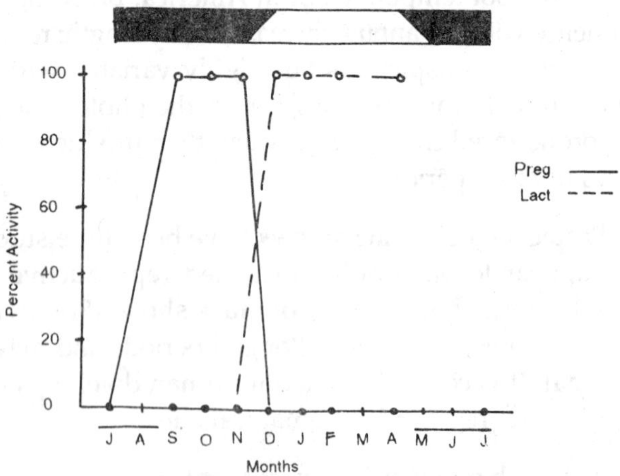

Fig. 12.13. Seasonal reproduction and environmental water balance of the Neotropical flat-headed bat (*Neoplatymops mattogrossensis*). The dark parts of the bat at the top indicate periods of water deficit, and the unshaded part is the period of water surplus. The solid line above the letters for the months indicates periods when testicular activity was apparent.

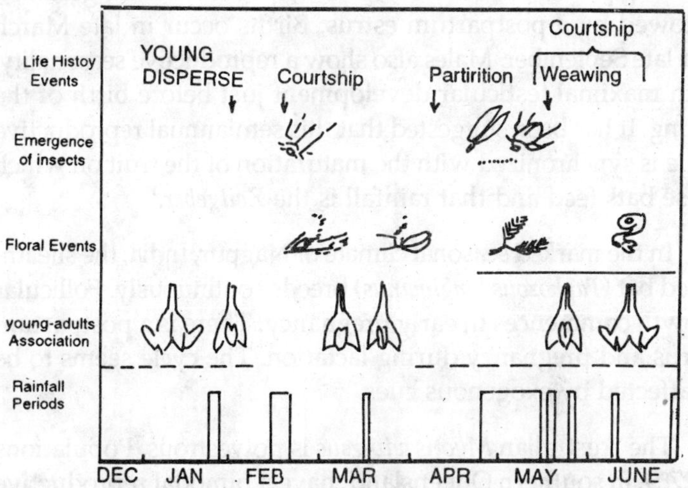

Fig. 12.14. Temporal relationships of reproduction in the yellow-winged bat (*Lavia frons*) of Africa, including rainfall, stages of *Acacia tortilis*, and emergence of insects. The dashed line indicates the period of association of the young with the mother: the solid line is the period when all acacias are in leaf; the dotted line is the time of nuptial flights of termites.

In Colombia *Artibeus jamaicensis* and *A. lituratus* appear to reproduce throughout the year; they are fruit eaters and their food is available at all seasons. In Panama, however, *A. jamaicensis* has two distinct periods of parturition, with births in the rainy season. The vampire bats (Desmodontidae) are found only in the New World tropics, where they feed on the blood of larger mammals. Their food is continuously available, and *Desmodus rotundus*, the most abundant species, breeds throughout the year.

Among a group of 10 species of bats in East Africa, reproductive seasonality is expressed in several patterns. These patterns include a distinct and restricted annual or semiannual reproductive period among some insectivorous species and a more extended, but synchronized, reproduction in fruit bats. In Uganda the fruit bat *Epomophorus anurus* has a semiannual breeding pattern, which includes a five- or six-months gestation followed by a postpartum estrus. Births occur in late March and late September. Males also show a reproductive seasonality, with maximal testicular development just before birth of the young. It has been suggested that this semiannual reproductive cycle is synchronized with the maturation of the fruit on which these bats feed and that rainfall is the *Zeitgeber*.

In the marked seasonal climate of Nagpur, India, the sheath-tailed bat (*Taphozous longimanus*) breeds continuously. Follicular growth commences in early pregnancy. There is a postpartum estrus and pregnancy during lactation. The cycle seems to be unaffected by exogenous cues.

The Australian *Myotis adversus* is polyestrous. Populations at 27°S, in southern Queensland, have a bimodal reproductive seasonality, but at 22°S, in central Queensland, the same species produces at least three litters. The litter size is one in both regions. Similarly, *M. nigricans* in Panama is polyestrous and appears to breed almost continuously.

In temperate regions both sperm storage and embryonic diapause are logical means of accelerating the onset of embryonic development in the spring, for subsequent weaning of young while insect food still abounds. These patterns persist, however, in some tropical and subtropical bats, which neither hibermate nor migrate. Delayed implantation is seen in tropical species of *Rhinolophus* and *Hipposideros*. Two species of bent-winged bats (*Miniopterus*) exhibit embryonic diapause both in the temperate climates of France, Japan, and southern Australia and in the tropical climate of southeast Asia. Sperm storage is well-known for some tropical species of vesper bats.

The cane rat (*Thyronomys swinderianus*) of tropical Africa is fairly aseasonal in it sexual activity. In Ghana most adult females are gravid at any given time, although there is a concentration of pregnancies at times of greatest rainfall. The mole rat (*Cryptomys hottentotus*) of Zimbabwe is nonreproductive for most of the year, and breeding is probably confined to the period of precipitation.

Many of the distinctive Neotropical caviomorph rodents have been discussed by Weir (1974), but most of her data are from captive colonies. Pearson (1949) studied wild populations of the viscacha (*Lagidium peruanum*) at 5000 ft in Peru. Breeding there occurs in the austral spring, and despite a long gestation, two broods are produced annually.

A discrete seasonality might seem to be complicated by a rather long gestation, from 21 to 22 months for the African elephant. Other large tropical mammals, which may be more or less aseasonal, also have long periods of pregnancy; gestation is some 13 months for tapirs and 15 months for rhinoceroses. A long gestation, however, does not preclude a favoured season for mating.

The two living species of elephants are said to exhibit little consistent seasonality. The African species (*Loxodonta africana*) may have annual peaks of pregnancy, with seasonality altered by population density. Thus seasonality would appear not to be associated with optimal conditions for the young. Separate populations of elephants, however, have fairly distinct periods of conception, such times following seasonal rains. This pattern probably reflects the stimulating effect of forage, which grows after the rains. There is, however, no seasonal fluctuation in gonadotrophins, gonadal steroids, or prolactin in the male African elephant or in non-pregnant females.

The two species of hippopotamuses are truly amphibious and do not stray from water, but their mating and parturition are associated with the seasonal cycle of rainfall. The large hippopotamus (*Hippopotamus amphibius*) mates at the close of the rainy season. After a gestation of eight months, the single young is born at the time of maximum precipitation, ground water and pond depth.

Temperate and Semiarid Environments

In higher latitudes the photocycle remains uniform from one year to the next, but other environmental features, especially rainfall, may change. Meteorological changes in themselves may alter reproductive seasonality and additionally alter vegetative characteristics from year to year. Although the onset of breeding is rather consistent from year to year, the end of reproduction may be extended in years of heavy mast crops.

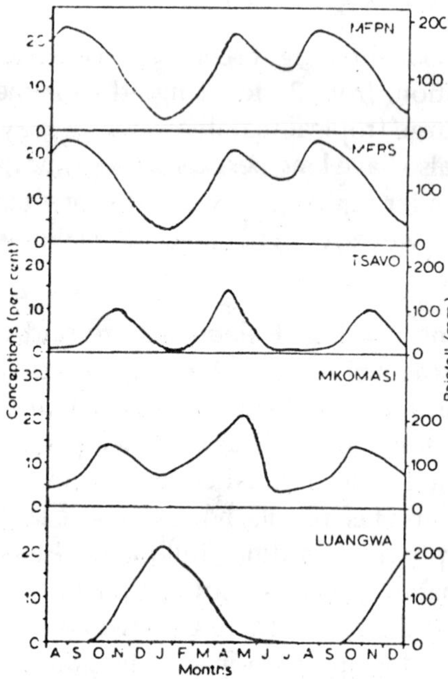

Fig. 12.15. Estimated monthly conception in percent (histograms) for five populations of elephants, plotted with mean monthly rainfall (solid line).

Autumnal and Winter Breeding. Numerous examples of small mice (*Apodemus* and *Peromyscus*) and voles (*Clethrionomys*) breed in late fall and winter. Such events demonstrate a partial independence from the annual photocycle. When autumnal breeding occurs, it involves most of the adult population and is frequently correlated with an unusually heavy crop of such favoured foods as pine seeds, acorns, and beechnuts. The bank vole (*Clethrionomys glareolus*) breeds through the spring and summer in the British Isles, and sexually active animals can be found in autumn or even into winter in years when mast is abundant. In a nearly pure stand of beech (*Fagus sylvaticus*) in Denmark, winter breeding of both the bank vole and the yellow-necked mouse (*Apodemus flavicollis*) followed the production of a very heavy crop to beechnuts. In southern Sweden the bank vole (*C. glareolus*), the wood mouse (*Apodemus sylvaticus*), and the yellow-necked mouse all breed in the winters in which there is a heavy crop of beechnuts. Similar patterns have been reported for many mice and voles.

In temperate region where rainfall is both sporadic and sparse, breeding of mammals may closely follow precipitation, which is similar to reproductive patterns in the tropics. Generally reproduction in many desert rodents in maximal during periods of rainfall, at which time there is almost always a flush of desert annuals. Seasonal fecundity end to be about as regular and predictable as the precipitation in any given region. Anecdotal accounts are numerous, and a few will illustrate this relationship.

In southwestern United States the round-tailed ground squirrel (*Spermophilus tereticaudus*) commence to breed after the winter rains have initiated the growth of annual plants. The rats vary from year to year, and for each increase of 13 mm over the mean, the breeding season is advanced about nine days.

Reproductive seasonairy of hares (*Lepus* spp.) is rather variable and subject to temporary changes in weather and habitat. In California the black-tailed jackrabbit (*Lerus californicus*) commences reproduction with the first appearance of green vegetation, and the duration of breeding varies with the growing season of forbs. The season of fecundity, moreover, is prolonged on irrigated land and rather short on dry cropland. Similar pattern have been noted for other species of leporids, and in desert-dwelling species breeding closely follows periods of substantial rainfall. In Arizona, for example, both *Lepus californicus* and *L. alleri* have semiannual breeding seasons that occur during the two annual rainy periods of the deserts.

The breeding season of the European rabbit (*Oryctolagus cuniculus*), introduced into Australia, closely parallels the pattern of sporadic rainfall. After rain, following and 15-month drought, pregnancies occurred within two weeks, before the growth of fresh vegetation. This suggests the perhaps humidity was a proximate cue, as is the case, experimentally, for the zebra finch.

Pocket gophers (Geomyitae) are abundant over much of western North America, especially in dry, sandy areas. They feed mostly on roots of herbaceous plants and are frequently pests on curvated land. Not surprisingly, they have been studied by many zoologists, and their reproductive cyclicity varies with the amount of water available for plant growth. The breeding season of the Botta pocket gopher (*Thomomys bottal*) is greatly prolonged in irrigated alfalfa fields in central California. In spring (March to May) pregnant females are equally numerous in both irrigated and non-irrigated alfalfa fields. By early summer (June), breeding cases on non-irrigated fields out continues where irrigated alfalfa is green and lush.

In the deserts of southwestern United States, germination and growth of winter annuals depend on adequate autumnal precipitation. In years with sparse rainfall, the subsequent development of annual forbs and grasses is poor. For five years the reproduction of desert moderns in southern Nevada paralleled the germination and growth of annual plants. With ample fall germination and sufficient subsequent rainfall the annuals retained greed leaves until February and March, by which time the rodents had begun to breed. Green leaves constitute an important vernal food for some species of heteromyid rodents (*Perognathus* and *Dipodomys*). In years when autumnal and winter rains are absent or short lived, these desert rodents fail to breed, and population decline. In the Mohave Desert heteromyid rodents breed only in years of ample winter rain. Also, in the Sonoran Desert, where rainfall is notoriously irregular, reproduction of the Merriam kangaroo rat (*Dipodomys merriami*) closely follows the rains. Within-year variations in pregnancies in two areas of contrasting precipitation were striking and highly significant. In the hopping mouse (*Notomys alexis*), a murid rodent of Australian deserts, breeding is facultative, apparently cued by irregular precipitation.

There is evidence that water per se may promote fecundity in desert rodents. Two species of gerbils (*Gerbillurus paeba* and *Rhabdomys pumilio*) of the Namib Desert (at 25°S) breed during periods of precipitation. When rains fail, reproduction is reduced. Artificial availability of drinking water was followed both by an increase in fecundity during the period of rains and by an extension of the reproductive period into the dry season, in contrast to the reproductive performance of these species in a control area. Sexual activity in the mole rat (*Heterocephalus glaber*), a fossorial rodent of the arid regions of southern Africa, is restricted to the season of rainfall. This rodent stores food, and rain provides for greater growth of plants that can be gathered. Increased body fat and fecundity follow a period of heavy rain, and breeding may not occur when rains fail.

There is distinct but different seasonality in the reproduction of two hyraxes in Kenya. In *Heterohyrax brucei* the young are born shortly prior to the period of heavy rains, at which time there is still some residual foliage on woody plants. This species is a browser and finds adequate food in the branches of trees and shrubs, even during the dry season. Lactation continues throughout the rainy period, when all vegetative growth resumes. In contrast, the young of the more terrestrial *Procavia habessinica* are born after the long rainy season, when there is a fresh growth of forbs and grasses upon which the lactating females subsist. Thus reproduction in both hyraxes is tied to the seasonal precipitation but clearly in different ways.

In Rwenzori National Park, which lies on the equator, there are two distinct rainy seasons, and native rodents breed only after the onset of the rains. It has been suggested that reproduction in the mice *Lemniscomys striatus* and *Praomys (Mastomys)* natalensis is stimulated by an increase of protein intake resulting from the inclusion of termites in their diet during the rains. At this time *P. natalensis* also feeds heavily on the fresh fruits of the shrub *Erythrococcus bongensis*. Both rodents experience a sharp rise in body fat after the onset of the rains, and thus the rains and the protein-rich food seem to relate to their seasonal pattern of reproduction.

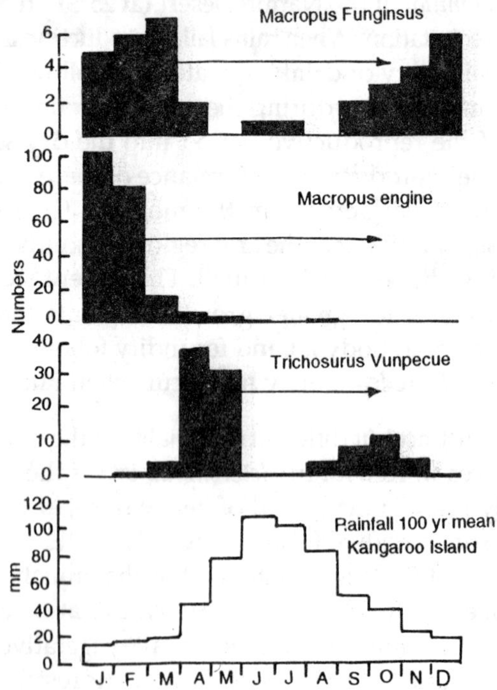

Fig. 12.16. Monthly occurrence (numbers) of births of macropod marsupials and the mean monthly rainfall (mm) on Kangaroo Island, Australia. The horizontal line with the arrow indicates the duration of pouch life of the joey.

Some ungulates in Africa are both gregarious and nomadic. They have long been a food source for aborigines, and consequently their movements and reproductive patterns are well-known. In the seasonally arid environments in which they occur, parturition is usually scheduled so that the young start life at a time when fresh forage is growing, which can be either annually or semiannually. The dik-dik (*Rhynchotragus kirki thomasi*) conceives at both the beginning and the end of the annual rainy season, and gestation lasts about six months. There is a postpartum conception, so that most females produce two young per year, and the young tend to be born when the grass is tall and fresh. Similarly, the impala (*Aepyceros melampus*) has two breeding periods annually, coinciding with the semiannual rainy periods. The springbok (*Aridorcas marsupialis*), in contrast, lives in an arid region of sporadic rainfall in southwest Africa and migrates in search of fresh herbaceous growth that develops after the nonseasonal rains. Reproduction, being dependent upon an extensive flush of plant growth, is also sporadic, although males are fecund throughout the year. Breeding can therefore occur in any month under favourable environmental conditions.

Many of the Australian marsupials dwell in arid regions where the rainfall is either seasonal and restricted or sporadic. Most Australian marsupials are seasonal in reproduction, responding to seasonal fluctuations in rainfall. In some species of macropod marsupials (such as kangaroos and wallables), breeding commences with the emergence of young from the pouch, which occurs during the period of annual precipitation. Some dasyurids breed in late winter, apparently as a result of photostimulation. *Isoodon obesulus*, a peramelid, exhibits a precise seasonality, suggesting photostimulation.

Migratory Marine Species

Among migratory seals, sea lions, and whales, parturition must be synchronized with the migratory schedule and the requirements of the newborn young. In some high arctic species these events must fit into a narrow growing season, and the environment does not allow much flexibility. Seasonal diapause is almost universal among these mammals, and there is either a postpartum or a postlactational estrus.

In the earless, or true, seals (Phocidae), lactation is rather brief. In most species mating occurs at the end of lactation, at the time the females abandon their young. In the eared seals or sealions (Otariidae), mating is postpartum, and lactation, even in polar regions, is more prolonged than in the phocids. Thus in both families mating must occur within the brief time of their terrestriality, at the end of their spring migration and just prior to their return to lower latitudes.

The walrus (*Odobenus rosmarus*) makes only local movements in the Arctic. Mating follows the birth of the young by several weeks, and gestation is about 11 months. Thus the single young is born at the onset of the brief period of arctic productivity; lactation is prolonged.

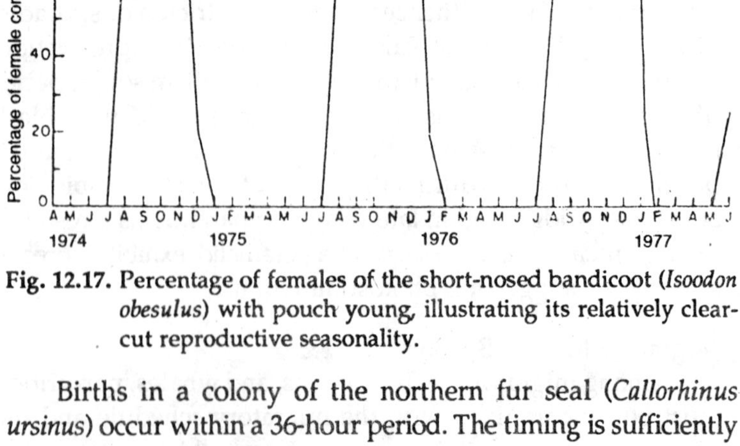

Fig. 12.17. Percentage of females of the short-nosed bandicoot (*Isoodon obesulus*) with pouch young, illustrating its relatively clear-cut reproductive seasonality.

Births in a colony of the northern fur seal (*Callorhinus ursinus*) occur within a 36-hour period. The timing is sufficiently late that low temperatures do not endanger the newborn pups and early enough to allow them to develop to the point of migrating some four months later.

The finback whales (Balaenopteridae) perform annual migrations to the high latitudes, where they accumulate thick layers of subcutaneous fat, and return to low latitudes as the day length shortens and the productivity of the polar seas declines. Birth of young (calving) takes place in shallow and relatively warm waters in the winter. Because of the alternation of the seasons globally, when the population of the Northern Hemisphere is calving, that of the Southern Hemisphere is in the high latitudes engorging on plankton. Seasonality is scheduled so that parturition occurs under the most auspicious environment for birth of the young, and lactation takes place at a time when energy-rich food is most abundant.

Hibernating Species

Insectivorous bats (Microchiroptera) are numerous and diverse in the tropics, and a few species occur at the higher latitudes, even dwelling in the subarctic. In the mid-and high latitudes hibernation is the rule and constitutes an important factor in the annual reproductive period of many bats. Additionally, many species are also migratory, and some species both migrate from their summer range and hibernate in their winter range.

In many Holarctic bats hibernation divides the reproductive season Mating occurs in the autumn, after which both sexes become torpid for several months. Hibernation is broken by periodic arousals throughout the winter, and active individuals may copulate also at those times. Such matings, however, are not seen in all species, for in some bats the sexes separate at the time of hibernation, and others have separate migratory patterns. In most bats of the temperate regions spermatozoa are stored in the oviducts, but in several kinds fertilization immediately follows mating, and there is either an embryonic diapause or a very slow rate of embryonic growth. In species that store spermatozoa, ovulation follows emergence from hibernation. In taxa that ovulate in the autumn, embryonic growth accelerates after hibernation.

Among the various groups of hibernating mammals, the timing of mating and parturition is closely bound to the period of dormancy. Hibernation is found in many species of ground-dwelling squirrels, including the marmots (*Marmota*), the prairie dogs (*Cynomys*), most of the ground squirrels (*Spermophilus* and *Ammospermophilus*), and most of the chipmunks (*Tamias*). One of the most thoroughly studied is the golden-mantled ground squirrel (*Spermophilus lateralis*) of western North America. In wild populations vitellogenesis commences in January, in the middle of the period of hibernation, and adult females mate immediately after emergence, or at least as soon as they appear aboveground. There is only a single brood, for the young must grow rapidly and accumulate an adequate supply of body fat before entrance into winter dormancy. Similar patterns are known for other species of ground squirrels. Reproductive cues are not apparent for these hibernators, and gonadal recrudescence may be bound, physiologically, to the cycle of hibernation. Among dormice pregnancy follows soon after emergence from hibernation, and usually there is only a single brood.

Several hibernating rodents are known to produce two or more broods a season, but they are the exception. Jumping mice (Zapodidae) may have two or even three litters annually. The gestation is only about 17 to 18 days, and there is a postpartum estrus. The eastern chipmunk (*Tamias stratus*), a hibernator, has at least two litters per year.

Hibernating bears, such as the North American black bear (*Ursus americanus*), mate in late summer, and the blastocyst experiences a diapause for several weeks. The young are born in mid-winter, long before actual emergence from hibernation.

Reproduction of hibernating mammals seems to be closely tied to the time of arousal from hibernation, for pregnant females are found shortly after emergence, and emergence is partly influenced by ambient temperature. Richardson's ground squirrel (*Spermophilus richardsonii*) breeds earlier in the year at lower latitudes, and in Saskatchewan (49^{0}N) this squirrel breeds approximately 10 days earlier in years with conspicuously warmer-than-usual days in late winter. It has also been suggested that the lemming (*Lemmus trimucronatus*) reproduces as minimum temperatures increase.

Primates

For a long time there was a prevailing notion that primates were characterized by year-long sexual activity, which seemed to provide a conceptual basis for their social cohesiveness. However, many accounts document alternating periods of mating and sexual inactivity in free-living populations of monkeys, baboons, and other primates. It is now widely accepted that wild populations of many species of primates are, indeed, quite seasonal in sexual activity.

Because of the comparatively tight social structure of many primate communities, some techniques that might be suitable for studying some kinds of mammals, such as wild mice, cannot be employed for primates. Most primates, however, are diurnal, and inasmuch as infants are frequently carried by a parent, times of birth can be estimated from visual observations.

As social activity becomes more complex, it assumes some control over sexual activity, and the latter seems to be less under environmental control. Many early studies of primate reproduction were made on captive colonies and sometimes under controlled conditions, so that environmental cues were not apparent.

In some of the rhesus monkeys or macaques (*Macaca* spp.), there is apparent seasonality of both copulation and birth. In the Japanese macaque (*M. fuscata*) kept under a 12L : 12D schedule with controlled temperature, menstrual cycles recur four to five times, commencing in December, but there is no clue as to a possible *Zeitgeber*. In a free-ranging population of the same species in Kyushu, mating occurs from October to April and births from April to October. When kept in large outdoor enclosures exposed to natural cues, a population of *Macaca mulatta* exhibited a clear-cut seasonality in mating. copulation was most frequent in October and November, at which time serum testosterone reached maximal levels. In other species of macaques, the schedule is slightly different, but there are generally discrete periods of copulation and parturition. Copulation usually occurs under a short or declining day length.

In contrast to macaques, baboons tend to be less seasonal in periodicity of mating and births. Some species have peaks, but there usually seems to be an extended period during which births occur.

The common langur (*Presbytis entellus*) in northern India does not show an obvious seasonality, and females in estrus can be found in any month of the year. Nevertheless, most births are in April and May. This may be an adaptation to food for the young, for fresh vegetation, on which the young feed, follows in a few weeks, when monsoons begin in June and July. On the other hand, the vervet monkey (*Cercopithecus aethiops*) of tropical western Cameroon, parturition in one population was synchronized with seasonal rains, while another group exhibited seasonality with the reverse pattern. Seasonality in this example is attributed to social, not environmental, factors. The lesser mouse lemur (*Microcebus murinus*) breeds seasonally both in the laboratory and in nature: mating is in the spring, and females are anestrous from September to February. The length of the estrous cycle, however, is increased by social contacts (olfactory, visual and tactile).

The ring-tailed lemur (*Lemur catta*) breeds in the winter in its native Madagascar, and its seasonality is under photoperiodic control.

Seasonal patterns in human conception, but probably not in coitus, have been reviewed by Cowgill (1969) and MacFarlane (1970). Although peaks in conception are not conspicuous, they do occur and reflect social rather than environmental factors. In some northern European societies, for example, conception arises at about the time of Christmas, perhaps as a consequence of the celebration of Teutonic fertility rites. In Utah, a region in which the prevailing religion proscribes the recreational use of alcoholic beverages, there is a conspicuous decline in conception rate from December to February. Another socially induced pattern has been noted among aborigines in western Abelam, New Guinea. In this equable climate, births reach a peak between October and March. The cultivation of yams is a major activity, and the social status of a man is elevated by his production of yams. Also, sexual activity is supposed to inhibit a man's capacity to grow yams, and there is therefore a strong incentive toward sexual abstinence during the yam-growing season, providing the basis for increased copulation and conception after the yams are harvested.

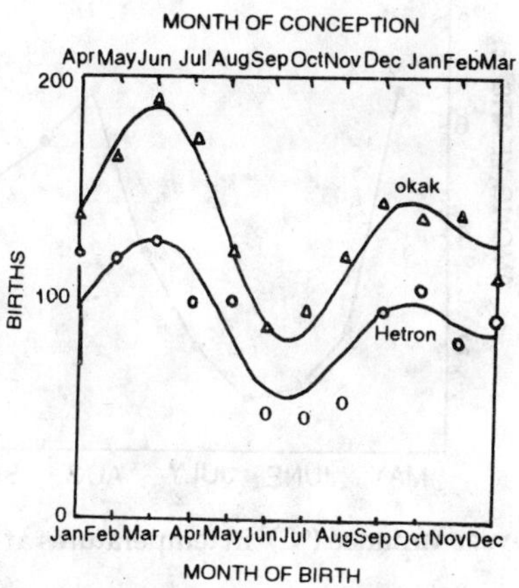

Fig. 12.18. Analysis of dates of conceptions and births from two Eskimo communities in Labrador. Data from Okak (1778-1918) and Hebron (1831-1940) are fitted by the summation of two sinusoidal functions with a 12-and 6-month periodicity.

Humans may not be entirely free from environmental influences. In upper socioeconomic strata of the eastern United States, seasonal rates of conception are independent of seasonal incidence of coitus, suggesting an exogenous influence on ovulation rate. The best case for exogenous influence on human conception exists for Eskimo villages in Labrador. Data from 100 or more years in each of two communities reveal a peak of conceptions in June and a peak of births in March. The amplititude of the changes in monthly means in birth dates is 75% and 84% from annual means. The factor inducing this seasonality is unidentified, but regular changes in food, photoperiod, and temperature are possibilities.

Temperate and Cold Environments

At the highest latitudes breeding seasons are short, and a high variance characterizes the modifying factors of the environmental variables. The annual photocycle, the changes

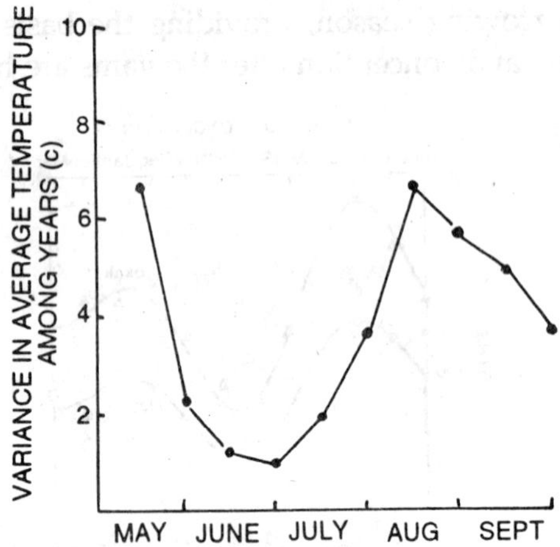

Fig. 12.19. Year-to-year variance (°C)^2 in temperatures at Barrow, Alaska.

in temperature, and the resurgence in plant growth are all of major magnitude and rather closely synchronized. Consequently, mammalian seasonality has little room for flexibility. In the Canadian arctic (Northwest Territories) the redback vole (*Clethrionomys rutulis*) breeds at the time of snowmelt, which varies by about 11 days from one year to the next. In years when the appearance of the first litter is delayed, sexual maturity of the young females in this litter is also delayed. If the initial litter is born earlier in the season, then those females may themselves reproduce in that summer.

In soricid shrews, which comprise the largest group of insectivores, breeding is usually vernal, and some species have postpartum estrus. In late winter and early spring, with the approach of maximal gonadal activity, there is a conspicuous increase in weight. There is also an annual turnover in population, so that no individual experiences more than a single breeding season. Thus the breeding season is delimited by sexual maturity in the spring and death in the summer.

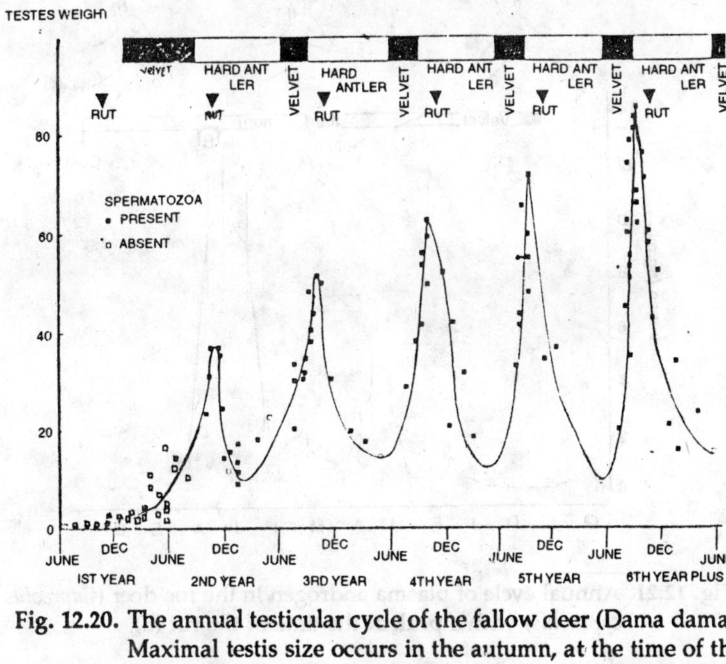

Fig. 12.20. The annual testicular cycle of the fallow deer (Dama dama). Maximal testis size occurs in the autumn, at the time of the rut, and declines to a minimum in the spring, at which time the antlers are lost.

Artiodactyles that dwell in seasonally cold regions mate so that the developing young make maximal use of the limited season of mild weather and abundant forage. Thus most species mate in the autumn, with parturition in late winter or spring, depending upon the species and the latitude. In the fallow deer (*Dama dama*), for example, by late winter and early spring the testes contain no spermatozoa, but sudden testicular growth occurs in midsummer, after the summer solstice, with concurrent spermatogenesis and growth of a fresh set of antlers. Similar patterns of seasonality occur in the most boreal artiodactyls, the caribou (*Rangifer tarandus*), the moose (*Alces alces*), and the muskox (*Ovibos moschatus*). The autumnal rut and the annual development of antlers of deer are under the seasonal output of androgens. There is direct development in all deer but the roe deer, with no embryonic diapause.

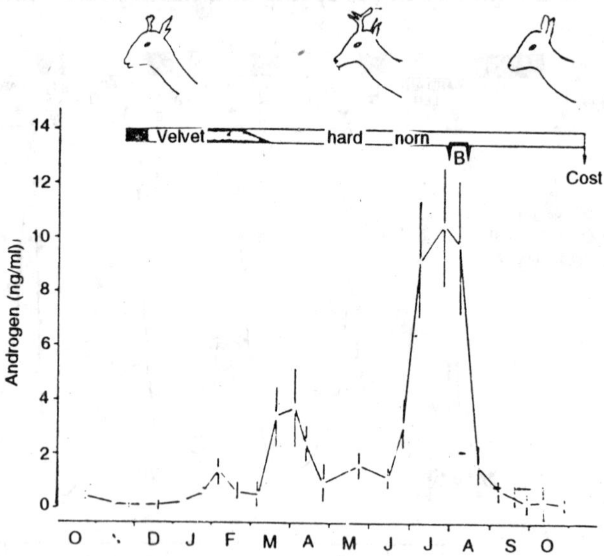

Fig. 12.21. Annual cycle of plasma androgen in the roe deer (*Capreolus capreolus*) with a peak at the time of the rut (b).

Adaptive Significance of Reproductive Seasonality

Reproductive seasonality is the association of (1) the series of physiological events from gonadal growth, steroidogenesis, gametogenesis, mating, and finally the appearance of the next generation with (2) a particular season. Although some species are now known to have strong endogenous elements to their annual reproductive cycles, environmental factors are generally responsible for synchronizing the physiological changes with annual climatic cycles.

Although the annual photocycle is regular from one year to the next, the onset of non-photoperiodic events varies slightly. In many vertebrates this results in year-to-year differences in the timing and duration of reproduction. Birds, which are *stimulated* by both the increase in day length and the occurrence of long days, are closely bound to the annual photocycle. Mammals, on the other hand, in which the hypothalamic-pituitary-gonadal axis is *repressed* by increasing darkness, or short days, and in which the gonads eventually become scotorefractory (and not photorefractory), sometimes respond to nonphotic factors of the environment. Thus seasonality is more predictable in birds than in mammals.

There are annual variations in the length of "summer" as measured in terms of heating degree days. Year-to-year variations in precipitation occur, and such variations result in both qualitative and quantitative changes in plant growth. These non-photic factors may alter the cyclicity of mammalian reproduction. Also, year-to-year differences in rainfall cause minor irregularities in stream flow and the formation and duration of temporary ponds, and such factors can be critical to the timing of spawning of amphibians and fish. Fluctuations in vernal warning can alter reproductive events in some polar vertebrtes, especially birds.

Climatic variations that we observe in a lifetime are relatively slight. Within the brief period of human history, however there have been fluctuations of greater magnitude and duration. The Little Ice Age (1600-1850) was characterized by America and Europe. Prior to that time, however, and up to about 1000 A.D., warmer and overall milder conditions permitted agricultural communities to survive in Greenland. Such climatic changes undoubtedly affected reproductive phenology of small mammals in the mid-and high latitudes. In addition, the survival of young lizards, some turtles, and voles, mice, and lemmings strongly suggests that reproductive seasonality is not characterized by a delicate, fine-tuned adjustment so that young are born precisely at the most optimal time of the year. Rather, reproductive seasonality is somewhat flexible, with the result that modern species have the capacity to make rapid adjustments in times of mating, oviposition, birth, and other such events within a given genotypic frame.

The result of these year-to-year variations is variability in the onset and termination of the reproductive calendars of many vertebrates. Whereas environmental factors determine seasons of gonadal activity for many forms, there are very few in which hatching of eggs or birth of young are set by exogenous factors. Some flexibility occurs in developmental rates of ectotherms, but in most groups thee onset and duration of mating of adults accounts for variation in the start and duration of natality. These variations are usually so minor that the young are nevertheless, produced when environmental conditions are usually optimal for their survival.

Whereas the *onset* of reproduction is subject to some year-to-year variation due to meteorological variations, there are frequently greater variations in *duration* of reproductive activity brought about by slight deviations in rainfall, temperature, and food supply. Consequently young born at the end of summer or in autumn do not face the same environment as young born in the spring do.

Ornithologists developed the concept that reproductive seasonality is adjusted so that young are produced at the time most favourable for their survival, which is certainly applicable to birds. If this concept is modified to be the time most favourable for maximal survival of the most vulnerable stage, it is more applicable to other vertebrate groups. For example, in many species of fish and reptiles the reproductive calendar provides for maximal survival of the eggs. In oviparous fish the oxygen dissolved in water dissolved in water appears to be more critical than conditions for the young, which have developed gill cover movement and pursue their own food. In some congeneric and sympatric species of lizards, oviparous species have a narrowly restricted breeding season in contrast to viviparous species, which may reproduce continuously. It was pointed out previously that oviposition in some tropical reptiles is cued to the humidity of the environment in the site for oviposition and that there is ample food for young throughout the year. The more-or-less continuous availability of food for neonate lizards at low latitudes is confirmed by the restricted seasonality of some oviparous species while viviparous species have continuous production of young.

The painted turtle (*Chrysemys picta*) lays its eggs in the late spring, and these eggs usually hatch in late summer. However, some eggs may overwinter and hatch the following spring, with the result that young may appear either in the autumn or in the spring. The same phenomenon has been reported for the snapping turtle (*Chelydra serpentina*) in New York (Hamilton, 1940) and is also known for the whiptail (*Cnemidophorus sexlineatus*) in Arkansas. The overwintering of reptilian eggs may be common, and survival of the young is probably as great in the spring as in the late fall.

Among the several species of small mammals for which reproductive phenology has been well studied, both vernal and autumnal production of young are not uncommon. It occurs in some species of Sciuridae, Muridae and Cricetidae. Because mammals do not usually become photorefractory, autumnal reproduction has no endogenous restriction. If it were deleterious for small mammals to produce young at the onset of winter, it would surely be very rare. It is significant that while wild mice breed in the autumn when food is abundant, autumnal breeding seems to be a regular occurrence in some species of midlatitude squirrels.

With birds the timing of nesting, incubation, and hatching is certainly very critical, and the reproductive chronology of birds is widely acknowledged to be adjusted so that young appear at the time most favourable for their rapid growth and maximal survival. Because the period of embryonic development is fairly constant for a given avian species, the schedule of oviposition would seem to be the mechanism that provides for hatching at the most advantageous time. The environmental conditions that provide the adults with the stimuli and energy for nesting are frequently the same as the environmental requirements of the young; this is especially true of ultimate cues. Moreover, inasmuch as incubation of most small birds is rather brief, major environmental (including nutritional) changes between egg laying and hatching are rare. In many passerine birds incubation requires only two weeks or less, and the first broods appear at times of maximal plant growth and insect populations.